The Significance of the Hypothetical in the Natural Sciences

The Significance of the Hypothetical in the Natural Sciences

Edited by
Michael Heidelberger and Gregor Schiemann

Walter de Gruyter · Berlin · New York

∞ Printed on acid-free paper which falls within the guidelines of the ANSI
to ensure permanence and durability.

ISBN 978-3-11-020694-4

Library of Congress Cataloging-in-Publication Data

A CIP catalogue record for this book is available from the Library of Congress.

Bibliographic information published by the Deutsche Nationalbibliothek

The Deutsche Nationalbibliothek lists this publication in the Deutsche Nationalbibliografie; detailed bibliographic data are available in the Internet at http://dnb.d-nb.de.

Printed in Germany

Cover design: Christopher Schneider, Laufen

Preface

This volume is to its greater part the outgrowth of an international conference held at the University of Tübingen in 2005 that brought together philosophers and historians of the philosophy of science. The conference was supported by a generous grant from the *Fritz Thyssen Stiftung* and a donation from the *Universitätsbund Tübingen*. We also wish to thank Dr. Gertrud Grünkorn for including the volume in the philosophy programme of de Gruyter and for her editorial support. Our final thanks go to two referees who have done an excellent job of carefully and critically reading the contributions. We hope that the interest in the volume will not remain hypothetical but become a living reality in the end.

Tübingen and Wuppertal
June 2009

Michael Heidelberger and
Gregor Schiemann

Contents

Introduction

Michael Heidelberger and *Gregor Schiemann*

Since early modern times, the significance of hypothesis in natural science has been judged in widely different ways and has become the source of many controversies. The purpose of this volume is to illuminate some general lines of development of those debates by treating cases from the history of science and philosophy. The case studies presented here deal especially with physics, astronomy, mechanics and chemistry as well as with problems posed for mathematical theories of natural science in general.

Taken together, these cases show that the role hypothesis played and plays for natural science is of central importance for the manner in which a science conceives of itself and its own methodology. Accordingly, different concepts of science entail different attitudes towards hypothesis, both in the history of science and in the discourses of the philosophy of science. The significance attributed to hypothesis is, so to say, a kind of a litmus-paper for the changing and diverging conceptions of science of the scientific actors themselves, as well as of the philosophers who reflect upon the sciences.

If we focus, though, on contemporary discussions, the concept of hypothesis seems to be taken almost as univocal. Historians and philosophers of science as well as scientists themselves seem more or less to agree on its meaning. A hypothesis is normally taken as a conjecture that is expedient for the gain of knowledge. Sometimes, this definition is accompanied by the conviction that the truth value of a hypothesis will finally be established with further research. We also find the view, however, that the hypothetical character of certain propositions will never be eliminated. These we can call "metaphysical hypotheses". Not only single propositions are called hypotheses, but also theories, clusters of them or even the whole of scientific knowledge. It is almost common sense in the philosophy of science to generally attribute a hypothetical character to empirical theories. According to this view, conjectures are not only useful for the production of knowledge, but scientific theories are nothing but a collection of conjectures. There are powerful and acknowledged arguments for this, both of a systematic and historical nature. In a systematic sense, the hypothetical nature of scientific theories relates to the insolubility of the problem of induction. The truth of general propositions a theory comprises can never be deduced from experience, the famous opposite opinions of Newton and Ampère notwithstanding. Yet experience and theory cannot be separated sharply from each other and their inseparability is a further reason for their insecure make-up. It seems to be impossible to identify the culprit for an error in the entangled net of experience and theory. This circumstance is in agreement with the historical claim that theories hardly ever lost their validity as a result of being conclusively refuted, but rather because their followers died out in the course of time – this being another support for the conclusion that the claim of scientific propositions to truth cannot be resolved.

It is now exactly this common sense in regard to the hypothetical character of science that can go with very different evaluations of its significance. It seems that the different views can be grouped in at least three ideal types. The first one takes the hypotheticity of science as being of highest significance. The conjectural character of scientific theories is taken as their hallmark. It follows that a true theory is impossible and that science must live with the perpetual susceptibility to error, and thus with permanent revision and a forever undecided truth-value. The high esteem of the provisional character of science leads to a revaluation of ignorance and uncertainty. The second type of evaluating hypotheticity to be distinguished from the first takes hypotheticity for an important and inevitable mark of science as well. Yet this type does not infer that we should renounce any claim to truth, but, on the contrary, to endorse it. According to this view, hypotheses can always come closer to the truth. The certainty of propositions can be improved, their domain can be increased and their grip on reality can become tighter. The position of the third type does not doubt the fundamental hypotheticity of science either, but denies its relevance. Science has not to be judged primarily according to its epistemic values but through its practical advantages. According to this view, science is taken as a context of action that aims to change the world in order to meet human needs. Theories can be helpful in reaching this goal, but this is not necessarily so. Their hypothetical status is therefore of lesser importance.

A closer look reveals that the present debate about the significance of the hypothetical is not only controversial, but also somewhat confusing. The different positions cannot always be categorized in the suggested ways and show some overlap. Yet the major fundamental positions remain visible. This is also exemplified by the case studies of this volume that refer to the present debate of the problem. Those positions that stress the practical dimension of science, like pragmatism in the follow-up of Charles Sanders Peirce or Bruno Latour's view, lead to a kind of downplay of the hypothetical aspect of scientific theories (third type). Hypotheses loose their immediate relevance for questions of truth and turn into productive anticipations of reality (Alfred Nordmann). If, however, one keeps to the epistemological attributes of theories, the hypothetical nature of science in general comes to the fore (second type). Two case studies discuss the retention of claims to the truth. The first deals with the relation between hypotheticity and scientific realism. The recognition of hypotheticity does not automatically exclude a realist point of view, provided it implies the possibility of approximate truth – on the contrary, such a view can even be based on hypotheticity if scientific realism is conceived as an empirical theory itself (Andreas Bartels). The other case study considers the metaphysics of nature as dependent on scientific hypotheses. One can understand metaphysics as being as hypothetical as the scientific theories from which it ultimately derives (Michael Esfeld). The esteem of hypotheticity can, however, also lead to a justification of the limited validity of scientific knowledge (first type). A further case study that can be counted as belonging to this position does not refer directly to the debate at present. The notion of a "closed theory" developed by Werner Heisenberg in the context of his quantum mechanics sees the reach of scientific theories as being limited through concepts and denies the possibility of a continuous progress of knowledge (Gregor Schiemann).

The different types of significance of the hypothetical in today's sciences have developed from specific historical constellations starting in the early modern era. The common origin is still present in the shared view that, in contrast to pre-modern conceptions from antiquity, hypothesis has a legitimate place and function in the process of knowledge. In Aristotelian science, which dominated the scene until the early modern period, the status of hypotheses is problematic. This is shown in a case study that deals with the development of astronomy of the middle-ages (Gad Freudenthal). One can even say that rejecting this aspect of Aristotelian science and incorporating hypothetical elements into the presuppositions and methods of modern science is among the decisive hallmarks of the Scientific Revolution (Ernan McMullin). That does not mean, of course, that this is enough for characterizing the fundamentals of modern science; suffice it to say that their changing relations to theology and religion as well as its new relevance for technology have to be taken into account.

Modern science's way of dealing with experience did not only imply the necessity of making room for hypotheses in science but also for limitations of scientific knowledge. There are reasons to suppose that even empiricist thinkers like Locke could not ground these limits exclusively in experience, but also in metaphysical presuppositions (Rainer Specht). All the new hypothetical elements notwithstanding, modern science still upheld truth as the goal of science in a largely comparable way as its predecessor. The second ideal type of significance of hypothesis that plays a role in the present debate originates from this historical constellation and is related to it with several different lines of development. One line is represented by the concept of induction that played a dominating role until deep into the 19th century (Laura Snyder; McMullin's section on Newton). Among the cases treated in this collection, Heinrich Hertz's hypothetico-deductivism provides an interesting further conception of the intricate and fragile balance of hypothesis and truth in science (Andreas Hüttemann).

In the 19th century, another process of change took place whose reverberations can still be noticed today in the first and third view on the significance of hypothesis. This was the growing conviction that all claims to final validity of scientific assertions about the world are hypothetical and with it all scientific theories. Instead of playing the role of useful and necessary conjectures on the way to truth that are finally overcome, hypotheses increasingly undermined the goal for which they were designed. In respect to considerations of validity, the traditional understanding of science began to change and was finally turned upside down: Instead of constituting irrefutable knowledge, science was increasingly seen as representing indemonstrable hypotheses. Science's immutable certainty was more and more disputed and taken over by refutability as a criterion of science; the trust in science's truth started to give way to a permanently effective suspicion of its possible failure. This development inaugurated a turn to a concept of science that eventually renounces any claim to truth and represents the first type of hypothesis as described above.

As in the Scientific Revolution, this process of change occurring in the 19th century has to be seen in the wider context of the changing social functions of science. Industrial development was based increasingly on scientifically produced technology (dye industry, electrical industry). Scientific results were now systematically incorporated into the production process of goods. Professional education in science and technology increasingly fell into the responsibility of the state which could in turn secure its influence by providing financial support for experimental research. State institutions had to watch over the implementation of large scale technology in science and industry. Developments like these, connected as they are with the applicational dimension of science, help to understand the course of increasing hypotheticity. In short, one can say that a key insight in the 19th century was the discovery that science can be socially useful even if epistemological questions, which had previously held priority, were left unanswered. Demand for the applicability of science overruled questions about its exact epistemic status.

The contributions of this volume do not deal with these more external relations of science to society but concentrate on the philosophical discussions that accompanied the developments described here and asked for the claim to truth. With the work of the mathematicians Carl G. J. Jacobi and Carl Neumann, the idea of axiomatic thought, to base knowledge of a field on self-evident assumptions, eroded "from above". The process of hypothesizing first principles likewise seized mathematics and mechanics (Helmut Pulte). The French philosopher of nature and science, Émile Boutroux, reached a hypothetical view of mathematics and science by making assumptions "from below", i. e. by admitting an irreducible variability and spontaneity on the micro-level of physical reality. The possibility of genuine novelty in nature as well as the hierarchical ordering of science into irreducible disciplines that build upon each other led to an insurmountable element of "contingency" in natural laws (Michael Heidelberger).

Both in Jacobi and Neumann as well as in Boutroux, one can discern considerations that come very close to the outlook on the foundations of mathematics and physics that was developed by Henri Poincaré at the turn to the 20th century. Poincaré's conventionalism can be regarded as one of the first and most effective formulations of a hypothetical conception of science. His epistemological analysis included a comprehensive classification of different meanings of hypothesis that are in use in mathematics (especially geometry), as well as in physics (Gerhard Heinzmann). To regard theories as hypotheses gives more possibilities in theory choice than if one remains wedded to a traditional conception that identifies science with truth and truth-seeking. One can decide only by convention between different theories of an object realm that are incompatible with each other but equally justified. This was one of the factors that made Poincaré keep to Galileo's conception of space-time against Albert Einstein's alternative (Scott Walter).

Poincaré's concept of hypothesis has interesting similarities with, but also differences to, Hans Vaihinger's concept of fiction. According to Vaihinger, a fiction is an idea that is blatantly false but nevertheless useful for dealing with reality. Together with hypotheses, fictions act as a basis for a pragmatic conception of science (Christophe Bouriau). To abstain from claims to truth concerning scientific assertions leads not only to a new appraisal of hypothesis, but also to a revaluation of the practical applications of science. The latter development is bound to advance the significance of hypothesis one step further and to completely cut the bond of hypothesis with its early modern origin.

Hypothesis in Early Modern Science

Ernan McMullin

Abstract:

My contention is that hypothesis gradually emerged from the shadows of natural inquiry in the period between Copernicus and Newton. Indeed, it might be argued that one perspicuous way to define the scientific revolution (an admittedly contentious term!) would be to detach just one theme, the conception of the sort of knowledge that constitutes the "science" of nature, and to note the profound shift in it that took place roughly between 1600 and 1700. Where the Aristotelian ideal of demonstration of a deductive sort from premises ultimately seen to be true in their own right could still claim wide authority among natural philosophers as the sixteenth century ended, it would have found little support among those pursuing systematic inquiry into the physical world a century later. And central to that profound change was the increasing visibility of hypothesis, both in the first-order role it played in the day-to-day activity of astronomers and natural philosophers and in the second-order way in which that role came to be discussed by them. What makes it worthwhile, I hope, to address in short space as broad a topic as this is to make good this claim.

Tracing the role of hypothesis in early modern science could prompt two quite different sorts of questions, what I will call first-order and second-order questions. One might investigate the methods, the strategies and the forms of reasoning implicit in the work of natural philosophers of the period from Copernicus to Newton with a view to determining to what extent and in what sort of contexts they had actually relied on hypothesis in their work. This first-order inquiry responds to a contemporary understanding of the two key terms in my title: Hypothesis in early modern science. But there is a second-order kind of inquiry which is also relevant: in their own reflective discussions of how systematic inquiry into the natural world should be carried on, what we might call their philosophy of science, what role did hypothesis play for the practitioners? I intend to ask both sorts of question and to note in some cases a striking dissonance between the answers. To do this necessitates, of course, careful attention to a third kind of question:

how does the contemporary sense of the term "hypothesis" relate to the sense that the agents of that earlier time would have attached to it?[1]

1. Prolegomenon

I realize that my use of such terms as "science", "hypothesis", and "philosophy of science", in speaking of the activities of agents long gone is likely to give rise to some worries.[2] Since the contemporary meanings of these terms are reasonably clear, however, to use them in a first-order context to describe the cognitive activities of that earlier time should be acceptable if carefully done, provided first, that it be made clear from the beginning that these are not necessarily the terms in which the past agents themselves would have described what they were doing, and second that a degree of analogy in the employment of the terms be permitted. In a second-order context, where one is involved with the categories the agents themselves employed, the matter is more complex and a developed sensitivity to the differences of context and of concept is called for. In particular, care must be taken not to impute to agents of the past philosophic concerns that would have been quite foreign to them.

Fortunately, the term of main concern to us here, "hypothesis", has maintained the fairly steady generic sense of *supposition* ever since its Greek origins as something proposed, put forward, postulated. As such, it was a common term in Greek. A computer search reveals that in Aristotle's works, for example, the term *"hupothesis"* (ὑπόθεσις) or its cognates occurs no less than 187 times. In the recent Oxford edition of those works, it is translated in nearly all those instances by "supposition" rather than by the more technical-sounding term "hypothesis" (Barnes 1984). Much could be said about Aristotle's second-order account of *epistêmê* (ἐπιστήμη), in which, despite his frequent use of the term, "hypothesis", there is no hint of the provisional. On the other hand, one could point to what *we* would certainly call "hypothesis" in Aristotle's own works on the physical world, in the *De caelo* and the *Meteorology,* and even in the *Posterior Analytics.*

The issue of how one could conclude to an *epistêmê* without any element of the provisional provoked a good deal of uneasiness in the later

---

1 I am grateful to an anonymous reviewer for several of his suggestions.
2 For a recent discussion, see Jardine 2003.

Aristotelian tradition, among Zabarella and his colleagues in the school of Padua in the sixteenth century, for instance. But it is striking in retrospect that the status of hypothesis in the pursuit of *epistêmê* did not provoke more discussion that it did, testimony perhaps to the undoubted attractiveness of the Aristotelian ideal of a knowledge-claim that has no element of the provisional, as well as, of course, to the enduring authority of Aristotle himself.

In the context of inquiry into nature, the context of concern to us here, the term "hypothesis" has most often been taken to refer to a supposition put forward in a provisional manner to account for something else, what we shall call a causal (or explanatory) hypothesis. There are two elements here: a hypothesis is provisional and therefore in need of further epistemic support, and it is presented as, in some sense, an explanation. But as we shall see, there have been some variations on that central sense along the way.

My contention is that hypothesis gradually emerged from the shadows in natural inquiry, both at first-order in the practice and at second-order in reflection on the practice, in the period between Copernicus and Newton, the period of interest to us here. Indeed, it might be argued that one perspicuous way to define the scientific revolution (another contentious term!) would be to detach just one theme, the conception of the sort of knowledge that constitutes *science*, that term understood as designating the ideal of knowing, and to note the profound shift that took place roughly between 1600 and 1700. Where the Aristotelian ideal of demonstration of a deductive sort from premises ultimately seen to be true in their own right could still claim wide authority among natural philosophers as the sixteenth century ended, it would have found little support among those pursuing systematic inquiry into the physical world a century later. And central to that profound change was the increasing visibility of hypothesis, both in the first-order role it played in the day-to-day activity of astronomers and natural philosophers and in the second-order way in which that role came to be discussed by them. What makes it worthwhile, I hope, to address in short space as broad a topic as this, given the superficiality that this limit necessarily imposes, is to make good this claim.

2. Copernicus: Different Senses of "Hypothesis"

Scanning the early modern period for significant developments in regard to the role of hypothesis in the natural sciences, I plan to focus on seven distinctive developments, each associated with a major figure. The first of these figures should obviously be Copernicus, whose heliocentric proposal was classified dismissively by many as a "hypothesis." But this was "hypothesis" in a rather special sense: a mathematical formalism serving as a useful means of practical astronomical prediction but nothing more. This had been the status commonly assigned to mathematical astronomy over the previous centuries, and the term "hypothesis" had been appropriated to convey this status.

What had led the Greek-derived term to take on this rather pejorative sense? There were two reasons in particular. First, from Apollonius's time onwards, Greek astronomers had been aware that two quite different formalisms, the epicycle and the eccentric, could deliver identical predictions under certain conditions. One could not, therefore, rely on the usual criterion of accurately saving the phenomena to discriminate between them and decide which of them yields the "true" motions. More disturbing, the Aristotelian system of concentric spheres offered a plausible physical explanation of the planetary motions; the Ptolemaic system did not. Yet Ptolemy's was much the more reliable in terms of predictive accuracy and virtually replaced its rival among the astronomers who had to produce such predictions. Since it seemed obvious that both criteria could not be satisfied together, which criterion then was to be the test of truth: explanatory power or predictive accuracy? Which of the two systems portrayed the orbits in which the planets *really* move? This issue was to attract the attention of Muslim and Christian philosophers throughout the medieval period. So strong, however, was the appeal of explanatory power as the testimony of truth that the commonest resolution of the dilemma was to limit mathematical astronomy in the tradition of Ptolemy to a merely practical role and to reserve the term "hypothesis", in an admittedly special sense, to signify this (McMullin 1984, 44–48).

This, as is well known, was the challenge that Copernicus had to face. His work was to all appearances in the tradition of mathematical astronomy. But he had become convinced along the way that "the appearance of daily revolution belongs in the heavens, but the reality belongs to the earth" (Copernicus 1543 [1984], I, 10, 6a = p. 15, line 11). He set out, therefore, to establish this in a way that would, as far as possible, satisfy both of the traditional criteria and to that extent bring mathematical astronomy and physical astronomy together in a single discipline. He did not abandon the term "hypothesis" as in these circumstances, he might have been expected to do. Rather, he expressly returned to its Greek roots, equating it to a principle or basic assumption taken to be true, just the contrary of the sense that had been for so long associated with mathematical astronomy.[3] When he describes the proposition that the earth moves around the sun as a "hypothesis", this word-choice of itself ought not to be understood as suggesting that the proposition is provisional: even the observations on which his theory depends are described as hypotheses. But he does make it clear that the truth of his assertion is to be tested by the consequences drawn from it. These latter are, he says, sufficient to "make it more probable that the earth moves than that it is at rest, especially in the case of the daily rotation, as that which is most proper to the earth" (Copernicus 1543 [1984], I, 8, 7a = p. 16, lines 29–30).

These consequences would be of two quite different kinds. First were the observed planetary motions that had for long been the domain of mathematical astronomy. But these, as he well knew, would of themselves scarcely be enough. On their account alone there was little to choose between his own system and that of Ptolemy. So in Book I, he turned instead to the superior explanatory ability of the heliocentric hypothesis. In such a scheme, it follows naturally that the outer planets are brightest and therefore (he assumes) nearest when in opposition; it does not have to be an extra postulate, as it was for Ptolemy. The situation is likewise with the retrograde motions of those planets and with the relative dimensions of those motions. These follow as part of the original heliocentric hypothesis; they have to be added on by means of cumbrous epicycles to the Ptolemaic model. Likewise again with the curious bond between Venus and Mercury on the one hand and the Sun on the other: they are never found in opposite quarters of the sky. This is natural, and thus explained, in a system where they are inner planets relative to the Earth: "All these things proceed from the same cause, which is the movement of the earth" (Copernicus 1543 [1984], I, 10, 10a = pp. 21, 15–16). Ptolemy could make no comparable claim. Copernicus was clearly ahead of his rival in the cat-

3 Even the observations on which his system depends are sometimes described by him as "hypotheses." See Rosen 1939, 22–33.

egory of cause, understood here as an explanatory factor, not as efficient cause.

Historians of science in the past, Thomas Kuhn among them, have been all too ready to dismiss considerations of the sort Copernicus adduces here as "merely" aesthetic or as relying overly on a subjectivist criterion of simplicity (for a fuller discussion see McMullin 1993, 71–75). And the fact that Copernicus uses terms like "bond of harmony", "commensurability", "order", is taken to support this dismissive construal. But it is clear that for Copernicus, at least, this form of argument, citing explanatory superiority as distinct from the single empiricist criterion of saving the phenomena of observed planetary positions, carried enough epistemic weight to outweigh the strong physical arguments that still existed against taking the earth to be in motion. He overlooked, of course, the possibility, later exploited by Tycho Brahe, that another system could, unlike Ptolemy's, call on the same explanatory considerations in its favor as he had proposed. But what is significant for our theme is that he clearly viewed the conclusion that his own hypothesis is the "more probable" one as a genuine epistemic claim even though it falls short of the traditional standard of demonstration.

3. Kepler: Saving the Phenomena May Not Be Enough

In 1599, when Kepler was seeking for his own reasons to be taken on by Tycho Brahe as his assistant, he found himself in the middle of an a acrimonious battle between Tycho and the Imperial Mathematician, Nicolai Baer, better remembered by the Latin version of his name, Ursus. The dispute focused around Tycho's planetary model for which Ursus claimed priority, to which Tycho responded by charging plagiarism. Tycho then, in effect, required Kepler (who had earlier shown some sympathy for Ursus) to distance himself from the offender by composing a critique of Ursus's *Tractatus*. The latter had defended the traditional view of mathematical astronomy as no more than a useful calculating device, in this way seeking to undermine the significance of Tycho's cosmological system. In late 1600 Kepler complied, but his *Apologia pro Tychone contra Ursum* remained unpublished; the death of both protagonists, Tycho and Ursus, led Kepler to set it aside. This is a pity, since in its own way it was one of the most perspicuous discussions of the epistemic status of hypothesis in astronomy to come down to us from this period. Jardine, indeed, titles his invaluable edition of the work: *The Birth of History and Philosophy of Science* (Jardine 1984)!

Kepler begins by distinguishing between two sorts of hypothesis. A *geometrical* hypothesis proposes a mathematical formalism as a way of summarizing a certain set of phenomena; a different formalism might, however, also serve the same purpose. (He is thinking here of the eccentric and the epicycle.) A successful *astronomical* hypothesis on the other hand should state what the real planetary motions are. But how is one to establish this, since different hypotheses could yield the same observed planetary motions?

Kepler has several suggestions to offer. One hypothesis may yield the cause of a phenomenon, the other not, giving the first a definite advantage. Following Copernicus's example, he notes that the heliocentric view can explain numerous features of the planetary motions that Ptolemy was forced to postulate in an ad hoc way: the exact yearly period of one of the two circular motions that Ptolemy associates with each planet, for instance. Alternatively, "related sciences" may favor one rather than the other. But his principal advice is to be patient and give the matter time to resolve itself:

> False hypotheses, which together yield the truth by chance, do not in the course of a demonstration in which they have been combined with many others retain this habit of yielding the truth but betray themselves (Jardine 1984, 140).

One must, then, follow the career of the hypothesis over time. Some years earlier, however, in his *Mysterium Cosmographicum* (1596), he had anticipated one possible outcome of such advice:

> That which is false by nature betrays itself as soon as it is considered in relation to other cognate matters: unless you would be willing to allow him who argues [from false premises] to adopt infinitely many other false propositions and never, as he goes backwards and forwards, to stand his ground (Kepler 1596, I, 15, quoted in Jardine 1979, 157).

One has to be quick, then, to question *ad hoc* moves designed to save a hypothesis when it encounters an anomaly.

Kepler also expresses some doubt about the original supposition that there could, in practice, be different physical hypotheses that would yield *exactly* the same predictions. He questions whether one would ever:

> come across any hypothesis, whether simple or complex, which will not turn out to have a conclusion peculiar to it and separate and different

> from all the others. Even if the conclusions of two hypotheses coincide in the geometrical realm [i.e. as mathematical formalism only], each hypothesis will have its own peculiar corollary in the physical realm (Jardine 1979, 141–142).

When he set out finally in his *Astronomia Nova* (1609) to construct his own world-system, embodying his ground-breaking discovery of the elliptical shape of the planetary orbits, he had foremost in mind the challenge posed for the astronomer who would wish to claim that his were the *real* orbits. He considers the three contenders: the Copernican (as significantly amended by Kepler himself), the Tychonic, and the Ptolemaic, calling each of them "hypotheses." His argument is carefully constructed and rests on two different sorts of consideration.

First, although all three of the rivals are roughly equivalent in terms of saving the phenomena of planetary positions, the Copernican-Keplerian one in the end displays some clear advantages over the other two even there. More important in his eyes, however, is that his own hypothesis is supported by an inquiry "into celestial physics and the natural causes of the motions" (Kepler 1609 [1992], 48). In the effort to overcome the long-standing debate between different astronomical hypotheses, "none other could succeed than the one founded upon the motions's physical causes themselves, which I establish in this work." He goes on: "The eventual result of this consideration [is to show] that only Copernicus's opinion concerning the world (with a few small changes) is true, that the other two are false ..." (*Ibid.*).

The causal theories he advances seem in our eyes somewhat speculative. He argues by means of what he calls a "physical conjecture" that "the source of the five planets's motions is the sun itself," calling on a vast emission of immaterial *species*, akin to light or to the agencies involved in magnetic action, and churned into a "whirlpool" by the postulated rotation of the sun itself (Kepler 1609 [1992], 52, 61). The details need not concern us here; in his later writings, Kepler kept working on them. What is important is the emphasis he gave to causal explanation and physical theory in making the case for his own hypothesis of the earth's double motion. If he were to prove its reality, a goal that the anti-realist of today would presumably have to question, it was not enough to predict planetary positions correctly: the epistemic force of explanatory success in a stronger sense had to be brought to bear. Kepler may have been the first to articulate a crucial difference that the admission of hypothesis to respectability in science made necessary. Unlike the combination of individual intuition and rule-governed deduction of the earlier ideal of demonstration, the new sciences would require *multiple* criteria for the assessment of hypothesis, criteria that were themselves not amenable to simple rule-like definition.

4. Bacon: The Ambiguity of "Induction"

There are those who regard Bacon as a pioneer of the hypothetico-deductive model of scientific inquiry, a champion of the role of hypothesis in the natural sciences.[4] But then there are others who emphasize the certainty that Bacon attaches to the products of his method. It "derives axioms from the senses and particulars, rising by a gradual and unbroken ascent, so that it arrives at the most general axioms last of all" (Bacon 1620 [1960], I, 19). In this way it provides "not pretty and probable conjectures but certain and demonstrable knowledge." Properly conducted, the method of induction thus "leads to an inevitable conclusion" (Bacon 1620 [1960], Preface, 36, 30). From this perspective, it would seem that hypothesis would have no real role to play as induction marches steadily on.

How are these two different construals to be reconciled? Bacon's claims for the certainty of the science his method can deliver occur for the most part in the introductory section of the *Novum Organum,* and can be discounted in some part as reassurance to the reader in the face of the inevitable objections by defenders of the traditional Aristotelian account. More fundamentally, however, the two construals treat different phases in inductive inquiry. The first describes the course of inquiry and the role played by hypothesis along the way; the second defines the end arrived at. Bacon is undoubtedly over-optimistic about the effective elimination of the hypothetical element when induction concludes, as he assumes it definitively will. But this still leaves open a role for hypothesis in the conduct of the inquiry itself.

It is true that he rarely uses the term "hypothesis" himself and when he does it is in a dismissive sense. But when he speaks, for example, of "axioms" that are put forward provisionally for "trial by fire", that is, for systematic testing, the term that would best convey what he has in mind would be "hypothesis", understood in our sense, not in his. He has a great deal to say about how this testing is to be carried out; while it is going on, the axiom (or "law" or "form") remains provisional but

4 See, for example, Horton 1973, 241–278; Ducasse 1960, 50–74.

is the central element in the process of inquiry itself. It is fair to say, then, that hypothesis plays a significant role in the plan of inquiry laid out in the *Novum Organum,* both at first-order and at second-order, even if he much too readily assumes that the process of assessment rapidly converges on a single answer.

But now we come to a fundamental ambiguity in his notion of induction, one that would remain a feature of the philosophy of science until the late nineteenth century and indeed in many quarters beyond that. Under that single label two quite different sorts of inference make their appearance. In the opening sections of Book I, induction appears as generalization over observed particulars, ascending from the senses in a gradual and unbroken ascent, as he puts it, leading finally to the formulation of the most general "axiom" or law-like proposition. The purpose here is the determination of "forms" or characteristic modes of behavior: "the laws ... that govern any simple nature as heat, light, and weight" (Bacon 1620 [1960], II, 17), working up systematically from specific instances of that sort of behavior to a general assertion.

Bacon has a good deal to say about how this procedure is to be carried out. His tables of "presence, absence, and degree" (later to be appropriated by J. S. Mill and renamed "sameness, difference, and concomitant variation") show how. The goal is to discover invariable (and thus significant) correlations among observable features of the physical world. The element of testing is provided, in particular, by the determination of absence (or difference) where presence might have been expected. In this way, in the celebrated but somewhat problematic example developed in such detail in Book II, heat is determined to be (and in that sense to be explained as) a form of motion.

But Bacon himself draws attention, if only implicitly, to a limitation of this kind of inference. Opening Book II he notes that bodies can be understood in two quite different ways: as "troops of natures" (and hence accessible to induction in the sense just defined), or as configurations of imperceptible particles and processes: "For seeing that every natural action depends on things infinitely small, or at least too small to strike the sense, no one can govern or hope to change nature until he has duly comprehended ... them" (Bacon 1620 [1960], II, 6). This is no longer a matter of noting correlations in observable features: one has to reach out imaginatively to natures that are *not* observed, thus requiring a very different form of inference. The emphasis here is on efficient or agent, rather than formal cause: the underlying configuration or latent process is postulated to be the agent cause of the observable feature or features under investigation.

Even the favored example of heat as the effect of motion has to call on the motion of imperceptible particles in many contexts, such as in the heating of an iron bar, where no perceptible motion appears. But Bacon has more accessible illustrations to offer, such as the cause of falling motion, where he supposes a choice between two possible explanations. One is that the earth attracts the body, the other is that bodies just naturally fall. And to illustrate his famous "instance of the fingerpost" (later known as a "crucial", i. e. "crossroads" experiment), he proposes to take two clocks, one spring-driven, the other moved by weights, up a high steeple or down a deep mine. In the event that attraction is the cause of falling motion, it should decrease or increase as one ascends or descends and thus the weight-driven clock should alter its timing relative to the other, thus allowing one to decide between the two possible causes of the motion (Bacon 1620 [1960], II, 36). This is, indeed, a textbook case of what came to be called hypothetico-deductive inference.

Bacon's practice of grouping these very different forms of inference under the single label "induction" had far-reaching consequences, not to be resolved finally until C. S. Peirce argued forcefully several centuries later that in the analysis of scientific inference the generic term "induction" had to be supplemented by at least one, and possibly several, others (McMullin 1992). Restricting "induction" to generalization over observables, and applying one or other of two new labels, "abduction" and "retroduction", to the inference from effect back to postulated effective cause, was one (only one) of the ways in which Peirce sought to resolve the issue.[5] The label I favor is "retroduction", since it conveys the direction of effect-to-cause inference and avoids the philosophical tangles associated by now with the (not quite equivalent) alternative label "hypothetico-deductive".

Corresponding to the two types of inference, induction and retroduction, there are two different types of hypothesis, the first *nomic* since it bears on the formulation of law-like statements, the other *causal* (or explanatory), since it involves the provisional identification of the efficient cause of a given natural feature, understood as an effect. The significance of the hypothetical status is quite different in the two cases. In

5 There is a growing literature on this topic among Peirce scholars; see, for example, Hintikka 1998.

the first, the proposed law-like generalization over observed instances is put forward with the understanding (whether expressed or not) that a wider or more accurate survey might lead to the modification, not the abandonment, of the generalization. In the other case, however, attributing hypothetical status is much more significant: if a retroductive inference singles out the wrong cause it may well have to be abandoned. Not surprisingly, it is to this form of inference and the provisionality it entails that the discussion of hypothesis in the seventeenth century is almost exclusively restricted.

5. Galileo: From Hypothesis to Demonstration?

Galileo plays a curiously ambivalent role in the early story of hypothesis in the sciences. In one respect, he faces forward: at first-order, his venturesome work on the nature of the objects his telescope revealed was hypothetical from beginning to end. But in another respect, at second-order, his relatively rare remarks on what he took the ideal in knowledge of the physical world to be show that he never really gave up on the ideal of science as demonstration that he had learnt as a young teacher commenting on the *Posterior Analytics* of Aristotle.[6] The obvious question one is led to ask is how these two poles could co-exist in the same person's work. Part of the answer lies in the kind of success he enjoyed in regard to the motion of falling bodies, where he could lay claim to having, in his own words, brought forward "a brand new science concerning a very old subject" (Galilei 1638 [1974], 147 = 1904, 190), a science that could plausibly lay claim to the sought-after mark of demonstration.

The other part of the answer lies in the domain where hypothesis seemed certain to remain the order of the day. Prior to Galileo's turning his telescope to the heavens in 1609, the science of nature had been confined to inference from what the unaided human senses could present as evidence. The one-step inference from properties to nature or essence that had been central to Aristotelian science had favored a phenomenalist rendering of natures in familiar terms. Thus, for example, the elements of which all bodies were composed were themselves said to be composed of mixtures of more fundamental elements character-

6 William Wallace has, in a number of his works, traced this story in some detail. See, for example, Wallace 1984.

ized by simple sense-properties: the hot and the cold, the dry and the moist. There was nothing there, it seemed, that could not be intuitively grasped. Even where the distant heavens were concerned, the planetary spheres could be directly inferred on the basis of principle to be solid, transparent, and with an eternally stable motion in the only figure that naturally returns on itself, the circle.

But the discoveries so dramatically announced in the *Sidereus Nuncius* and the potential they opened up for further surprises were a different matter. The dark shadows on the lunar surface were now revealed in sufficient detail to suggest as a plausible hypothesis that that surface was similar to that of earth. And in his *Dialogue on Two Chief World Systems,* Galileo explicitly used the consequences of that hypothesis, some of which were spectacularly verified, as grounds for the hypothesis itself. It was likewise in the case of sunspots, where analogies with the clouds that drift over the terrestrial sky served him well. The method followed here was a two-step one. It called first on imagination and analogy (not the intuition of the Aristotelian model), and then on the systematic testing of the hypothesis by the observational consequences derivable from it. And of course, the most celebrated instance of this form of inference in Galileo's work – the one with which the *Dialogue* concludes – is where he labored hard to give a causal explanation of the earth's proposed dual motions by an appeal to its tides.

There was, of course, nothing new about this form of inference. But what was strikingly new, more effective here than in Bacon's less concrete discussion of the microworld, was the opening up of a domain of nature that challenged the familiar identifications of everyday human perception. Here only hypothesis could serve the purposes of the inquiry at hand. Only some sort of inference from observed effect to unobserved (or at least unrecognized) cause could hope to bring this distant domain into focus. And there was always the possibility, as Galileo's debate with Grassi about the nature of comets amply illustrated, that a different cause might account for the same effects as well or better. This is the epistemic challenge that has come to be called "underdetermination" in recent philosophy of science, where the known effects are insufficient of themselves to determine a single cause.

Could the new mechanics escape this challenge? On one level, yes, but only for a reason that marked this mechanics off as a special case: namely that it did not, in fact, involve causal hypothesis. Galileo himself may not have altogether appreciated how far-reaching the consequences are of the move he makes right at the beginning of his inquiry:

> The present does not seem to me to be an opportune time to enter into the cause of the acceleration of natural motion, concerning which various philosophers have produced various opinions … Such fantasies [he elsewhere includes among them the notion of an attraction between the earth and the falling body] would have to be examined and resolved, with little gain (Galilei 1638 [1974], 159 = 1904, 202).

Even though he leaves aside the *cause* of falling motion, Galileo found that he could still make a foundational contribution to kinematics, significant enough to justify his claim to have made of it "a brand new science." By applying to the real motion of fall the geometrical formalism associated with uniform acceleration, a formalism that had been known for several centuries since its first formulation by the Merton School in Oxford, he could provide an elegant formula relating space fallen with time taken. But there was an important proviso: causal "impediments", as he called them, like friction and air resistance, would have to be experimentally minimized. So his law of fall (to use a later label) was an idealization, in the sense of its not being a description of fall as it would be encountered in unadjusted nature. It would, however, be in one respect universal in scope, in the sense that it would be held to apply to all material bodies, whatever their composition.

What was the epistemic status of this law and on what sort of inference did it rest? At an early point in the discussion, Galileo has this to say of it:

> Let us take this for the present as a postulate, whose absolute truth will later be established for us by our seeing that other conclusions, built on this hypothesis, do indeed correspond with and exactly conform to experience (Galilei 1638 [1974], 164 = 1904, 208).

It is to be established in the manner common to hypothesis, he says, here and elsewhere, by the success of the observational consequences drawn from it, the consequences here being no more than further instances of the motions postulated by the law, filling in gaps in the law's coverage, as it were. His reason, he says, for saying that uniformly accelerated motion of the kind described in the formalism is in fact the motion found in the world is "the very powerful reason that the essentials successively demonstrated by us correspond to … that which physical experiments show forth to the senses" (Galilei 1638 [1974], 153 = 1904, 197).

But is this argument from consequences back to universal claim sufficient to ensure the sort of "absolute truth" he clearly craves? He hesitates at this point. When speaking of this mode of validation, he keeps slipping in cautious qualifiers like "falling just a little short of necessary demonstration," or "worthy of being conceded as if it had been proved" (Galilei 1638 [1974], 162, 164 = 1904, 205, 207). On one occasion, therefore, he has recourse to a very different line of argument that could lead to a stronger conclusion. Uniform acceleration, he suggests, is "the simplest and most evident rule" that could govern a body's fall and thus the one most likely to be adopted by Nature (*ibid.* See McMullin 1978a, 225–240). He evidently felt the need of an epistemic boost to close the gap between the sort of support that hypothesis can derive from its consequences and the status of demonstration that he clearly desires to attach to it. And there is a further potential challenge of which he shows himself to be aware, putting it in the mouth of Simplicio. How can one be sure in any given context that the causal impediments that could mar the application of the law in this particular case are in fact absent? Galileo concedes that in the end "conclusions demonstrated in the abstract are altered in the concrete," but pleads that demonstration in the abstract ought still in the circumstances be sufficient for the purposes of science (Galilei 1638 [1974], 224 = 1904, 274).

It seems clear that Galileo does derive some of his assurance (or perhaps it would be more correct to call it his hope) that his two laws of motion can claim demonstrative status, and that they cease to be merely hypothetical, from the fact that they *do* function as geometry does in the abstract. There is an implicit transfer of warrant from the operation of demonstration within the formalism to the justification of the formalism itself as a representation of the physical order. He leaves out of account the possibility that setting aside the *cause* of falling motion may well affect the adequacy of the description he gives of that motion, considered as physics and not just as geometrical formalism. So at this fundamental level, causal hypothesis is not, and cannot be, entirely overcome. In the end, a less constrained account will be needed; the problem of the cause of motion will return, and when addressed, will give rise to a very different mechanics where Galileo's laws will hold only as approximation.

6. Descartes: Hypothesis Rampant but Largely Disowned

Descartes is another transitional figure in our story, claimant on the one hand to a deductivist system based on premises that are held themselves to rest ultimately on the nature of God, but on the other hand, the imaginative proponent of a whole panoply of unobserved mechanisms that are supposed to explain how the world in all its variety actually works. At one level, he leaves no room, no need, for hypothesis; at another, he seemingly calls on it at every turn while minimizing the departure that this entails from his announced epistemic ideal (McMullin 2008a).

His earliest work, the *Regulae* of 1628, left unfinished and unpublished, came closer than did any other of that period to defending an ideal of natural knowledge that left no room for the provisional or the probable. The aim was to deal only with objects "which admit of as much certainty as the demonstrations of arithmetic and geometry," so as to arrive at a form of knowledge "incapable of being doubted" (Descartes 1628 [1985], I, Rule 3, pp. 13, 10). Only two mental operations, intuition and deduction, would be required, an echo of Aristotle though with his own very different construals of both those terms. It was a wildly optimistic vision, adequate for the geometry that had suggested it, but altogether unrealizable, of course, in the face of the complexities of the physical world.

Ten years later, in his *Discourse on the Method* (1637) this optimism was tempered but still very much in evidence:

> I showed what the laws of nature were, and without basing my arguments on any principle other than the infinite perfections of God, I tried to demonstrate all those laws about which we could have any doubt, and to show that they are such that, even if God created many worlds, there could not be any in which [these laws] fail to be observed. After that, I showed how, in consequence of those laws, the greater part of the chaos had to become disposed and arranged in a certain way which made it resemble our heavens, and how at the same time, some of its parts had to form an earth, some planets and comets, and others a sun and fixed stars (Descartes 1637 [1985], part 5, 132).

Thus, from the initial hypothesis of a chaos of particles in motion, and relying only on laws of mechanics themselves arrived at on the basis of a claimed intuitive insight into God's nature, he could set out to construct an entire physics "by demonstrating effects from causes and showing from what seeds and in what manner nature must produce them" (Descartes 1637 [1985], Part 5, 134).

The only hypothetical element Descartes allowed here was the supposition that God would have acted in this way instead of directly creating the variety of kinds, non-living and living, as the *Genesis* account of origins, taken literally, implied. This was a prudent reservation on his part, but the reader would have been in no doubt which way he himself leant. In any event, aside from that proviso, there is no room for hypothesis in this account, nor in his later discussion of the varieties of the living world. All is supposed to be deducible from the operation of the laws of mechanics over time on an unspecified initial distribution of matter.

He gives no clue how, in fact, planets, comets, and the rest *are* to be arrived at. Something else enters in at this point: what he is really saying is that all this profusion of physical variety is, in principle, *capable* of being explained by the laws of mechanics. The emphasis is on the explanatory potential of mechanics. There is no hint of testing *specific* explanatory hypotheses by consequences drawn from them. It is as though only one very general explanatory hypothesis based on mechanics could possibly fit what we know of comet, planet, or for that matter, starfish ... Remember Aristotle and the implicit assumption underlying the illustration of demonstration that he gives in the *Posterior Analytics*: nearness is not only a plausible explanation of the planet's not twinkling, one can assume it to be the *only* such.

Part Six, with which the *Discourse* ends, was written several years after Part Five, from which the quotation above was taken. By then, Descartes had evidently been worrying a good deal about this very issue. He returns to this program of deducing all physical variety from an unspecified material starting point but now recognizes what he is up against:

> Reviewing in mind all the objects that have ever been present to my senses, I venture to say that I have never noticed anything in them which I could not explain quite easily by the principles I had discovered. But I must also admit that the power of nature is so ample and so vast, and these principles so simple and so general, that I notice hardly any particular effect of which I do not know at once that it can be deduced from the principles in many different ways and my greatest difficulty is usually to discover in which of these ways it depends on them. I know of no other means to discover this then by seeking further observations whose outcomes vary according to which of these ways provides the correct explanation (Descartes 1637 [1985], Part 6, 144).

What he is saying here is that even though he is confident that he can explain in a general way the origin of such features of the universe as comets or planets by means of the laws of mechanics, he has to admit that a variety of different explanations consistent with those laws would still be open. The theme of underdetermination has once again emerged. Accordingly, the only way to select the *correct* explanation is to call on further observation in order to discriminate between the alternatives. Implicit in this is the admission that prior to this testing, all he will have is hypothesis in regard to the intermediate-level causal configurations, though the ultimate causes, the laws of mechanics, are of course taken to be known with certainty. But he remains confident that at the end of this period of testing the correct causal explanation will be found and the resultant explanatory theory will be in no way hypothetical.

Others were not so easily persuaded of this. Several of his correspondents pressed him on the issue, notably Marin Mersenne, who insisted that the hypothetical character of the explanations Descartes offered in his account of the physical world could not be overcome: probability was the best that could be hoped for. In response, Descartes showed no sign of taking his own advice about observational testing, but rather extolled the persuasive quality of the explanations originally offered, the variety of aspects of the given nature that they accounted for, and the assurance in many instances that the explanation offered is the only one that could fit the case (McMullin 2008a, 94–96). One can see here the genesis of the sort of causal hypothesis that Newton would excoriate later: a hypothesis that appealed only to a broad sort of explanatory plausibility, without any real attempt to derive from it precise enough predictions to admit of genuine test.[7]

Six years later, in 1644, Descartes produced his major work on the natural world, *The Principles of Philosophy*. Undaunted, we find him still prefacing that work with the claim that from the principles he is about to lay down, all the other things he is about to propose can be deduced (Preface to the French ed. of Descartes 1644 [1985], 183), assimilating the explanatory relation of effect-to-cause to the deductive relation of cause-to-effect, a move with which we are by now familiar.

Here, however, he tries out a new argument:

> Suppose, then, that we use only principles which we see to be utterly evident, and that all our subsequent deductions follow by mathematical reasoning: if it turns out that the results of such deductions agree accurately with all natural phenomena, we would seem to be doing God an injustice if we suspected that the causal explanations discovered in this way were false. For this would imply that God had endowed us with such an imperfect nature that even the proper use of our powers of reasoning allowed us to go wrong (Descartes 1644 [1985], III, sec. 43, p. 225).

---

7 This is the main reason why Descartes cannot plausibly be taken to be an originator of the hypothetico-deductive method, as has sometimes been suggested.

Ingenious, but his critics could still of course point to two flawed assumptions in this argument: first, that agreement with *all* natural phenomena has in fact been reached, and second, that this manner of overcoming the hypothetical character of the explanations he proposes *is* indeed "a proper use of our powers of reasoning." At the end of the work he struggles on page after page with this same issue, made all the more urgent by his frequent appeal to imperceptible particles in explaining the properties of material bodies. And then there is what amounts to a complete concession: "With regard to the things which cannot be perceived by the senses, it is enough to explain their *possible* nature, even though their actual nature may be different" (Descartes 1644 [1985], IV, sec. 204, p. 289).

On the last page of the work, however, he makes a plea to his readers: "Perhaps even these results of mine will be allowed into the class of absolute certainties, if people consider how they have been deduced [here is that term again] in an unbroken chain from the first and simplest principles of human knowledge" (Descartes 1644 [1985], IV, sec. 206, p. 290). He never really gave up! In no other writer of that century, perhaps, did the discrepancy between second-order reflection and first-order practice in the matter of hypothesis cause so prolonged an interior struggle.

As the seventeenth century wore on, later natural philosophers in the Cartesian tradition tended to draw a sharp distinction between the science of mechanics, still perceived as having a priori and demonstrative status, and a "physics" that postulates the hidden configurations of ether or of unseen particles responsible for the phenomena one actually observes.[8] The latter in its tentative mode of inferring from effects to causes can, they admitted, never be better than probable. Hypothesis is thus acknowledged to be a permanent feature of physics, though not in mechanics; it is not just a temporary expedient to be tolerated only insofar as demonstration is perceived to be on the way.

Christiaan Huygens struggled mightily with the issue of hypothesis bequeathed by Descartes. He was willing to retain the Cartesian idea of a set of principles that allowed a special epistemic status to mechanics. But the complexity of the physical process of light led him to allow that hypothesis was the best that could be hoped for in determining the underlying physical configurations responsible for that process, and that the way to assess the epistemic credentials of a hypothesis

8 See, for example, Régis 1690: 1: 277, cited in Clarke 1980, 298–301.

was through its observational consequences. In a celebrated passage in the Preface to his *Traité de la lumière* (1690), he remarks that in physics, as distinct from mechanics, this is the best one can hope for and it affords, in favorable cases, "a degree of probability which very often is scarcely less than complete proof." And he notes two criteria that, if satisfied, afford "strong confirmation" – first, the variety of the experimental consequences accounted for, and second, the ability of the hypothesis to "imagine and foresee new phenomena" that are then verified. Here, once again, is the hint we already found in Kepler: what lends particular epistemic weight to a theory is its success in opening up new domains, domains that were not part of the original evidence-base.

7. Boyle: The Requisites of a Good Hypothesis

This leads naturally to consideration of Robert Boyle, the person who perhaps saw most clearly that a new form of non-deductive (what we have called retroductive) inference was now becoming widely used, if as yet not so widely discussed at second-order, and that some account of how it should be regulated was called for. His own work in pneumatics involved testing causal hypotheses about the "spring" of the air, and his extensive studies of chemical reactions, in the light of the corpuscular hypothesis he shared with a growing majority of natural philosophers, invited speculative guesses as to what underlying configurations at the corpuscular level might be responsible.

In a short unpublished paper, "The requisites of a good hypothesis," Boyle listed six criteria that should govern what he called a "good" hypothesis, and added four more that would, if satisfied, qualify a hypothesis in natural philosophy as "excellent" (Westfall 1956). A *good* hypothesis should be internally consistent, should accord with the phenomena under consideration, and should not contradict other phenomena. An excellent hypothesis should, besides, be simple, not forced, enable the making of new predictions that can be tested, and should be either the only hypothesis that can explain the phenomena or at least explains them better than any other.

In the "experimental philosophy" that he aims to define in terms of his own practice, the purpose of a hypothesis is

> to render an intelligible account of the causes of the effects ... without crossing the laws of nature. The more numerous and the more various the particles are whereof some are explicable by the assigned hypothesis ... the more valuable is the hypothesis and the more likely to be true. For it is much more difficult to find an hypothesis that is not true which will suit with many phaenomena, especially if they be of various kinds, than but with a few (Boyle 1772 [1965], 4: 234, cited in Sargent 1995, 58).

Boyle makes it quite clear that in the science of nature, as he sees it, hypothesis has an honored place. Demonstration is no longer the norm: it is now acceptable to speak in terms of likelihood.

It was one thing to test a hypothesis in pneumatics like the one of the "sea of air" that weighs down on the mercury surface. But as Boyle himself emphasizes, tracing chemical properties of the sort that his experiments enabled him to classify back to specific underlying configurations of unseen corpuscles is a very different matter. It was inherently speculative since it did not, as yet at least, permit serious test. Yet this was, in effect, the only sort of causal hypothesis permitted by the mechanical philosophy in the realm of chemistry. In his work as a "skeptical chemist", his hypotheses, he must have known, would never reach the rating of "excellent."

As an inheritor of the Baconian tradition, Boyle focused his researches largely on the sort of inductive generalization that Bacon had so clearly envisaged. But he was also impressed by the potential of the sort of causal hypothesis that the corpuscular philosophy called for and the need to regulate it in order to avoid Cartesian excess (Sargent 1995). This increasingly accepted view of Nature offered the challenge of devising hypotheses that would stretch back from observed properties and processes to the underlying shapes, sizes and motions of the corpuscles that were deemed to be their cause. But it would be many a year before that challenge would begin to be met. Indeed, John Locke, one of the most perceptive commentators on the natural philosophy of his day, was convinced that the barriers facing hypothesis-testing of this sort were insurmountable, and that what he called "a science of bodies", in the classical sense of the term "science", was in consequence forever out of reach in this realm of inquiry. One can fall back on analogy in guessing at appropriate corpuscular causes for given observed properties but the best that could ever be achieved in this regard could never escape what he called "the twilight of probability" (Locke 1690, IV, xiv, 2).

But probability or likelihood is already an achievement, especially if this is the best that can be done in this sort of hypothesis-driven inquiry into causes themselves unobserved. Hypothesis here is not just a question posed, it is an answer given, though admittedly not an answer that would pass muster were demonstration or finality to be insisted on. What is emerging here is a new idea of what constitutes knowledge or science appropriate to a new and evidently fruitful sort of inquiry. Because of the vulnerability of the effect-to-unobserved-cause form of retroductive inference, it has to be recognized as provisional yet the source of genuine insight all the same. As the protagonists of this sort of inference, from Kepler to Huygens, had seen, it has not the logical simplicity of either deduction or inductive generalization. Nor is it simply hypothetico-deductive if this be taken to mean that a single form of validation suffices, namely testing deductions from the hypothesis against observation. The criteria that hypothesis must satisfy in this arena are much more diverse and, it must be said, less easy to specify (McMullin 2008b).

A hypothesis that goes an acceptable way towards satisfying these criteria has at this point evidently moved up a level in epistemic terms. While remaining hypothetical, it is no longer *just* a hypothesis. The term "theory", with its root meaning "understanding", was already available and at this time was gradually beginning to acquire the additional overtone of "provisional", something less than demonstrated, to fit the new recognition of what was achievable in those forms of inquiry that sought to go beyond the immediately observable to the causal structures responsible for what was observed.

8. Newton: From Hypothesis Back to Query

With the support of Boyle, Locke and Huygens, it should have seemed that the place of hypothesis as an integral element in science proper would have been secure. But that was not to be. As the seventeenth century ended, the great authority of Isaac Newton was responsible for a surprising reversal. The celebrated passage in the General Scholium added to the second edition of his *Principia* (1713) comes immediately to mind:

> For whatever is not deduced from the phenomena is to be called an hypothesis, and hypotheses, whether metaphysical or physical, whether of oc-

> cult qualities or of mechanical, have no place in experimental philosophy. In this philosophy, particular propositions are inferred from the phenomena, and afterwards rendered general by induction (Newton 1687 [1972], General Scholium).

What can this limiting definition of the "experimental philosophy" mean? Newton could not have been unaware of the second-order acceptance of hypothesis on the part of some of his most respected predecessors and contemporaries. Partial explanations come readily to mind. His early exposure to the profusion of vaguely plausible but untested and often untestable causal hypotheses in the Cartesian tradition certainly played a part. He explicitly sought to separate his own work as sharply as possible from the ether vortices and screwlike pores of Descartes's cosmology. Another reason sometimes advanced for his antipathy to the admission of hypothesis into science proper was his extreme sensitivity to the slightest suggestion of epistemic defect in his own work. To him, "hypothesis" implied "provisional," so that to characterize any part of his work as hypothesis he tended to find offensive. Allied to that was his strong aversion to what he once called "troublesome and insignificant disputes" – the sort of disputes that the admission of hypothesis into respectable science would inevitably encourage (Cohen 1958, 179).

But something much deeper than this was at work here, and in concluding it is important to ask what that may have been in order to understand what happened to the philosophy of science in the century that followed (McMullin 2001). The full title of his great work, *The Mathematical Principles of Natural Philosophy*, gives us a clue. He will confine himself to *mathematical* principles only, he asserts. And everyone, of course, knew that in mathematics strict demonstration was possible and hypothesis had no place other than as a starting-point. In the opening pages of the *Principia*, he emphasizes that he is not attempting to provide a *physical* explanation of motion that would (as he knew) inevitably involve causal hypothesis: "I here design only to give a mathematical notion of those forces, without considering their physical causes and seats" (Newton 1687 [1972], Def. 8). And a few lines later, he insists that the "forces" he is talking about are to be understood "mathematically," not "physically." The reader is not then to imagine, he warns, that he is taking on himself the task of specifying "the causes or the physical reason" of the motions he discusses, or that forces are to be attributed "in a true and physical sense" to centers that he describes as "endued with attractive powers."

What enabled Newton to set aside the search for agent-causes of a traditional sort was a peculiarity of the new mechanics he was proposing. Its key concepts, *force*, *attraction*, *gravity*, were significantly ambiguous. They could be understood in a descriptive sense as specifying no more than a *disposition* to motion of a specific kind in a designated material context: "With the sun at rest here and Mars in stable motion there, Mars will move in a certain way." But the terms themselves suggest agency: the Sun is said to *attract* the planets; force is said to *cause* motion to be accelerated. So the impression is given of a regular causal explanation, of something more than just dispositional description. Leibniz picked up on the apparent equivocation and argued that no real explanation had been given, so that Newton's mechanics did not, in his view, qualify as science proper.

What led Newton to restrict himself to the "mathematical" (and thus non-causal) aspects of motion, thus leaving himself open to Leibniz's reproach, was that the way to a conventional causal explanation of gravitational motion appeared to be blocked (McMullin 1978, 75–106). He had himself shown that a mechanical ether could be ruled out because of the drag it would cause on planetary motion. And he was persuaded, as virtually every natural philosopher since Aristotle's time had been, that action at a distance could be excluded a priori. The sun could not act on a planet without some sort of causal connection through the surrounding medium. Might God be acting directly, perhaps, without any physical intermediary? But that would lead physics out of the realm of testable claim. So what was left? The only alternative left him at that point was to do just as he in fact did, and simply bracket out the "physical" issue of causal agency entirely, leaving that puzzle to a later generation. Yet he was convinced that he had hit upon a new science of motion, something much more than a mere Ptolemaic description that merely fitted the observations.

Had he left himself without an explanation of any sort, as his critics charged? He still sometimes described what he had achieved in the *Principia* as deducing the "causes" of motion from their effects, even though he had admitted elsewhere that "the cause of gravity I know not." What he was implicitly developing here was in fact a novel form of quasi-causal explanation that one might term "dynamic" explanation, somewhere between traditional causal explanation which specified the nature of the agency involved and inductive description of the motion in noncommittal terms.[9] Terms like "force" and "attraction" denote an agency of a common kind at work in all these varied instances of motion, terrestrial and celestial, thus making a unified and quantitative treatment of motion possible. How the agency operates is left unspecified, but there is an implicit promise that whatever the nature of the agency is, it will bind earth and sky in a common pattern.

So, *pace* Leibniz, it *does* qualify as explanation, admittedly of a weak, but nonetheless real kind. And by being weak it had, in Newton's eyes, the inestimable advantage of avoiding the need for a more specific causal hypothesis. And so, in all good faith, he could exclaim on the closing page of the *Principia*, "I feign no hypotheses," no causal/explanatory hypotheses, that is. There was nothing provisional, in his own eyes at least, about the axiomatic structure composing the eight Definitions and the three "Laws or Axioms" that preface the work. The specific laws of force governing particular material contexts were another matter. As a product of inductive generalization "made general by analogy," he would eventually allow of such a law that: "If no exception occurs from phenomena, the conclusion may be pronounced generally. But if at any time afterwards any exception shall occur from experiments, it may then begin to be pronounced with such exceptions as occur" (Newton 1717 [1952], 404). This was, of course, to concede in effect their status as hypothesis, but only (in our terms) as nomic hypothesis, not as the kind of causal hypothesis that was on no account to be "feigned."

What about optics, the other area to which so much of Newton's experimental effort had been devoted? Here the contrast between second-order rejection and first-order utilization of hypothesis is once again as remarkable as it had earlier been in Descartes's work in optics. But Newton's way of handling it was quite different from that of his predecessor. The *Opticks* (1704, with its later editions in 1706 and 1717), where he finally gathered together the results of his earlier optical inquiries, divides rather sharply into two parts. The first contains fairly straightforward arguments from experimental results to inductive laws, but when the phenomena of color make their appearance, Newton was forced to appeal to explanatory hypotheses of all sorts to account for some of the phenomena he had discovered: bodies bending light-rays at a distance, vibrations in the light-rays themselves, the famous "fits of easy reflection and easy transmission," and more.

9 See "Dynamic explanation" in McMullin 2001, 295–299.

The contrast between the *Principia* and the *Opticks* in regard to causal hypothesis could not have been sharper. The reason is clear: the sort of bracketing out of underlying causal structure that had been possible in the *Principia* while still permitting substantive results to be achieved, was simply not available in optics once the phenomena of color came under scrutiny. The physical phenomena of color are themselves too causally complex in their generation to permit hypothesis-less treatment. And, of course, these were the sorts of complex phenomena that Huygens and Boyle had earlier had in mind. In short, in the basic mechanics of the *Principia* Newton had created the only science in which the banishment of hypothesis that he proclaimed with such force could be made at all plausible.[10]

Newton's mode of handling the apparent second-order discrepancy of allowing the generous use of hypothesis in optics was to qualify these hypotheses as "queries," questions to be asked but not to be understood as in any real sense answers to be given, with their own appropriate epistemic criteria. From the perspective of mechanics as he had circumscribed it, all explanation elsewhere that appealed to unobserved causes was to be labeled "query". Only when hypothesis was understood as a query worth following up, but no more than a query, was it to be admissible to temporary status in science proper. But a query has no epistemic standing in its own right whereas a hypothesis may well have. Boyle's concern to define criteria that would qualify a hypothesis as having a degree of likelihood, and the insistence on the part of Huygens and Locke on the genuineness of the probability that in favorable cases could be assigned to hypothesis, would be effectively challenged by Newton's use of the devaluing term "query."

Newton's legacy in regard to the notion of hypothesis was thus a troublesomely ambivalent one. Part of the problem was his own ambiguous use of the term "hypothesis" itself.[11] But there can be no mistaking the growing hostility to what he called "hypothesis" in his later work (Cohen 1966) at the very time that "queries" were making up an increasing part of the successive editions of the *Opticks*. And yet the "queries" were clearly hypotheses in at least the broad sense of questions

10 That its axiomatic preface might itself qualify one day as provisional and in that sense, at least, as hypothetical would scarcely have occurred to him (McMullin 2002).

11 Koyré distinguishes between five different uses in Newton's writings (Koyré 1965, 40).

posed, but more significantly in the sense of theories for which there was already at least some amount of evidence, notably in the case of his corpuscular theory of light. Which advice was one to follow? Was one to exclude hypothesis from "experimental philosophy" entirely, as the author of the *Principia* had ruled? Or was one to ignore this ruling, as the author of the *Opticks* had evidently done? This was the issue that would surface over and over in the century that followed.

9. Afterword

A brief indication of how the next generation responded to this situation will have to suffice in closing. Those, like Euler and Lagrange, who carried on the work of the *Principia* turned to a mathematical reworking of the original framework, to a "rational mechanics," as it came to be called. Their criteria were primarily simplicity and elegance, and in their efforts they were eminently successful. There was little suggestion of hypothesis about this work; it was primarily understood as exploration of a mathematically expressed system that already carried complete conviction as an account of physical motion. The ideal of science that it embodied suggested closure, an absence of provisionality. Applying it to concrete situations might encounter the sorts of limitations that one had to expect when inductive generalization had to be relied on. But the framework itself rapidly came to be regarded as science in the fullest sense.

To illustrate how seriously Newton's ban on hypothesis was taken by some, there would be no clearer example than Thomas Reid who in his extensive writings on the nature of perception and of the "intellectual powers," as well as in his philosophy of science generally, advocated an absolute ban on "conjecture," "hypothesis," and "theory" (he tended to equate the three). A "query" could be tolerated but only observation and inductive generalization should govern belief. Citing Newton as his authority, he denied that even a single agent cause can be discovered in natural philosophy; all that can be known is an invariable succession of observed events. Hypothesizing hidden causes as explanations for what is observed is to invoke fictions.[12]

12 Reid's fulminations against hypothesis, which were prompted in part by undoubted excess in that respect in the psychology of his day, may not have had much influence among those of his contemporaries who were engaged

Hume's critique of induction raised doubts about the purportedly inductively based Newtonian system, doubts that Immanuel Kant set out to allay later in the century in his astonishingly original metaphysical system. There is considerable disagreement among Kant scholars as to the importance of this motive to Kant's transcendental project generally. Whatever of this, it is fair to say that Kant powerfully reinforced the image of the Newtonian system as definitive, as in no danger from further empirical discovery. In his *Metaphysical Foundations of Natural Science* (1786), he claimed to derive Newton's laws of motion a priori from an analysis of the conditions required for the possibility of experience, given merely the empirical concept of matter as the spatially movable. Newtonian mechanics thus in his eyes could now claim the status of *real* science, setting it firmly apart from epistemically lesser efforts elsewhere in natural inquiry. In his later years, Kant came to recognize (as we now know from his *Opus Postumum*) that these latter efforts, in chemistry particularly, could scarcely be excluded any longer from epistemic status entirely. But his discussions of how they might be converted from empirically-based hypothesis into what he could accept as science proper, the conversion that his transcendental method had made possible in the case of Newtonian mechanics, never came to anything.

So much, then, for the legacy of the *Principia*. But, turning back in time, a quite different sort of inquiry looked instead to the experimental character of the *Opticks* and its tolerance for query (understood now as testable hypothesis), as well as back to a rich tradition dating back to the like of Boyle and Huygens. The queries of the *Opticks* themselves prompted further developments in chemistry, attempting (not very successfully) to explain, for example, chemical affinities in Newtonian terms as due to attractions between clusters of attracting particles. Georg Stahl explained combustion and eventually a range of other phenomena as the emission of a substance he named phlogiston. After the invention of the Leyden jar, static electricity was widely studied, giving rise to a profusion of theories about its nature and transmission. Daniel Bernoulli derived Boyle's Law from the hypothesis that gases consist of enormous numbers of tiny corpuscles, the first empirical verification of that venerable hypothesis. Leonhard Euler rejected Newton's corpuscular account of the propagation of light despite all of the verified conse-

in physical inquiry, but the distrust of theory that was a prominent part of his "commonsense philosophy" found a long-lasting resonance in some segments of the general public, particularly in the U.S.

quences that Newton had drawn from it, formulating instead a wave theory involving vibrations of an ether. Hypotheses began to be formulated about the rock layers that lie bare at the earth's surface, leading finally to the far-reaching theories of James Hutton.

The eighteenth century was a time of intense practical activity also in other fields like mining, metallurgy, medicine, pharmacy, leading to the formulation of innumerable inductive generalizations about the characteristics of the physical world at every level. Inevitably, this would prompt hypotheses about the sources of these characteristics and about their link with one another. With the advent of a new century, and the triumphs of Lavoisier and Dalton in chemistry as well as of Young and Fresnel in optics, hypothesis in the form of explanatory theory had finally come into its own as an indispensable part of the sciences, even if this, in effect, entailed a redefinition of what counted as science.

Bibliography

Bacon, Francis (1620 [1960]), *The New Organon and Related Writings*, transl. J. Spedding et al. Fulton H. Anderson (ed.), New York: Bobbs-Merrill.

Barnes, Jonathan (ed.) (1984), *The Complete Works of Aristotle: The Revised Oxford Translation*, vol. 2. Princeton: Princeton University Press.

Boyle, Robert (1772 [1965]), *The Works of the Honourable Robert Boyle*, 6 vols. Thomas Birch (ed.). Repr. Hildesheim: Olms.

Clarke, Desmond M. (1980), "Pierre-Sylvain Régis: A paradigm of Cartesian methodology", *Archiv für Geschichte der Philosophie* 62: 289–310.

Cohen, Bernard (ed.) (1958), *Isaac Newton's Papers and Letters on Natural Philosophy and Related Documents*. Cambridge, Mass.: Harvard University Press.

— (1966), "Hypotheses in Newton's philosophy", *Physis* 8: 163–184.

Copernicus, Nicolaus (1543 [1984]), *De revolutionibus libri sex*. Critical ed. H. M. Nobis and B. Sticker (eds.) (Vol. 2 of the *Gesamtausgabe*) Hildesheim: Gerstenberg.

Descartes, René (1628 [1985]), *Rules for the Direction of Mind*, transl. D. Murdock, in J. Cottingham, R. Stoothoff and D. Murdock (eds.), *The Philosophical Writings of Descartes*, vol. 1. Cambridge: Cambridge University Press, 9–78.

— (1637 [1985]), *Discourse on the Method*, transl. R. Stoothoff, in J. Cottingham, R. Stoothoff and D. Murdock (eds.), *The Philosophical Writings of Descartes*, vol. 1. Cambridge: Cambridge University Press, 111–151.

— (1644 [1985]), *The Principles of Philosophy*, transl. J. Cottingham, in J. Cottingham, R. Stoothoff and D. Murdock (eds.), *The Philosophical Writings of Descartes*, vol. 1. Cambridge: Cambridge University Press, 174–294.

Ducasse, Curt J. (1960), "Francis Bacon's philosophy of science", in R. Blake, C. J. Ducasse and E. H. Madden (eds.), *Theories of Scientific Method: The Renaissance through the 19th Century*. Seattle: University of Washington Press, 50–74.

Galilei, Galileo (1638 [1974]), *Two New Sciences: Including Center of Gravity* and *Force of Percussion*, transl. S. Drake. Madison, Wisc.: University of Wisconsin Press.

—— (1904), *Le opere di Galileo Galilei: edizione nazionale*. 20 vols. 1890–1909. Florence: Barbera. Vol. 8: 1904.

Hintikka, J. (1998), "What is abduction? The fundamental problem of contemporary epistemology", *Transactions of the Charles S. Peirce Society*, 74: 503–533.

Horton, Mary (1973), "In defence of Francis Bacon", *Studies in the History and Philosophy of Science* 4: 241–278.

Huygens, Christiaan (1690 [1912]), *Treatise on Light*, transl. S. P. Thompson. London: Macmillan.

Jardine, Nicholas (1979), "The forging of modern realism: Clavius and Kepler against the Sceptics", *Studies in the History and Philosophy of Science* 10: 141–173.

—— (1984), *The Birth of History and Philosophy of Science: Kepler's A Defence of Tycho against Ursus with Essays on its Provenance and Significance*. Cambridge: Cambridge University Press.

—— (2003), "Whigs and stories: Herbert Butterfield and the historiography of sciences", *History of Science* 41: 125–140.

Kepler, Johannes (1596), *Mysterium Cosmographicum*. Tübingen: Georg Gruppenbach.

—— (1609 [1992]), *New Astronomy*, transl. W. H. Donahue. Cambridge: Cambridge University Press.

Koyré, Alexandre (1965), *Newtonian Studies*. Chicago: University of Chicago Press.

Locke, John (1690), *An Essay Concerning Human Understanding*. London.

McMullin, Ernan (1978), *Newton on Matter and Activity*. Notre Dame, Ind.: University of Notre Dame Press.

—— (1978a), "The conception of science in Galileo's work", in R. E. Butts and J. C. Pitt (eds.), *New Perspectives on Galileo*. Dordrecht: Reidel, 209–257.

—— (1984), "The goals of natural science", *Proceedings and Addresses of the American Philosophical Association* 58: 37–64.

—— (1990), "Conceptions of science in the Scientific Revolution", in D. C. Lindberg and R. S. Westman (eds.), *Reappraisals of the Scientific Revolution*. Cambridge: Cambridge University Press, 27–92.

—— (1992), *The Inference That Makes Science*. Milwaukee: Marquette University Press.

—— (1993), "Rationality and paradigm change in science", in P. Horwich (ed.), *World Changes: Thomas Kuhn and the Nature of Science*. Cambridge, Mass.: MIT Press, 55–78.

—— (2001), "The impact of Newton's *Principia* on the philosophy of science", *Philosophy of Science* 68: 279–310.

— (2002), "The significance of Newton's Principia for empiricism", in M. J. Osler and P. L. Farber (eds.), *Religion, Science, and Worldview: Essays in Honor of Richard S. Westfall.* Cambridge: Cambridge University Press, 33–59.
— (2008a), "Explanation as confirmation in Descartes's natural philosophy", in J. Broughton and J. Carriero (eds.), *Blackwell Companion to Descartes.* London: Blackwell, 84–102.
— (2008b), "The virtues of a good theory", in *The Routledge Companion to the Philosophy of Science,* S. Psillos and M. Curd (eds.), London: Routledge, 498–508.
Newton, Isaac (1687 [1972]), *Philosophiae Naturalis Principia Mathematica*, 3rd ed. A. Koyré and I. B. Cohen (eds.) Cambridge, Mass.: Harvard University Press.
— (1717 [1952]), *Opticks: or, A Treatise of the Refractions, Inflections & Colours of Light.* Based on the 4. ed., London 1730. New York: Dover.
Régis, Pierre-Sylvain (1690), *Système de Philosophie,* 3 vols. Paris: D. Thierry.
Rosen, Edward (1939), "The views of Copernicus concerning the nature of astronomical hypothesis", in *Three Copernican Treatises*, transl. with introd. and notes E. Rosen. New York: Dover, 22–33.
Sargent, Rose-Mary (1995), *The Diffident Naturalist: Robert Boyle and the Philosophy of Experiment.* Chicago: University of Chicago Press.
Wallace, William (1984), *Galileo and His Sources.* Princeton: Princeton University Press.
Westfall, R. S. (1956), "Unpublished Boyle papers relating to scientific method", *Annals of Science* 12: 63–73, 103–117.

Experience and Hypotheses: Opinions within Locke's Realm

Rainer Specht[1]

Abstract:

It is a wide-spread opinion in Locke and his circle that men are able to perceive sensible alterations in bodies, whereas the essences and procedures of natural causes and effects are concealed to them. Spirits cannot be perceived by us and visible bodies are systems of smallest particles whose figures and movements are inaccessible to our senses. It is therefore reasonable to be content with observations, experiments and classifications relating to appearances and to deal with speculative systems of natural philosophy only insofar as they suggest new observations and experiments. It is true that natural histories do not procure any general and necessary truths, but they allow useful inventions for the improvement of human life. Especially in post-Kantian Germany, Locke used to pass for a simple psychologist who confined himself to the presentation of facts without being interested in the philosophical quaestio juris. One may doubt, however, if this was right. For there are some statements in Locke's "Essay" concerning the situation of man in his 'state of pilgrimage' and about the reasons why God dimensioned man's senses so scantily. Modern interpreters of Locke have tended to disregard these statements, because they cannot be justified empirically. But for Locke, they seem to have had the function to show that it is reasonable and in accordance with God's will to be content with natural history and to distrust speculative systems. So these texts can be seen as constituting an effort by Locke to take the quaestio juris into consideration.

The concepts of "empiricism" and "empiricist" only came into use during Kant's time. He characterised an empiricist as a philosopher who thinks that the modes of 'knowledge through pure reason' (*Vernunfterkenntnis*) are derived from experience.[2] Since then it has become a commonplace to characterize Locke with a term that he would most probably not have understood. After Kant many things have changed and a term that can be used both for Locke and Carnap has acquired some vagueness. Locke and his friends believed that our knowledge of nature – of the corporeal and the spiritual kind – comes in two parts: The first

1 Abridged and revised version of an article in German: "Erfahrung und Hypothesen. Meinungen im Umkreis Lockes." *Philosophisches Jahrbuch 88* (1), 1981, 20–49.

2 Kant, *Critique of Pure Reason*, A 854, B 882.

is *natural history* which serves for the description and classification of phenomena. The second is *natural philosophy* which is used for their systematic explanation. Both they together form what we today call natural science. Since good natural philosophy is based upon experience, it is also called *experimental philosophy*. The version of experimental philosophy that is discussed in the following consists, besides empirical generalisations and general explanations of phenomena, of propositions about God and man – that is, propositions which we would classify today as metaphysical.

Thomas Sydenham, a great physician of the 17th century, said, for example, that the deeper causes of things are hidden from us through divine resolution. It is true that God has ordered nature in its tiniest aspects,[3] but we do not know the inner essence of things and have no intuitive knowledge of them. It is true that we can observe how nature acts through visible external causes, but we will never be able to grasp her effects.[4] For this reason, nobody will ever become a philosopher in the strict sense.[5] Sydenham uses these metaphysical convictions in order to justify the view that we will not advance through speculation but only with sensory experience. Systems resting on fantasies are not only nonsensical,[6] but also detrimental: they divert from diseases, exaggerate or fabricate symptoms and are not geared to truth but to wool-

---

3 "... supremum illud Numen, cujus vi producta sunt omnia, et a cujus nutu dependent, infinita sua sapientia sic disponit omnia, ut ad opera destinata se certo quodam ordine atque methodo accingant; neque frustra quidquam molita, neque nisi, quod optimum est, ac toti Rerum fabricae, suisque privatis naturis maxime accomodatum, exsequentia ...". (Thomas Sydenham, *Pestis Annorum 1665 et 1666*, in Sydenham 1741, 129)

4 "But proud man, not content with that knowledge he was capable of and was useful to him, would needs penetrate into the hidden causes of things, lay downe principles and establish maximes to him self about the operations of nature, and then vainely expect that Nature, or in truth God him self, should proceede according to those laws his maximes had prescribed him. Whereas his narrow weake facultys could reach noe farther then the observation and memory of some few effects produced by visible and externall causes but in a way utterly out of the reach of his apprehension ..." (from Sydenham's *De arte medica*, as quoted in Dewhurst 1966, 81–82).

5 "*Philosophus* vero, saltem pro augustiori hujus nominis majestate, nemo Mortalium evadet umquam" (Th. Sydenham, *Tractatus de hydrope*, in Sydenham 1741, 504).

6 "His Natura cum ingenii modum largita est, quo de illa docte nugari queant; at prudentiam non dedit, qua intelligant, se eam haud aliter scire posse, nisi experientia Indice ...." (*Ibid.*, 503–504)

gathering.[7] Speculations about natural bodies that do not improve anything, nor make it faster or less difficult, or which do not lead to new beneficial inventions are a waste of time.[8] Accurate histories of diseases and therapies, however, enable precise classifications, because they are not based on speculative systems, but on observable signs and characteristics.[9] Those signs have to be recorded painstakingly, but all systems must go, because systems only cause prejudice.[10] One has also to distinguish between stable phenomena and occasional ones that have to be attributed to special circumstances like temperament and age of the patient or to the chosen therapies.[11] Occasionally, Sydenham surmises that if there were mature natural histories one could have a system that would guide practice. Such systems would, however, not rest on speculation but on assured phenomena and be ultimately derived from practical experience.[12]

7 "... utpote [auctores] quorum scripta fere omnia *Hypothesibus* innituntur, quas peperit lascivientis ingenii et phantasiae luxuria; ipsaque adeo Morborum Phaenomena (in quibus eorumdem Historia vertitur) prout ab iis describuntur, ex eadem Hypotheseωs officina prodeunt; quin et ipsa Praxis, qua Morbos aggrediuntur (quod humani generis Pestis est certissima et pernicies) ad hujusmodi Postulata componitur, non ad rei veritatem" (Th. Sydenham, *Epistola* II, in Sydenham 1741, 326).

8 "... all speculations in this subject [the knowledg of naturall bodys] however curious or refined or seeming profound and solid, if they teach not their followers to doe something either better or in a shorter and easier way then otherwise they could, or else lead them to the discovery of some new and usefull invention, deserve not the name of knowledg, or soe much as the wast time of our idle howers to be throwne away upon such empty idle phylosophy" (Th. Sydenham, *De arte medica*, as quoted in Dewhurst 1966, 83).

9 Th. Sydenham, *Praefatio*, in Sydenham 1741, 13–14.

10 "Porro autem in scribenda Morborum Historia, seponatur tantisper oportet quaecumque *Hypothesis Philosophica*, quae scriptoris judicium praeoccupaverit; quo facto tum demum Morborum *Phaenomena* clara ac *naturalia*, quantumvis minuta, per se adcuratissime adnotentur ..." (Th. Sydenham, *Praefatio*, in Sydenham 1741, 14–15).

11 Sydenham, *Praefatio*, in Sydenham 1741, 15–16.

12 "Quamvis autem *Hypotheses* speculationibus Philosophicis innixae futiles sint prorsus, cum nemo hominum scientia intuitiva praeditus sit, qua fretus Principia queat substernere, quibus mox superstruat; attamen si Hypotheses ab rebus ipsis fluant, ex eis tantum observationibus natae, quas Phaenomena Practica et Naturalia suggerunt, stabiles manent et inconcussae: ita ut: licet Praxis Medica, si scribendi ordinem respicias, ex Hypothesibus orta videatur; nihilominus ipsae Hypotheses, si modo solidae fuerint ac genuinae, Praxi quadamtenus originem debeant" (Th. Sydenham, *De Hydrope*, in Sydenham 1741, 493).

Sydenham is important for Locke's development. Locke had already occupied himself for ten years with medicine when he met Sydenham in 1677 and established a narrow professional and personal relationship with him. Several of Sydenham's texts are handed down in Locke's handwriting, e.g. *De arte medica, Anatomia*, and parts of former sketches of the *Observationes medicae*[13] whose authorship has been contested. In the *Essay* Locke calls Boyle and Sydenham *master-builders.*[14]

Some of Sydenham's views on man's ignorance recall the opinions of Pierre Gassendi who became increasingly important for British scientists during the second part of the century.[15] Gassendi's philosophy contains not only metaphysical justifications for the adequacy of the method of natural history that are several years older than Sydenham's, but also reasons that draw on atomism, about which humans will for ever remain ignorant of the essences of phenomena. Sydenham would have called these reasons speculative. Gassendi's theory of the *tabula rasa* assumes that ideas are the only objects of our thought. Our understanding is like a blank piece of paper[16] and can receive ideas and knowledge only from the senses.[17] Since the singular is always known before the general, we can reach general propositions only through the processing of singular sense perceptions.[18] One should, however, not be satisfied with occasional observations, because things are not always the way

13 See Dewhurst 1966, 73 and 75.

14 Locke 1690, *The epistle to the reader*, p. 9, lines 33–37.

15 See Puster 1991, especially "Skizze der britischen Gassendi-Rezeption im 17. Jahrhundert", 46–94.

16 "Videtur ergo potius concipi non male quasi charta munda, seu papyri purissimi folium" (Gassendi 1658, *Physica* 3/2.8.3; II 406a, 59–61).

17 "*Omnis Idea aut per Sensum transit, aut ex iis, quae transeunt per Sensum, formatur.*" – This is in accordance with Gassendi's demand for empirical confirmation of statements about facts (see Gassendi 1658, *Institutio logica* I, can. 3; I 92b, 48–50). Cp. also "... cum constet notitiam omnem quae in nobis est, vel sensuum esse, vel manare a sensibus, ideo constare etiam videtur non posse aliquod de ulla re iudicium ferri nisi cui sensus ferat testimonium" (Gassendi 1658, *Exercitationes paradoxicae* 2.6.2; III 192b, 29–33).

18 "... quod obiter dixi, nihil a nobis generatim, nisi singularibus prius notis intelligi, modum attingit, quo Intellectus in cognoscendo progreditur. Siquidem tametsi plerumque ex generalioribus ad specialiora argumentetur; omnino tamen debemus prius a singularibus incoepisse, ut generaliora colligeremus, ex quibus possemus deinceps ad specialiora et ad usque singularia procedere" (Gassendi 1658, *Physica* 3/2.9.4; II 458a, 68–459b, 9).

they appear under given circumstances[19] – observations need a methodical examination.[20]

Natural philosophy has to be true to the phenomena, thus be a *scientia experimentalis, apparentialis* or *historica.*[21] Scholastic school philosophy has left natural history to artisans and hucksters, concentrated on speculations, and only dealt with what is supposedly in accordance with philosophy.[22] This, alongside the habit to prefer authorities to experiment and observation, led to its eventual failure.[23] Gassendi justified the dependency of successful natural philosophy on experience with his doctrine of the *tabula rasa.* He used his atomic theory, however, in order to show that human knowledge of the essence of bodies is strictly limited in principle. The reception of his atomic theory by Newton and Boyle made it the most successful corpuscular theory of the 17th century. Wolfgang Detel has shown that Gassendi's atomic theory is much

19 "Plerumque videlicet ea, quae percipiuntur sensibus, alia, aliove modo apparent, quam in se reipsave sint. ... Nimirum licet experientia Sensibus peracta sit regula summa, ad quam, dum de re quidpiam dubitatur, confugiendum sit; non quaevis tamen talis est habenda; sed ea solum, quae est purgata omni instantia, omnique dubio; quaeque adeo est evidens, ut expensis omnibus, contradici iure non possit" (Gassendi 1658, *Institutio logica* I, can. 11; I 96a, 56–96b, 5, and 96b, 12–21).

20 "... quoties ambigitur de re, quae sensu probari potest, sit-ne, an-non sit; talisne, an alia; idcirco ad sensum recurrendum est, et ab evidentia, quae per ipsum fit, standum. Evidentia, inquam, quae habetur, quoties impedimentum nullum est, aut si est, postquam est sublatum. Impedimentum autem appello, v. c. distantiam, ob quam res magna videtur parva; quadrata, teres, et caetera ..." (Gassendi 1658, *Institutio logica* IV, can. 4; I 122a, 17–26).

21 "... scientiae tamen experimentalis, et ut sic dicam apparentialis ..." (Gassendi 1658, *Exercitationes paradoxicae* 2.6.7; III 207a, 54–55). – Cp. also "... scientiam, quam dicere historicam, seu experimentalem soleo ..." (Gassendi 1658, *Ad librum Herberti de veritate*; III 413a, 20–21).

22 "Quam iuvaret enim nosse historiam lapidum, metallorum, plantarum, animalium, caeterorumque huiusmodi, quorum est adeo iucunda cognitu varietas! Et ista tamen, inquiunt, noverint lapidarii, aurifices, herbarii, venatores; Flocci nempe faciunt, quod nimis vulgaria sint: iactantque interea seligere se, quae proprie spectant ad Philosophiam" (Gassendi 1658, *Exercitationes paradoxicae* 1.1.7; III 107b, 25–33).

23 "Desinunt videlicet suum adhibere iudicium: dum id ratum habent, quod ab eo, quem probant, iudicatum vident. Haec [authoritas] ipsa est, quae quasi onus Aethna gravius imminet, quaeque efficiet ut si veritas occultata iampridem sit, periculum non leve sit, ne in aeternum iaceat consepulta" (Gassendi 1658, *Exercitationes paradoxicae* 1.1.9; III 114a, 63–69).

more than just a revival of the atomic theory of Epicurus.[24] We can never, Gassendi held, reach to the marrow of nature, and have to be content with the façade and the bark of things, that is, with their external effects and perceptual qualities.[25] These are the result of unobservable atoms which determine the essence or inner nature of objects in their shape, size, movement and arrangement, and from which follow perceptual qualities and forces with necessity.[26] We have neither ideas nor knowledge of the intimate nature of reality because our senses are not suitable to the perception of atoms.[27] It is true that we can make a guess with regard to phenomena, but we have to count ourselves lucky if they give us an impression only similar to the truth.[28] This is also valid for the principles of experimental philosophy: We scent the occult essence under the sensual qualities as their fountain and their cause.[29]

Individual and particular statements about correctly observed phenomena are true and their set could be called science in a wide

24 See Detel 1978.

25 "… quod tales species referant solum quasi externos cortices rerum, non vero aut intimam naturam aut quae ex illa profluunt, aut necessitatem, modumque profluxus" (Gassendi 1658, *Physica* 3/2.9; II 456b, 19–23).

26 "… acceptis coniunctim proprietatibus Atomorum, et commistis praesertim, variatisque iis, de quibus dictum hactenus, raritate videlicet, densitate, et caeteris, oriuntur rerum facultates, quae actiuae et motiuae cum sint, id a pondere, seu mobilitate Atomorum habent; et cum hae hoc modo, illae illo agant, id habeant oportet, tum a speciali Atomorum magnitudine, et figura, tum a vario inter ipsas ordine atque positu; tum a laxitate, compressione, connexione, seiunctione, etc." (Gassendi 1658, *Philosophiae Epicuri syntagma*, 2.15; III 22b, 36–47).

27 "Et vndenam esse putemus, cur non cognoscamus naturas rerum intimas, intima actionum principia, intimosque agendi modos; nisi quia Intellectus noster sensibus, quasi ducibus caret, quibus omnia intimius, quam hisce crassis rudibusque scrutetur?" (Gassendi 1658, *Physica* 3/2.6.1; II 333a, 26–32).

28 "Nobis, vt, quod res est, fateamur, nulla facultas, aut Qualitas, dum causa rogatur vrgeturque, non Occulta est. Nam et quae hucvsque disseruimus, beati simus, si vel quamdam speciem probabilitatis obtineant; et quantumcumque nonnullae causae non prorsus remotae, sed aliquanto propinquiores afferantur; proximae tamen, et quas mens nosse peruelit, semper latent" (Gassendi 1658, *Physica* 1.6.14; I 449b, 31–39).

29 "… non posse nos tamen habere latentis illius naturae aliquam in Phantasia speciem; atque idcirco posse nos quidem subolfacere et quasi suspicari esse universe aliquam; at qua speciatim facie, seu cuiusmodi illa sit, neque intelligi, neque dici a nobis posse" (Gassendi 1658, *Physica* 3/2.9.5; II 463a, 20–26).

sense.[30] If, however, the word "science" is used in the strict sense and one reserves it only for general, necessary and proven knowledge, then experimental philosophy can make no claim to "science" for several reasons.[31] First of all, it lacks necessity because we cannot know the inner constitution of corporeal causes nor the forces and qualities that follow from them.[32] Secondly, it lacks true generality, for universal statements are never certain. We can never know ourselves to have run through all their singular instances.[33] However, the inaptness of induction and conjecture to give us any certainty should not make us despair.[34] Fortunately, we do not depend on knowing the innermost nature of objects or on deriving their properties and forces, because God has supplied us with everything we need to lead a good life by furnishing things with perceptual qualities and employing us with five

---

30 "... admittere posses appellandam Scientiam esse notitiam quandam experimentalem et rerum apparentium, ut cum dicor scire me sedere, me nunc potiusquam stare, interdiu potiusquam noctu, impransum potiusquam saturum, domi potiusquam in foro ... Ne igitur quispiam hic nobis facere inuidiam conetur, quod perspectas adeo et illustres res pernegemus, et insequamur; idcirco praemonendum est eiusmodi genus scientiae a nobis hic non impugnari ..." (Gassendi 1658, *Exercitationes paradoxicae* 2.6.1; III 192a, 33–49).

31 "Si satis constanter tueretis Scientiam esse alicuius rei certam, euidentem et per necessariam causam, seu Demonstrationem habitam notitiam; hac enim ratione illa experimentalis seu apparentium notitia nomine *Scientiae* non veniret" (Gassendi 1658, *Exercitationes paradoxicae* 2.6.1; III 192a, 49–54).

32 "... si ex me quaeras an sciam mel mihi apparere dulce vel me experiri degustando dulcedinem mellis, respondeo scire et hoc modo Scientiam huiusce rei haberi concedam. At vero tamen si quaeras deinde an sciam mel esse ex natura sua, secundum se, et reuera dulce, hoc est quod tum demum me nescire fateor, quippe necessariam causam aut Demonstrationem non habens cur res ita se habeat ..." (Gassendi 1658, *Exercitationes paradoxicae* 2.6.1; III 192b, 6–15).

33 "... ostendendum est nullam tuto colligi aut haberi posse Propositionem vniuersalem. Primum igitur si sit aliqua propositio vniuersalis illa non potest alio modo quam inductione colligi ... Atqui Inductione colligi non potest Vniuersalis propositio, siquidem percurri prius et enumerari non possunt omnia singularia ratione quorum propositio dicenda sit vniuersalis. Ratio est quia singularia innumera sunt ..." (Gassendi 1658, *Exercitationes paradoxicae* 2.5.5; III 187b, 58–188a, 2).

34 "... neque enim desperatio propterea inducitur iis, qui philosophari volunt, quod videant magnos philosophos profiteri nihil sciri posse, intelligo quod ad naturas rerum intimas attinet, siquidem cum hac in parte ignorantes sese agnoscant; at aliunde tamen agnoscuntur scientissimi, quod earum rerum, quae sciri possunt, nihil prope ipsos lateat" (Gassendi 1658, *Exercitationes paradoxicae* 2.6.6; III 207b, 33–41).

senses and the understanding. He obviously intended to hide the intimate nature of things from us, and if we pretend to know something about it we have lost our temperance.[35] With this, Gassendi provided a rational theological justification for the deficiency of our senses.

Gassendi attracted the attention of the English not only with his atomic theory, but also with his opinions on human understanding.[36] His works became widely accessible through the Lyon edition of 1658. From these, his objections to Descartes's *Meditationes* stand out which had already appeared earlier, in the *Disquisitio Metaphysica* of 1644.[37] Two of his works had appeared in London: the *Institutio Astronomica* in 1654[38] and the *Philosophiae Epicuri Syntagma* in 1660.[39] The latter work was known to Boyle who recommended it to his readers.[40] In 1654, Walter Charleton, the king's personal physician and later member of the Royal Society, wrote a clear and very original presentation of Gassendi's philosophy, the *Physiologia Epicuro-Gassendo-Charltoniana.*[41] There, pithy views on natural philosophy can be found such as the following: "Nor is Physiology, indeed, more then the larger Descant of Reason upon the short Text of Sense [...]."[42] One can also find there pronouncements on the certainty of an atomic hypothesis concerning the nature of colours: "... the foundation of it is not layed in the rock of absolute Demonstration, or desumed *a Priori*, but in the soft-

35 "... quicquid fuit nobis de re vnaquaque nosse necessarium, illud nobis [Deus] apertum fecit, tribuendo rebus proprietates, per quas innotescerent, et nobis sensus varios, quibus illas apprehenderemus, ac facultatem interiorem, qua de iisdem iudicaremus. Quod ad internam vero naturam et quasi scaturiginem, illam vt nobis cognitu non necessariam occultam voluit; et nos, cum nosse affectamus aut praesumimus, intemperantia laboramus" (Gassendi 1658, *Disquisitio metaphysica* 2.8.1; III 312b, 28–38).

36 Similar statements can be found e.g. in Newton 1726: "Videmus tantum corporum figuras et colores, audimus tantum sonos, tangimus tantum superficies externas, olfacimus odores solos, et gustamus sapores: intimas substantias nullo sensu, nulla actione reflexa cognoscimus; et multo minus ideam habemus substantiae dei" (Third Book, *Scholium generale*, p. 763).

37 Gassendi 1658; III, 269–410. – See also the French translation in Gassendi 1644.

38 Gassendi 1658; IV, 1–65.

39 Gassendi 1658; III, 1–94.

40 Together with Descartes's *Principia*! See Boyle 1772, *Considerations touching Experimental Essays in General*; I 302, 6–8.

41 See Charleton 1654.

42 Charleton 1654, I, 3.1.5; 18.

er mould of meer *Conjecture*, and not deeper than *a Posteriori*."[43] Finally, from 1655–1662 appeared Thomas Stanley's *History of Philosophy*,[44] whose part XIII conveyed the philosophy of Epicurus in close dependence on Gassendi.

Also, some of the metaphysical hypotheses used by Boyle resemble those of Gassendi. Boyle entertained ideas regarding the benefits of natural philosophy, and the relation of knowledge and power in a quite resolute way.[45] He wrote, for example: "... I shall not dare to think myself a true naturalist, till my skill can make my garden yield better herbs and flowers, or my orchard better fruit, or my field better corn, or my dairy better cheese, than theirs that are strangers to physiology."[46] ("Physiology" here means "natural science".) It is difficult to assign Boyle to a definite school. He was clearly influenced by Descartes and Gassendi and did not care to settle any disagreements arising between the two. He thought that the phenomena can be explained by both systems and that most of the controversies can be attributed to definitional problems.[47] The most important thing for him was that both sides worked with corpuscular explanations.[48] In his writings concerning

---

43 Charleton, III, 4.3.10; 196.

44 See Stanley 1701. The part on Epicurus can be found on pp. 533–633.

45 E. g. "... that natural philosophy, wherein his [man's] dominion over the creatures chiefly consists ..." (Boyle 1772, *The Usefulness of Natural Philosophy* I, 2; II 18, 48–49). – Cp. also: "The advantages of the knowledge of nature towards the increasing the power of man ..." (Boyle 1772, *The Usefulness of Natural Philosophy* II, 1; II 64, marginal note). – "For in man's knowledge of the nature of the creatures, doth principally consist his empire over them, (his knowledge and his power having generally the same limit)" (Boyle 1772, *The Usefulness of Natural Philosophy* II, 1; II 65, 4–6).

46 Boyle 1772, *The Usefulness of Natural Philosophy* II, 1; II 64, 18–21.

47 "... the reason why there cannot be a void, being taken, not from any experiments, or phaenomena of nature, that clearly and particularly prove their hypothesis, but from their notion of a body, whose nature, according to them, consisting only in extension (which indeed seems the property most essential to, because inseparable from a body), to say a space devoid of body is, to speak in the schoolmen's phrase, a contradiction *in adiecto*. This reason, I say, being thus desumed, seems to make the controversy about the vacuum rather a metaphysical, than a physiological question; which therefore we shall here no longer debate, finding it very difficult either to satisfy Naturalists with this Cartesian notion of a body, or to manifest wherein it is erroneous, and substitute a better in its stead" (Boyle 1772, *New Experiments Physico-Mechanical*; I 37, 48–38,8).

48 "... I considered, that the Atomical and Cartesian hypotheses, though they differed in some material points from one another, yet in opposition to the Peri-

the doctrines of substances and qualities, however, Boyle proceeded more like a follower of Gassendi than of Descartes. In controversies connected with innate ideas and principles, he tended to follow Descartes and foster some reserves against Gassendi's doctrine of the *tabula rasa*.[49] In justifying the historical and descriptive methods, he preferred to refer to their usefulness in explanation and to God's intention – something which becomes evident from the provision of our sensory equipment.[50] He also held that one has to pay a price for the usefulness of the historical method: anyone practicing experimental philosophy will more often experience disappointments than successes and needs not only brains but also money.[51]

---

patetic and other vulgar doctrines they might be looked upon as one philosophy ... both the Cartesians and the Atomists explicate phaenomena by little bodies variously figured and moved" (Boyle 1772, *Some Specimens of an Attempt to make Chymical Experiments useful to illustrate the notions of the Corpuscular Philosophy*; I 355, 33–37, and also pp. 42–43).

49 "... I see not, why we may not reasonably think that God, who, as themselves confess, has been pleased to take care men should acknowledge Him, may also have provided for the securing of a truth of so great consequence, by stamping characters, or living impresses, that men may know his wisdom and goodness by, as well without, upon the world, as whithin, upon the mind" (Boyle 1772, *A Disquisition about the Final Causes of Natural Things*, sect. II; V 402, 2–7).

50 "We must not suppose that, at least in our present state, our reason and other faculties are given us, to reach all that are knowable, even to corporeal creatures; but only things that are in such a sphere of intelligibility, that they are proportioned to our present faculties, and convenient for our notice in our present state or condition. As our eyes are not given us to see all that is visible, and might be discovered by us, in case they were framed, sometimes like telescopes, and sometimes like microscopes; but to discover those visible objects which are not so very minute, or so remote from us, but that it may concern us in point of safety or welfare, to distinguish and discern their bulk, distance, figure, colour, etc." (Boyle 1772, *Appendix to the First Part of the Christian Virtuoso*; VI 696, 12–20).

51 "For though the knowledge of nature be preferable by odds to those other idols, which we have mentioned, as inferior to it, yet we here attain that knowledge but very imperfectly, and our acquisitions of it cost us so dear, and the pleasure in them is so allayed with the disquieting curiosity they are wont to excite ..." (Boyle 1772, *The Usefulness of Natural Philosophy* I 5; II 60, 28–32). – Cp. also "... the advancement of that experimental philosophy, the effectual pursuit of which requires as well a purse as a brain ..." (Boyle 1772, *New Experiments Physico-Mechanical, touching the Spring of the Air*; I 6, 4–5).

Boyle agreed with Gassendi and Locke on several points when he referred to the uncertainty of hypotheses in natural philosophy. First of all, it is the size, shape, arrangement and contexture of the invisible atoms that play an important role for the properties and activities of objects.[52] Secondly, we can never be certain whether our generalizations are also valid for hitherto unobserved specimens of a kind.[53] Boyle's pronouncements upon the necessity of the confirmation of sensory information were more detailed than those of Gassendi and Locke, and put more weight on the state of the observer and his instruments.[54] His assertions were, however, twenty years of stormy progress younger than those of Gassendi. Boyle also agrees with Gassendi that sensibility and understanding have to work together in natural history and philosophy: "... experience is but an assistant to reason, since it doth indeed supply informations to the understanding; but the understanding remains still the judge, and has a power or right to examine and make use of the testimonies, that are presented to it."[55]

Boyle's opinions on hypotheses in natural philosophy basically correspond to those of Gassendi but are based more extensively on exper-

52 "When I consider how much most of the faculties of bodies, and consequently their operations, depend upon the structure of their minute, and singly invisible particles, and that to this latent contexture, the bigness, the figure, and the collocation of the intervals and pores do necessarily concur with the size, shape and disposition or contrivance, of the substantial parts ..." (Boyle 1772, *Experiments and Considerations about the Porosity of Bodies*; IV 759, 30–34).

53 "And I very much question, whether *Physeophilus* do know ... how incompleat the history of nature we yet have, is, and how difficult it is to build an accurate hypothesis upon an incompleat history of the phaenomena it is to be fitted to; especially considering, that (as I was saying) many things may be discovered in after-times by industry or chance, which are not now so much as dreamed of, and which may yet overthrow doctrines specially enough accommodated to the observations, that have been hitherto made" (Boyle 1772, *The* Excellency *of* Theology; IV 59, 36–43).

54 E.g. "1. That by reason of the various and unheeded pre-dispositions of our bodies, the single and immediate informations of our senses are not always to be trusted. 2. That though common weather-glasses are useful instruments, and the informations they give us are in most cases preferable to those of our sense of touching, in regard of their not being so subject to unheeded mutations; yet even these instruments being subject to be wrought upon by the different weights of the atmosphere, as well as by heat and cold, may (upon that and perhaps some other accounts) easily misinform us in several cases, unless in such cases we observed by other instruments the present weight of the atmosphere" (Boyle 1772, *Deficiencies of Weather-glasses, etc.*; II 498, 43–51).

55 Boyle 1772, *The Christian Virtuoso*; V 539, 21–24.

imental experience. Hypotheses in natural philosophy, including those of the atomic one, are uncertain,[56] depend on risky presuppositions and are sometimes obstructive and short-lived.[57] Since almost every day brings a new discovery, so Boyle contends, no method can be correct for a longer period of time and will soon rightfully be replaced through a better one.[58] There is a steady growth of knowledge and philosophical insight: "You can as little take stable measure of it, as a taylor can take such measures of a child of seven years old, as will continue to fit him during his whole life".[59] A reasonable scientist will, therefore, alter his assumptions as soon as he encounters incongruous phenomena or better alternatives.[60] Even corpuscular philosophy – and Boyle uses this term almost like a unificatory formula for Cartesian and atomic views[61] – has to be disposed with where required, although one has

---

56 See "... our experiment affords us a considerable argument in favour of that part of the corpuscular or mechanical hypothesis ..." (Boyle 1772, *A Chymical Paradox*; IV 501, 11–12). Principally philosophy "is a thing of a more noble nature, and of greater extent, than the hypothesis of any one sect of philosophers ..." (Boyle 1772, *Appendix to the First Part of the Christian Virtuoso*; VI 700, 14–15).

57 For hypotheses in the sense of systems see "... it has long seemed to me none of the least impediments of the real advancement of true natural philosophy, that men have been so forward to write systems of it, and have thought themselves obliged either to be altogether silent, or not to write less than an entire body of physiology; for, from hence seem to have ensued not a few inconveniences," as, for example, that authors made up observations which they had no knowledge of and that the impression arose that nature had already been investigated completely (Boyle 1772, *Some Considerations touching Experimental Essays in general*; I 300, 20–25).

58 "... natural philosophy, being so vast and pregnant a subject, that (especially in so inquisitive an age as this) almost every day discovers some new thing or other about it, it is scarce possible for a method, that is adapted but to what is already known, to continue long the most proper; as the same clothes will not long fit a child, whose age will make him quickly out-grow them. And therefore succeeding writers will have a fair pretence to compile new systems, that may be more adequate to philosophy, improved since the publication of the former" (Boyle 1772, *The Excellency of Theology*; IV 55, 12–19).

59 Boyle 1772, *Appendix to the First Part of the Christian Virtuoso*; VI 708, 10–13.

60 "... rational philosophers scruple not to alter or renounce the opinions, that specious reasons had suggested to them, when once they either find those opinions contradicted by experience, or meet with other opinions more conformable to experience" (Boyle 1772, *The Christian Virtuoso;* V 538, 26–29).

61 In the "*Origin*," Boyle wants to write neither in the atomistic nor in the Cartesian way but "rather for the Corpuscularians in genere, than any party of

to defend it presently because it commands all the characteristics of a good system. It contains perspicuous principles, in fact as few and simple ones as possible; it can explain all hitherto known phenomena[62] and is in this respect superior to all competing systems, e.g. the Peripatetic and Paracelsian ones.[63] The concepts and hypotheses of the corpuscular system are instruments[64] constructed by reason itself, which not only help us in our work, but also lead us to new knowledge and knowhow.[65] These views also correspond to a principle of rational theology: The

---

them." (Boyle 1772, *The Origin of Forms and Qualities*; III 7, 22). – According to Boyle, the two parties can be regarded as representatives of one single philosophy: Cp. *Some Specimens of an Attempt to make Chymical Experiments useful to illustrate the notions of the Corpuscular Philosophy*; III 355, 34; 356, 7.

62 "... there cannot be fewer principles than the two grand ones of mechanical philosophy, matter and motion ... Nor can we conceive any principles more primary, than matter and motion ... Neither can there be any physical principles more simple than matter and motion ... The next thing I shall name to recommend the corpuscular principles, is their great comprehension" (Boyle 1772, *Of the Excellency and Grounds of the Corpuscular or Mechanical Philosophy*; IV 70, 3–18).

63 "... that the four peripatetick elements and the three chemical principles are so insufficient to give a good account of anything near all the differing phaenomena of nature, that we must seek for some more catholick principles, and that those of the corpuscularian philosophy have a great advantage of the other in being far more fertile and comprehensive than they" (Boyle 1772, *The History of particular Qualities*; III 296, 31–35). – Cp. also "... in about two thousand years since Aristotle's time the adorers of his physicks, at least by virtue of his peculiar principles, seem to have done little more than wrangle, without clearing up (that I know of) any mystery of nature, or producing any useful or noble experiment: whereas the cultivators of the Particularian philosophy ... must often make and vary experiments; by which means nature comes to be much more diligently, and industriously studied, and innumerable particulars are discovered and observed which in the lazy Aristotelian way of philosophy would not be heeded" (Boyle 1772, *The Origin of Forms and Qualities*; III 75, 15–28).

64 Reason "manages a frame or system of ideas and propositions, wherewith she is furnished by sciences and arts; which, though as it were tools or instruments of her own framing, yet are such, as being once made, her operations are regulated by them; as when the eye and the hand make rulers, compasses, and telescopes; but are afterwards in their operations guided, as well as assisted, by these instruments ..." (Boyle 1772, *Appendix to the First Part of the Christian Virtuoso*; VI 713, 25–29).

65 "The advantage of the knowledge of nature towards the increasing the power of man; and its uses as to health of the body and goods of fortune" (Boyle 1772, *The Usefulness of Natural Philosophy* II, 1; II 64, in the margins).

revelation of God's magnificence and the well-being of man as God's most noble creature were part of the main goals of God's Creation.[66]

As we mentioned already, Locke respectfully called Boyle a *master-builder* in his letter to the reader of the *Essay*.[67] This might have been a friendly reaction to an early preface where Boyle assesses himself as "under-builder".[68] It is this context, in which Locke calls himself an "under-labourer".[69] When he published this, he had been a friend of Boyle's for twenty-five years already. He had conducted experiments under Boyle's guidance, discussed with him the nature of substances, methods and principles, collected observations for him, accompanied his writings, and edited his *History of the Air* posthumously. He was also one of Boyle's executors. Locke's *Essay* is a work that belongs to Boyle's circle and shows a mixture of interests in natural history and metaphysical assumptions very similar to Boyle's writings. Locke's metaphysical views about the order of the universe, the inability of human understanding to adequately grasp it and the wisdom and benevolence of God's providence in assessing our human sensibility and understanding are in general very close to those of Gassendi and Boyle. As regards content, Locke's *Essay* is a natural history of human understanding as becomes evident from Locke's explicit commitment to "historical, plain method"[70] and from the repeated remark that the *Essay* does not treat any natural philosophy whose subject would have been body and soul.[71] The *Essay* differs from natural philosophy à la Boyle through,

66 "... two of God's principal aims in the creation were the manifestation of his own glorious attributes, and the welfare of his noblest visible creature, man ..." (Boyle 1772, T*he Usefulness of Natural Philosophy* I, 2; II 18).

67 Locke 1690, *The epistle to the reader*, p. 9, lines 34–37.

68 "... by the way of writing, to which I have condemned myself, I can hope for little better among the more daring and less considerate sort of men, should you shew them these papers, than to pass for a drudge of greater industry than reason, and fit for little more, than to collect experiments for more rational and philosophical heads to explicate and make use of. But I am content, provided experimental learning be really promoted, to contribute even in the least plausible way to the advancement of it; and I had rather not only be an under-builder, but even dig in the quarries for materials towards so useful a structure, as a solid body of natural philosophy, than not to do something towards the erection of it" (Boyle 1772; *Some Considerations touching Experimental Essays in general*; I 307).

69 Locke 1690, *The epistle to the reader*, p. 10, line 3.

70 Locke 1690, I.i.2.

71 For example Locke 1690, II.viii.4 and 22.

among other things, the distinctiveness of its subject matter[72] and the right classification of the pertinent phenomena, ideas and operations of the mind. Gassendist elements also emerge powerfully, as already Leibniz remarked.[73] This applies not only for the so-called empiricism of ideas, for parts of the views on certainty and for the background physics. The fact that Locke does not explicitly refer to Gassendi is not a plagiarist omission as it was generally known that some of Gassendi's views had become opinions of Locke's and Boyle's circles. In the *Essay*, Locke ostentatiously refuses to cite authorities or erudites generally, but only authors of travelogues.

If one disregards Charleton's influence, there are at least three ways through which Gassendi's thought is transmitted to Locke. There is first of all Locke's own reading of Gassendi, which is certified for the second half of the sixteen-sixties.[74] And secondly, there is, most probably in 1677, the Paris encounter with Gilles de Launay who gave lectures on Gassendi's philosophy, and four of whose philosophical works Locke carried around with him in his baggage.[75] In 1677 he became also acquainted with François Bernier who was one of the best experts on Asia at the time, having extensively travelled through Africa and Asia, served as personal physician to the Great Mogul for thirteen years, and had written a number of then famous travelogues. Locke had a soft spot for reports of this kind, as he reveals in the *Essay*, and had discussions with Bernier on travelling, as his notes show.[76] It is not known whether he also discussed medicine and philosophy with him, but it is very probable, since Bernier had published at about the same time a handy *Abregé de la Philosophie de Gassendi*, which Locke bought for himself and which most probably left its traces in the *Essay*.[77] And finally, Locke came into contact with Gassendi's theories through Boyle and his circle.

Locke is generally seen as the progenitor of modern empiricism. If one understands the *Essay* as a natural history of the human understanding trained in Boyle's views, then a critique that was especially common in Germany does not seem too implausible: Locke is something like a

72 Essay 1975, I.i.1.
73 Leibniz 1704/ 1765, 63.
74 Cranston 1985, 102–103.
75 Cranston 1985, 169–170.
76 Cranston 1985, 170.
77 Bernier 1678.

"psychologist" who deals with the *quaestio facti* and tries to make clear what happens when someone makes use of his or her understanding. It is exactly the function of natural history to clear up facts like these. And, it is held, Locke greatly overdoes his psychologism in overlooking the decisive *quaestio juris*. Kant had already written about Locke's theory of "pure concepts" (in Kant's terms): "Since this attempted physiological derivation concerns a *quaestio facti*, it cannot strictly be called deduction; and I shall therefore entitle it the explanation of the *possession* of pure knowledge."[78] It is, indeed, Locke's intention to show "how he comes by them [i.e. his ideas]."[79] But this is not the only aspect with which he is interested. The *Essay* does not only show that understanding proceeds in a certain way, but also that it is unable to proceed otherwise. In so far Locke does deal with the *quaestio juris* – yet he does so in his own way, rather than following a Kantian route.

In spite of several invectives against speculation, Locke's *Essay* also contains propositions about God and man that cannot be justified through experiment or observation. They are usually read over in a friendly way or regretted as rationalist atavisms. Locke explains, for example, that God created the universe as a system of corpuscles but that He, at the same time, denied us its adequate cognition in benevolent intention, and endowed us instead with five senses adjusted to medium size objects and phenomena, and with an understanding that can reappraise the information conveyed by these senses. He has, however, as Locke contends, connected certain sensory ideas with certain movements of atoms in an infinitely purposeful way. This connection has given them their factual reference and made us capable of presuming the causes of the phenomena, which does not lead us to scientific certainty, but to a greater admiration of God and to useful inventions. Such statements, which today's empiricists take note of at best with amazement, correspond to metaphysical assumptions of Gassendi, Sydenham and Boyle. Similar opinions can also be found in Newton. Tullio Gregory writes in his preface to the reprint of the Gassendi edition of 1658 that the same is true for many other empiricists of the 16th century.[80]

It would be of little use to assume that all these authors were so unintelligent as to take the mentioned assumptions for empirical generalizations or experimentally founded explanations of phenomena. Because

78 Kant, *Critique of Pure Reason*, A 86–87/B 119.
79 Locke 1690, II.i.1.
80 Gassendi 1658, I, xix.

of their perseverant appearance we can surmise that these suppositions had to fulfil a function that could not have been accomplished by empirical generalizations or hypotheses of natural philosophy. Gassendi, Sydenham, Boyle and Locke used them in order to argue that natural philosophy is in need of sense-based natural history. They also tried to explain with them why every human philosophy of nature will always suffer from a lack of necessity and universality. Locke never regarded such metaphysical statements as empirical generalizations nor as explanatory hypotheses of natural philosophy, but neither did he introduce them through an oversight. It is true that they are metaphysical hypotheses, but Locke uses them in order to justify the experimental method and not for the deduction of a philosophical system. Such hypotheses cannot be justified within the bounds of the same experimental philosophy for whose justification they were created in the first place. The modern reader is amazed by these seemingly foreign remnants of rationalism and feels deceived in the expectation of Locke the modern empiricist. Yet in reality these remnants made this special brand of Franco-British empiricism appear reasonably justifiable and facilitated its reception for the average 17th century reader who was not yet used to an empiricist strain of thought. One could forego this justificatory metaphysics only when empirical methods had become so self-evident that nobody cared anymore for their justification except perhaps philosophers like Kant and Hegel.[81]

Only after these early empiricist metaphysics were set aside did the gulf between rationalists and so-called empiricists look as wide as it appears today. It made sense to distinguish between these two traditions only after one could refer to different ways of using metaphysical hypotheses, to different assumptions about the origin of human concepts and to different senses of "decency," so to say, in communicating hypotheses. Only then could the rationalist appear as a philosopher who thinks it useful and admissible to form hypotheses for the explanation of phenomena and to claim that they can be deduced from highest metaphysical principles. An empiricist could then appear as someone who only values those hypotheses from natural philosophy that can explain the phenomena as directly and parsimoniously as possible. Both tendencies are justified by metaphysical hypotheses, albeit of different content and function: on one hand, the efficiency of the natural light of reason is

81 Perhaps Hegel's best known criticism is in the *Enzyklopädie der philosophischen Wissenschaften im Grundrisse* (1830), §§ 37–39; cp. especially the note to § 38.

stressed, and, on the other, the deficiency of human understanding. The first tendency corresponds to a special interest in theories and ways of constructing philosophical systems, whereas the other tendency has, firstly, a special interest in hypotheses of natural philosophy explaining phenomena as unswervingly and squarely as possible and, secondly, a disinterest in hypotheses connecting statements close to the observational basis with highest metaphysical principles, perhaps even in an *ad hoc* manner.

This difference does not prevent empiricists and rationalists from agreeing with each other on the significance and production of the empirical basis. It tells, however, something about the different role of the metaphysics employed. For Cartesians, there is a direct route from metaphysics to physics and vice versa. An empiricist like Locke is sometimes inspired by the results of experimental philosophy to meditate on God and man in a way that can be justified from a rational point of view, albeit not empirically. Yet for him metaphysics stands on a different level from the Cartesians: it does not form the basis for physics but serves to validate a philosophy of experience that does not only wish to be justified by its success, but also theoretically. The early British empiricists were content with the metaphysics Gassendi had used for giving good reasons for empiricism. Locke used it to treat the *quaestio juris* in his own way. It is not convincing to accuse him of having overlooked this question. Those commentators who raised this charge regarded Locke's metaphysical hypotheses as short-lived lapses of an otherwise deserving empiricist and were sometimes generous enough to ignore them. If, however, one perceives them as metaphysical hypotheses answering the *quaestio juris* and justifying science and its method, one can only be astonished at the fact that they are mentioned in the *Essay* only incidentally. If one takes, however, the goal of the *Essay* seriously, it becomes clear that the literary genre of a natural history allows only for occasional parentheses, but not for a complete tract of metaphysics. Boyle had already proceeded in a similar way to Locke: his natural histories are intermingled with short metaphysical contemplations every now and then, but do not contain any lengthy and complete considerations.

The early empiricist postulate to carry on empirically and not by speculation refers to the subject-matter of natural philosophy and natural history, and demands to leave metaphysics and rational theology out of the question. The authors discussed above advocate an empiricist metaphysics which goes back to Gassendi and whose traces can still be found in David Hume. There were, however, also experimental philosophers who did not uphold such a dire empiricist metaphysics but more blooming versions. There was, for example, Francis Glisson who admitted to be an experimentalist in natural philosophy but a follower of Suarez in matters metaphysical, although not a slavish one.[82] Some time later the metaphysics of the empiricists, as sparse as it was, went so out of fashion that modern readers, who now refused to take man as principally ignorant or science as principally uncertain, could only perceive it as a strange oddity. What originally seemed to be indispensable for a theoretical accreditation of empiricism with the educated public could, after the impressive success of empirical science, be dropped. Science was now seen as providing enough justification for empiricism by itself. What remained were the empirical generalizations of natural history and the explanatory hypotheses of natural philosophy, both purified of metaphysics and metaphysical hypotheses of justification. These remnants correspond best to what one called "empiricism" during the 19th century when this term came into wider currency. The strange character of this philosophy as a leftover of earlier developments makes it conceivable why it got into such formidable difficulties when, for several different reasons, the question of how empiricism is to be justified was posed again in the 20th century.[83]

Translated by Michael Heidelberger

Bibliography

Bernier, François (1678), *Abregé de la Philosophie de Gassendi*. Lyon 1678. Reprint of the edition of 1684 ed. Sylvia Murr and Geneviève Stefanie. Paris 1992.

Boyle, Robert (1772), *The Works*, 6 vols. Ed. Thomas Birch, 2nd ed. London. Reprint Hildesheim 1965. (After the title of the work follows the volume number [in Roman numerals] and the page).

Charleton, Walter (1654), *Physiologia Epicuro-Gassendo-Charltoniana: or A Fabrick of Science Natural upon the Hypothesis of Atoms*. London. Reprint with indices, ed. and introd. R. H. Kargon. (The Sources of Science, vol. 31) New York/London 1966 (cited by book number (in Roman numerals), chapter, section and article, followed by page number).

82 Scorraille 1913, 437 quotes Glisson as saying: "Suarius, quem prae aliis mihi ducem in rebus metaphysicis elegi, sed non iuratus in verba Magistri."

83 The interpretation of the history of rationalism has led to similar conclusions; so e.g. in École 1979 and Poser 1979.

Cranston, Maurice (1985), *John Locke. A biography*. Oxford 1985.
Detel, Wolfgang (1978), *Scientia rerum occultarum. Methodologische Studien zur Physik Pierre Gassendis*. (Quellen und Studien zur Philosophie, Bd. 14) Berlin.
Dewhurst, K. (1966), *Dr. Thomas Sydenham. His life and original writings*. London.
École, Jean (1979), "En quels sens peut-on dire que Wolff est un rationaliste?". *Studia Leibnitiana 11:* 45–61.
Gassendi, Pierre (1658), *Opera Omnia*. 6 vols. Lyon. Reprint Stuttgart-Bad Cannstatt 1964 (after the title of the work with chapters and §§ follows the volume number (in Roman numerals), column and line).
Gassendi, Pierre (1644), *Disquisitio metaphysica seu Dubitationes et instantiae adversus Renati Cartesii metaphysicam responsa*. Amsterdam. Reprint with French translation ed. Bernard Rochot. Paris 1962.
Leibniz, Gottfried Wilhelm (1704/ 1765), *Nouveaux Essais sur l'entendement humain*. Amsterdam and Leipzig. Reprint in *Philosophische Schriften von Leibniz*, vol. 5. Ed. C. J. Gerhardt. Berlin 1882. Reprint Hildesheim 1965.
Locke, John (1690), *An Essay concerning human understanding*. (The Clarendon edition of the works of John Locke) Ed. Peter H. Nidditch, Oxford 1975 (cited by book, chapter and paragraph).
Newton, Isaac (1726), *Philosophiae naturalis principia mathematica*. 3rd ed. London. Reprint ed. A. Koyré and J. B. Cohen. Cambridge 1972 (1st ed. 1687).
Poser, Hans (1979), "Die Bedeutung des Begriffes 'Ähnlichkeit' in der Metaphysik Christian Wolffs". *Studia Leibnitiana 11:* 62–81.
Puster, Rolf W. (1991), *Britische Gassendi-Rezeption am Beispiel John Lockes*. (Quaestiones 3) Stuttgart-Bad Cannstatt.
Stanley, Thomas (1701), *The History of Philosophy*. 3rd ed. London. Reprint ed. B. Fabian et al. Hildesheim 1975.
Scorraille, Raoul de (1913), *François Suarez, de la Compagnie de Jésus*, vol. 2. Paris.
Sydenham, Thomas (1741), *Opera universa*, Leiden.

Hypotheses in 19th Century British Philosophy of Science: Herschel, Whewell, Mill

Laura J. Snyder[1]

Abstract:

In nineteenth-century Britain, Francis Bacon was often seen as the "hero" of inductive science. Most writers on science claimed to be followers of Bacon. Yet modern commentators on the period have tended to see it as dominated by the "method of hypothesis," pointing to William Whewell and John Herschel as the main proponents and popularizers of that position (for example, Laudan 1980, Butts 1987, Yeo 1993). The method of hypothesis is described as similar to twentieth-century hypothetico-deductivism and opposed to inductivism. In this paper I explore this paradox. I argue that, like Whewell, Herschel endorsed an inductive method of science. He believed he was continuing the legacy of Bacon. Indeed, Herschel argued with Whewell over which man was the most faithful to Bacon's inductivism. After showing that the nineteenth century was an age of inductivism, rather than hypothetico-deductivism, I suggest reasons for the common but mistaken view.

Previous to the publication of the *Novum Organum* of Bacon, natural philosophy, in any legitimate sense of the word, could hardly be said to exist. ... [Bacon] will, therefore, justly be looked upon in all future ages as the great reformer of philosophy.

J. F. W. Herschel, 1830

If we must select some one philosopher as the Hero of the revolution in scientific method, beyond all doubt Francis Bacon must occupy the place of honor.

William Whewell, 1840

It was ... by pointing out the insufficiency of [the older] rude and loose conception of Induction, that Bacon merited the title so generally awarded to him, of Founder of the Inductive Philosophy.

J. S. Mill, 1843

1 I thank Michael Heidelberger and Gregor Schiemann for useful comments on this paper. I am grateful to the other conference participants for stimulating discussion of the issue of hypotheses in science.

1. Introduction

Nineteenth-century Britain saw a revival of interest in the writings of Francis Bacon, and in his inductive philosophy. The major writers on scientific methodology all pointed to Bacon as the "hero" of the inductive method, claiming – with greater or lesser degrees of accuracy – that they were following in his footsteps. It is therefore notable that modern writers on the history of the philosophy of science have seen early to mid-nineteenth-century Britain as an age dominated by the "method of hypothesis," a methodology similar to that endorsed by recent proponents of hypothetico-deductivism, and one opposed to the inductivism of Bacon (see, for instance, Laudan 1980 and Butts 1987). In my paper I will explore this paradox.

The method of hypothesis endorses the view that scientists invent theories by postulating a hypothesis; no reasoning is required at this stage of the process, and the invention of the hypothesis is often described as a matter of guesswork or conjecture. These hypotheses are then tested by drawing out the consequences that would follow if the theory were true, and seeing if these deductive consequences obtain. In contrast, the inductive method proposes that scientists use reasoning from empirical data even at the stage of inventing their hypotheses (generally speaking, inductivists suppose that these hypotheses must also be tested by drawing out deductive consequences).

The major writers on methodology in nineteenth-century Britain were J. S. Mill, William Whewell, and John Herschel. Hardly anyone would deny that Mill was an inductivist (the only exception to this that I know of is Jacobs 1991a and 1991b; see Snyder 2006, chapter two for an argument against his position). Those who claim that 19th century British methodology was dominated by the method of hypothesis point to Whewell and Herschel as the proponents of this view, arguing that both writers rejected the use of reasoning at the stage of inventing hypotheses. I have argued extensively that this is a mistaken characterization of Whewell's position (see Snyder 1997a, 1997b, 1999, 2006). Here I will argue additionally that even Herschel presents a complex case. On the one hand, he certainly allowed a more liberal use of "bold leaps" in science than did Mill or Whewell, claiming that one legitimate method for discovering a law was "by forming at once a bold hypothesis, particularizing the law, and trying the truth of it by following out its consequences and comparing them with facts."

Yet at the same time Herschel, unlike twentieth-century proponents of hypothetico-deductivism such as Karl Popper or Carl Hempel, placed inductive constraints on these "bold hypotheses;" they were not intended to be the result of mere guesswork on the part of the scientist. By examining the case of Herschel, and the disagreement between Herschel and Whewell over who was the legitimate heir to Bacon, we will gain insight into how the role of hypotheses was perceived in early to mid-nineteenth-century British philosophy of science. I will show that the method of hypothesis was clearly not the dominant methodology of the time. Although different writers had diverse notions of what inductivism entailed (and, indeed, Whewell, Herschel and Mill each saw his task as being that of defining induction), the major writers on the topic all proposed inductive methods for the natural sciences. I will end the paper by discussing some reasons for the mistaken view that this period was dominated by the method of hypothesis.

2. Disciples of Bacon

As undergraduates, Whewell, Herschel, and their friend the future political economist Richard Jones were drawn together by their shared interest in the works of Bacon. Whewell later recalled that their favorite reading was the *Novum Organum*, and that they spent many a Sunday morning discussing it together (Whewell 1859). The three of them agreed with Bacon that a proper inductive method must be established for every area of thought, not only physical science. Even after receiving their degrees, when they were no longer gathered together at Cambridge, Whewell, Herschel and Jones continued to express this as their shared goal. In 1831, Jones claimed to be seeking an "outline of reasoning ... inductively on almost all subjects."[2] During the time he was engaged in his astronomical observations in the mid-30s at the Cape of Good Hope, Herschel told Whewell that he was working on "an enquiry into the moral nature of men," which he hoped would culminate in "the inductive construction of a system of Ethicks."[3] In his *Preliminary Discourse on the Study of Natural Philosophy,* Herschel noted that the methods of the physical sciences could be applied fruitfully to

2 Richard Jones to William Whewell, 25 February 1831, Whewell Papers (Hereafter WP) Add.Ms.c.52 folio 21.

3 John Herschel to William Whewell, 9 May 1835, WP Add.Ms.a.207 f. 26.

problems in moral and social philosophy; indeed, he presented this as one of the motivations for writing the book (Herschel 1830, 72–4).

Whewell himself thought that the "hyperphysical branches of knowledge ... art, language, political economy, morals and the like," as well as the physical branches, could be improved by reforming induction.[4] The three set for themselves the task of carrying out this reform, which required first that they define a "true idea of induction." In an early notebook entry, Whewell lamented "that the true idea of induction has not been generally fixed and agreed upon must I think be very obvious."[5] Bacon's own view, though a useful starting point, was itself flawed; it needed to be renovated (thus, in later years, Whewell gave the title *Novum Organon Renovatum* to one volume of the third edition of his *Philosophy* [Whewell 1858]).

Both Whewell and Herschel wrote books on scientific method, and each began their works with aphorisms of Bacon. Herschel even placed the bust of Bacon on the title-page of his book. Yet although each man wished to renovate Bacon's inductivism, they had very different views of what Baconian induction entailed. This difference is reflected in the aphorisms with which each chose to open his book. Whewell began with Aphorism 18 from Book I of the *Novum Organum:* "The discoveries which have hitherto been made in the sciences are such as lie close to vulgar notions, scarcely beneath the surface. In order to penetrate into the inner and further recesses of nature, it is necessary that both notions and axioms be derived from things by a more sure and guarded way; and that a method of intellectual operation be introduced altogether better and more certain." Herschel chose the first aphorism: "Man, as the minister and interpreter of nature, is limited in act and understanding by his observation of the order of nature: neither his knowledge nor his power extends farther." Not surprisingly, these aphorisms highlight the different elements of Bacon's thought they emphasized: Whewell the slow, careful, rational method which can penetrate to unobservable parts of nature; Herschel the empiricist epistemology.

These different interpretations of Bacon are also apparent in the reviews each man wrote of his friend's work. In his review of Herschel's *Preliminary Discourse*, Whewell accused his friend of having omitted to relay Bacon's "condemnation of the method of anticipation, as opposed to that of gradual induction; a judgment indeed which of itself almost

4 William Whewell to Richard Jones, 5 August 1834, WP Add.Ms.c.51 f. 174.

5 See notebook dated 28 June 1830, WP R.18.17 f. 12, pp. v–ix.

conveys the whole spirit and character of his philosophy" (Whewell 1831b, 399). What Whewell considered here to be expressive of "the whole spirit and character" of Bacon's inductive method was his injunction against hasty "anticipations" or leaps to hypotheses; instead, Whewell stressed, Bacon prescribed gradual generalizations to laws. And as Whewell noticed, Herschel, in his *Preliminary Discourse*, did ignore or at least underplay the importance of this aspect of Bacon's method of "interpretation of nature."

In the book Herschel characterized John Dalton's discovery of the law of multiple proportions as involving only "the contemplation of a few instances, without passing through subordinate stages of painful inductive ascent by the intermedium of subordinate laws" (Herschel 1830, 303–305). Herschel seems to have been influenced by Thomson's view of Dalton's discovery, according to which, after experiments analyzing only methane and ethylene, Dalton leapt to his conclusion that all elements combine in simple integral ratios and in fixed proportions (see Nash 1956, 101–2; Thomson 1825, I: 11–12; and Herschel 1830, 305n).

On Whewell's view, in contrast, Dalton drew his conclusion from a number of experiments with different chemical elements. Thus he admonished Herschel, "such language we cannot but think is liable to be mistaken. If Mr. Dalton had guessed the law to be true from a few instances, and done no more, he would have been the first person whose name has been permanently connected with the history of science in virtue of such an unexamined simplification." He then quoted from their friend Adam Sedgwick's recent address to the Geological Society: "'the records of mankind offer no single instance of any great physical truth anticipated by mere guesses and conjectures'" (Whewell 1830, 401). Whewell was worried that Herschel might seem to be encouraging a spirit of "gratuitous theorizing," by not cautioning against anticipatory leaps to hypotheses. But it is important to note that Whewell did not believe Herschel was endorsing guesswork as a way of inventing hypotheses. He saw that Herschel was allowing a rational inference to a law from only one or two (strong) instances, and this is what Whewell characterized as contrary to Bacon's inductivism.

Nine years later, Herschel had the chance to return the charge of anti-Baconianism. When Whewell's *Philosophy of the Inductive Sciences* was first published, Herschel wrote to him: "Your book is a tough one – when I ruminate it chapter by chapter I chew the cud of both sweet and bitter fancies – you are too a priori rather for me – as soon as one has worked one's way up to a general law you come cranking in and tell me it is a Fundamental Idea innate in everybody's mind ..."[6] In his lengthy joint review of the *Philosophy* and Whewell's *History of the Inductive Sciences,* Herschel claimed that Whewell was the one who had abandoned Bacon. What Herschel objected to in Whewell's book was the ideal or conceptual element of Whewell's epistemology. Whewell argued that every act of knowledge required an ideal as well as an empirical element, an idea or conception of the mind as well as sensations or perceptions from outside the mind. Whewell's view was not a purely rationalist or ideal one, as some have claimed, but was "antithetical," in the sense of combining empirical and rational components (for a more detailed discussion of Whewell's epistemology, see Snyder 2006, chapter 1).

Whewell believed that this position was consistent with Bacon's own epistemology. He claimed that Bacon did not ignore the conceptual side of knowledge altogether; the problem is that Bacon never was able to complete his task of reforming philosophy: Whewell explained that "if he had completed his scheme, Bacon would probably have given due attention to Ideas, no less than to Facts, as an element of our knowledge" (Whewell 1860, 136). Although it has been argued that Whewell more or less invented this reading of Bacon in order to "detach" the British inductive tradition from French positivism, this conceptual element *can* be found in Bacon's writings in various ways, though not as explicitly as it is developed in Whewell's epistemology (see Yeo 1993, 247). To take just one example, Bacon claimed that he "established for ever a true and lasting marriage between the empirical and the rational faculty" (Bacon 1877–89, IV: 19). He elaborated on this marriage in his famous aphorism urging the scientist to emulate the bee:

> Those who have handled sciences have been either men of experiment or men of dogmas. The men of experiment are like the ant; they only collect and use: the reasoners resemble spiders, who make cobwebs out of their own substance. But the bee takes the middle course; it gathers its material from the flowers of the garden and of the field, but transforms and digests it by a power of its own. Not unlike this is the true business of philosophy; for it neither relies solely or chiefly on the powers of the mind, nor does it take the matter which it gathers from natural history and mechanical experiments and lay it up in the memory whole, as it finds it; but lays it up in the understanding altered and digested. Therefore from a closer

6 John Herschel to William Whewell, 6 August 1840, WP Add.Ms.a.207 f. 45.

> and purer league between these two faculties, the experimental and the rational, (such as has never yet been made) much may be hoped. (Bacon 1877–89, IV: 92–93; see also Rossi 1984, 255)

Moreover, Bacon cautioned that the forms of simple nature that science seeks cannot be discovered until conceptions or "notions" are clarified (Bacon 1877–89, IV: 49–50, 61–62). And he further claimed that the work of fact-collecting could not be conducted blindly, without some theoretical guidance (*ibid.*, 94–95, 81).

Herschel, however, did not interpret Bacon as having attempted to marry the conceptual and the empirical in this way; he thus criticized Whewell for having strayed from Bacon's epistemology. In his review Herschel described Whewell's position as that of the "high a priori Pegasus," in contrast to the inductive philosopher, a "plain matter of fact roadster":

> The high a priori Pegasus … is a noble and generous steed who bounds over obstacles which confine the plain matter of fact roadster to tardier paths and a longer circuit. There is no denying to this philosophy, for one of its distinguishing characters, a verve and energy which a merely tentative and empirical one must draw from foreign sources, from a solemn and earnest feeling of duty and devotion, in its followers, and a firm reliance on the ultimate sufficiency of its resources to accomplish every purpose which Providence has destined it to attain. (Whewell 1841, 223)

So, ironically, Herschel accused Whewell of allowing "bold leaps" epistemologically, while Whewell accused Herschel of allowing "bold leaps" methodologically. The disagreement between the two was sharp enough that Herschel was moved to write: "we begin at diametrically opposite points and meet only in our central love of truth … We are like two staunch politicians Tory and Radical who agree in love of country and whom a thousand delightful associations keeps from tearing each others eyes out."[7]

This paper is not the place to argue over who was the legitimate heir to Bacon (though I think that Whewell was the more perspicuous reader of Bacon – for more on this see Snyder 1999 and 2006). The important point is that both Herschel and Whewell saw themselves as following in the path of the "inductive master" (as they referred to Bacon) by endorsing the use of inductive method in the sciences. This is why Herschel and Whewell could each consider the other as devoted to the same general project, even as they accused each other of diverging from Ba-

---

7 John Herschel to William Whewell, 17 April 1841, WP Add.Ms.a.207 f.46.

con's view. The shared task was that of supporting the reform of scientific method, and more specifically promoting inductive method over deductive views, such as that proposed by Richard Whately and his followers at Oriel College, Oxford (see Snyder 2006, chapter 1). Thus, even after reviewing a book that seemed to promote "hasty anticipations" as opposed to gradual generalizations, Whewell could still dedicate his *History of the Inductive Sciences* to its author, saying "[these volumes] are the result of trains of thought which have often been the subject of our conversation, and of which the origin goes back to the period of our early companionship at the University. And if I had ever wavered in my purpose of combining such reflections and researches into a whole, I should have derived a renewed impulse and increased animation from your delightful Discourse on a kindred subject ..." (Whewell 1837 [1857], vol. I). And of the philosopher whom he considered an "a priori Pegasus," as bad as some of the "later German metaphysicians," Herschel could still describe his work as "among the most important contributions which have ever been made to the philosophy of mind" (Herschel 1841, 183). The disagreement between Whewell and Herschel, though based on very real differences, was seen by them as a family argument, a dispute between brothers devoted to each other and to their father, but in conflict over which one was most deserving of his legacy.

3. Herschel and Hypotheses

Numerous commentators have claimed that Herschel was a proponent of the method of hypothesis (in addition to those cited earlier, see Yeo 1993, Schweber 1981, and Laudan 1969). Yet Herschel did not endorse a method allowing any hypothesis to be tested, no matter how it was invented. Herschel, unlike twentieth-century proponents of hypothetico-deductivism such as Popper or Hempel, placed inductive constraints on hypotheses at the stage of initial invention. He explained that

> It must not be ... supposed that, in the formation of theories, we are abandoned to the unrestrained exercise of imagination, or at liberty to lay down arbitrary principles, or assume the existence of mere fanciful causes. The liberty of speculation which we possess in the domains of theory is not like the wild license of the slave broke loose from his fetters, but rather

> like that of the freeman who has learned the lessons of self-restraint in the school of just subordination.[8] (Herschel 1830, 190–1)

Even the "leap" to a "bold hypothesis," Herschel believed, requires "self-restraint." As is apparent in his use of the example of Dalton, Herschel – unlike Bacon (and Whewell) – allowed the inference to a hypothesis from what he called "two or three strongly impressive facts, rather than by affording the whole mass of cases a regular consideration" (Herschel 1830, 182). But the way the hypothesis is reached is through inference. Herschel thus described a "well-imagined hypothesis" as one which has "been suggested by a fair inductive consideration of general laws" (Herschel 1830, 196).

Herschel continually stressed that both induction and deduction are used in inventing and testing hypotheses, using Bacon's image of an ascending and descending ladder.[9] It is especially important the hypotheses reached on the basis of "leaps" from few phenomena be confirmed by drawing out their deductive consequences and testing them. Indeed, since there are so many difficulties involved in drawing out the deductive consequences of hypotheses that are the result of "bold leaps," Herschel claimed that this "effectively precludes this method from being commonly resorted to as a means of discovery, unless we have some good reason, *from analogy or otherwise*, for believing that the attempt will prove successful, or have been first led by partial inductions to particular laws" (Herschel 1830, 200; emphasis added).

As this quote suggests, Herschel especially emphasized the use of analogical inference in reaching these "well-imagined hypotheses."[10] The

8 Although Herschel is here speaking of the formation of theories, rather than "inductions" i.e. to hypotheses, he does note that "The ultimate objects we pursue in the highest theories are the same as those of the lowest inductions; and the means by which we can most securely attain them bear a close analogy to those which we have found successful in such inferior cases" (p. 191).

9 Herschel 1830, 174–5; see also 181. Following Bacon, Herschel noted that the tendency to leap to a generalization after few instances is a part of our psychology; yet, unlike Bacon, Herschel allowed that this tendency was often useful: "such is the tendency of the human mind to speculation, that on the least idea of an analogy between a few phenomena, it leaps forward, as it were, to a cause or law, to the temporary neglect of all the rest; so that, in fact, almost all our principal inductions must be regarded as a series of ascents and descents, and of conclusions from a few cases, verified by trial on many" (1830, 165).

10 In allowing leaps to hypotheses as long as these were justified by analogical inference, Herschel seems to be following Dugald Stewart's interpretation of Bacon's inductivism. Although Stewart had claimed that Bacon required hypoth-

emphasis on analogical inference is clear in Herschel's use of Newton's principle of the *vera causa* in order to explain how scientists should infer causal laws. In his *Principia*, Newton had expressed four "Rules of Reasoning in Philosophy." The first of these methodological precepts read, in part, "We are to admit no more causes of natural things than such as are both true and sufficient to explain their appearance." By requiring causes that are "true causes," Newton was rejecting an instrumentalist view of theories, in which causes may be postulated as long as they "save the appearances," regardless of whether they are actually true. Further, the rule has seemed to some commentators to suggest that there must be empirical evidence for the existence of a cause independent of its explanatory power in a particular case. Thus, for instance, Newton's eighteenth-century supporter Thomas Reid interpreted Newton's methodological injunction in this way, claiming that the requirement for true causes meant that causes are "not to be conjectured to exist without proof."[11]

In his *Preliminary Discourse* Herschel gave much the same characterization of true causes as had Reid, explaining that scientists seek "causes recognized as having a real existence in nature, and not being mere hypotheses or figments of the mind." He used the following example:

> The phenomenon of shells found in rocks, at a great height above the sea, has been attributed to several causes. By some it has been ascribed to a plastic virtue in the soil; by some, to fermentation; by some, to the influence of the celestial bodies; by some, to the casual passage of pilgrims with their scallops; by some, to birds feeding on shell-fish; and by all modern geologists ... to the life and death of real mollusca at the bottom of the sea, and a subsequent alteration of the relative level of the land and sea. Of these, the plastic virtue and celestial influence belong to the class of figments of fancy. Casual transport by pilgrims is a real cause, and might account for a few shells here and there dropped on frequent passes, but is not extensive enough for the purpose of explanation. Fermentation, generally, is a real cause, so far as that there is such a thing; but it is not a real cause of the production of a shell in a rock, since no such thing was ever witnessed

eses in arriving at laws, Stewart's definition of hypotheses differs from that of the modern-day hypothetical-deductivist. Stewart claimed that the term "hypothesis" denotes "not fictions altogether arbitrary," but rather suppositions "supported by strong analogy" (see Stewart, 1792–1827, II: 403; and Rashid 1985, 255). It is noteworthy that Stewart presented a copy of his *Elements of the Philosophy of the Human Minds* to the astronomer William Herschel, who bestowed it to his son John (I owe this point to Pietro Corsi).

11 See Reid 1785. The concept of the *vera causa* is thoroughly discussed in Kavaloski 1974.

> as one of its effects, and rocks and stones do not ferment. On the other hand, for a shell-fish to die in the mud, where it becomes silted over and imbedded, happens daily; and the elevation of the bottom of the sea to become dry land has really been witnessed so often, and on such a scale, as to qualify it for a *vera causa* available in sound philosophy. (Herschel 1830, 144–45)

We see, then, that there are two requirements for a *vera causa* on Herschel's view. First, it must be a cause whose existence in other cases is already known. Secondly, it must be a cause whose ability to produce a similar – or analogous – effect is known independently of its putative responsibility for causing the phenomenon in this case. As Herschel explained, *verae causae*, when not directly observable in the case at hand, "are not to be arbitrarily assumed; they must be such as we have good inductive grounds do exist in nature, and do perform a part in phenomena *analogous to those we would render an account of*" (Herschel 1830, 197; emphasis added). The alleged causes of "plasticity" and celestial influence fail on the first condition, namely by not being causes whose existence is known in other cases. Fermentation is known to be a cause that exists, but only in cases quite different from cases of rocks and shells; thus it fails to be a *vera causa* in this case by the second condition. The dropping of shells by pilgrims or birds is a cause known to exist, and is also known to be responsible for the appearance of shells in other instances, but not in cases with such a vast number of shells. Thus it too fails as a cause whose ability to produce a similar effect is known independently.

Note that a "true cause" might not be true, in the sense of being the cause that actually is at work in a particular case. A *vera causa* is true in the sense that it is a causally efficacious agent at work in analogous instances. Another example used by Herschel can illustrate this point: Lyell's theory of climatic change. It was recognized by geologists at this time – from evidence of fossil remains – that the climate of the earth, or at least large portions of it, had become cooler. One theory claimed that the earth's past warmer climate was due to the immensely greater action of volcanoes in the past, bringing internal heat to the surface of the earth. Yet, Herschel noted, there was no independent evidence for greater volcanic activity in the past (Herschel 1830, 146; see also Ruse 1976, 123).

On the other hand, Herschel explained that Lyell's proposed cause – the varying influence of the distribution of the land and the sea over the globe – relied upon the action of phenomena similar to that which we know "by actual observation." As Herschel explained, "a change of such distribution, in the lapse of the ages, by the degradation of old continents, and the elevation of the new, [is] a demonstrated fact; and the influence of such a change on the climate of particular regions, if not of the whole globe, [is] a perfectly fair conclusion, from what we know … by actual observation" (Herschel, 1830, 146–7). That is, we know by observation that changes in the distribution of land and sea have occurred in particular regions of the earth (so it passes the first condition). Moreover, we know that climate differs in "continental, insular, and oceanic" regions, and seems to be caused by features of these different regions. Thus it is reasonable to infer that climate change over the whole globe, or at least large regions of the globe, could be caused by changes in the geographic structure of the regions themselves. Accordingly, Herschel could claim that Lyell's theory was worthy of consideration, even though he was unwilling to commit fully to its truth.

Another important aspect of Herschel's use of the *vera causa* principle is that this principle allows for the inference to theoretical or unobservable causes, as long as these are analogous to other known causes. This is why Herschel was able to accept that the wave theory of light offered a *vera causa* of optical phenomena, even though the cause postulated was a theoretical cause: the motion of unobservable waves in an unobservable ether. The motion of waves in a medium was known empirically to have certain effects in other cases (such as water waves), and very often wave theorists of light did describe optical effects in analogy with these cases, strengthening their case for inferring the action of undulations in an unseen ether. Herschel certainly drew attention to this analogy in his discussion of optics in the *Preliminary Discourse.* (See Herschel 1841, 234, 251, 260 and Kavaloski 1974, 59).[12] Herschel's view of true causes does not in fact rule out theoretical or unobservable causes, as long as they are analogous to other known causes; indeed, this is the way in which Herschel allowed for theoretical science while still endorsing an empirical, inductive method.

Thus, far from allowing a scientist to postulate any cause hypothetically, as modern commentators would have it, Herschel required that theoretical causes be strongly analogous to already-known causes. Even Whewell, who believed that Herschel was too licentious in the

12 Thus I disagree with Ruse's claim that Herschel's acceptance of the wave theory of light forced him to abandon his notion of the *vera causa* and to accept Whewell's less-stringent interpretation of it (see Ruse 1979, 59).

use of hypotheses, thought that this requirement was overly-restrictive.[13]

4. A Historiographic Puzzle

This brief examination is enough to show that, like Whewell, Herschel cannot plausibly be considered the philosophical ancestor of Popper or other writers who reject the use of induction in the invention of hypotheses and who allow non-inductive "guesses" to be confirmed or found "not yet falsified" by the testing of their deductive consequences. Yet commentators persist in claiming that early to mid-nineteenth century Britain was dominated by the method of hypothesis, pointing to the major writers on methodology, Whewell and Herschel, as exemplifications. Since neither Whewell nor Herschel proposed views similar to modern-day proponents of hypothetical deductivism, a historiographic question is raised: what accounts for this interpretation and its predominance?

One reason is the mistake of reading certain terms used in the nineteenth century as if they held the same meaning as they did in the twentieth century or do today. Thus, since Whewell spoke at times of "conjectures" and "guesses," we are told that he shared Popper's methodology of "conjectures and refutation," even though he used these terms in ways that differ quite radically from the way they were used by Popper. Whewell often used these terms in a way connoting a conclusion that is simply not yet confirmed; other times he suggested that they were an-

13 Whewell agreed with the first condition, namely that causes are not to be "arbitrarily assumed," and that we must have "good inductive grounds" for believing in their existence. Thus he explained that verae causae are "those which are justly and rigorously inferred" (see 1840, II: 189). However, he disagreed with Herschel's second condition, which requires an analogous connection to known causes. Were we to limit ourselves in this way, Whewell argued, science could never make progress. This type of interpretation "forbids us to look for a cause, except among the causes with which we are already familiar. But if we follow this rule, how shall we ever become acquainted with any new cause?" (Whewell, 1840 II: 442). On these grounds Whewell criticized Lyell (and uniformitarian geology in general) for ruling out the possibility of catastrophic causes of geological change; i. e., for not allowing that causes of unknown kinds and intensity may have been at work in the past (see Whewell, 1831b and 1832. For more on the relation between Lyell, Herschel and Whewell see Ruse 1976 and 1979).

ticipations rather than slow and careful generalizations. For instance, in speaking of Kepler's move from his discovery of Mars's elliptical orbit to his first law of planetary motion, Whewell claimed that "When he had established his premise, that 'Mars does describe an Ellipse around the Sun,' he does not hesitate to guess at least that … 'All the Planets do what Mars does.'"[14] But this was no mere guess on Kepler's part, but was, rather, an inductive generalization of a property found to exist in one member of the class of planets to all its members. Although such an inference may be a rather weak one, because he was inferring from a single case to a universal generalization, in this instance Kepler surely had additional rational support for inferring that all the planets shared the property of the elliptical orbit from the premise that one of them had this property; there would be strong analogical and causal reasons for thinking that the planetary orbits would all lie on the same type of curve (because it was reasonable to suppose that each orbit is caused by the same physical mechanism).

Whewell often used the term "conjecture" in a similar way.[15] His use of these terms does not imply the absence of any rational inference. Rather, his usage is consistent with ways that the term was used prior to the twentieth century. In the nineteenth century and earlier, "conjecture" was often used to connote not a hypothesis reached by non-rational means, but rather one which is "unverified," or which is "a conclusion as to what is likely or probable" (as opposed to the results of demonstration). Writers whose work was well-known to Whewell, including Bacon, Kepler, Newton and Stewart, used the term in this way. This common usage of the term explains why Whewell was not interpreted by nineteenth-century reviewers as advocating the method of hypothesis. Indeed, his sometime-nemesis David Brewster criticized him for *not* proposing such a view (see Brewster 1842).

Another reason for misinterpreting Whewell and Herschel as hypothetical-deductivists arises from reading paragraphs out of the larger contexts of the work at hand and the general philosophy and intentions of its author. Since Herschel spoke of making "bold leaps" to hypotheses, it is supposed that this summarizes his methodology. But if we look, as we must, to the broader context of the work as a whole, where he makes clear the need for inductive restraints, it is clear that Herschel's view is rather different from the modern hypothetical deduc-

14 Whewell 1858, 75.
15 See, e.g., Whewell 1837 [1857], I: 299).

tive position. This is even more obvious if we consider also the argumentative context of his controversy with Whewell over the proper way to interpret Bacon. Herschel wanted to emphasize that Bacon's most distinctive characteristic was his empirical epistemology, rather than his gradualist methodology. By situating his comments about leaps to hypotheses in the context of a book sealed with Bacon's bust, prefaced with quotes from Bacon, and filled throughout with explications of Bacon's methods, Herschel was signaling his inductive view loudly and clearly.

One explanation for the view that Whewell and Herschel proposed a non-inductive methodology similar to twentieth-century hypothetical deductivism has to do with the influence of Mill's *System of Logic.* Thus, ironically, Mill is partly responsible for the misconception that his century was dominated by a methodology opposed to his own. Mill's *System of Logic* promoted, and popularized, a view of induction as a rather narrow logical operation, involving only eliminative and enumerative forms of reasoning as found in Mill's famous "Methods of Experimental Inquiry." His work was incredibly successful, becoming the standard work in logic at both Cambridge and Oxford and, eventually, American universities as well. In the twentieth century, the acceptance of Mill's view of induction led to a false dichotomy: between a narrow Millian inductivism and hypothetico-deductivism. It was argued by some proponents of hypothetico-deductivism that scientific discovery could not be a matter of merely calculating enumerations of observed instances, perhaps together with some eliminative process; discovery could not proceed by a logical "rule-book" (recall that Mill himself claimed to be giving rules for inductive science analogous to the syllogism in deductive logic). Further, it was often noted that theoretical entities were an important part of modern science, and, since Mill's methods could not reach theoretical entities, these methods could not be the proper path to discovery. In this way it was concluded that scientific discovery was not, and could not be, inductive. Instead, it must consist in a non-inferential process, one involving no "logic." Mill's influence in redefining induction helped lead to the view that scientific discovery must be hypothetical rather than inductive (for a further development of this claim see Snyder 2006, 331–2).

Yet before Mill, the term "induction" often referred to a broader logical operation, one involving more than just enumerative and eliminative reasoning. Mill's inductivism involved a break from earlier meanings of the term induction, and there is no reason why we should use Mill's categories to characterize the methodologies of Herschel or Whewell. Like Mill, Herschel and Whewell wanted to define inductive methods for science (and indeed for all areas of thought). The three men had very different ideas about how to go about doing this. Nevertheless, it is clear that theirs was predominantly an age of inductivism, not of hypothetical-deductivism.

Bibliography

Bacon, Francis (1857–61 [1877–1889]), *The Works of Francis Bacon.* Collected and Edited by J. Spedding, R. L. Ellis and D. D. Heath, in 14 volumes. London: Longman and Co.

Brewster, David (1842), "On the History of the Inductive Sciences", *Edinburgh Review* 74: 139–161.

Butts, Robert E. (1987), "Pragmatism in Theories of Induction in the Victorian Era: Herschel, Whewell, Mach and Mill", in: Herbert Stachowiak, ed., *Pragmatik: Handbuch Pragmatischen Denkens.* Hamburg: F. Meiner, 40–58.

Herschel, John F. W. (1830), *Preliminary Discourse on the Study of Natural Philosophy.* London: Longman, Rees, Orme, Brown and Green and John Taylor.

— (1841), "Whewell on Inductive Sciences", *Quarterly Review* 68: 177–223.

Jacobs, Struan (1991), *Science and British Liberalism: Locke, Bentham, Mill and Popper.* Aldershot: Avebury.

— (1991), "John Stuart Mill on Induction and Hypotheses", *Journal of the History of Ideas* 29: 69–83.

Kavaloski, Vincent C. (1994), "The Vera Causa Principle: A Historico-Philosophical Study of a Metatheoretical Concept from Newton Through Darwin". Ph.D. diss., University of Chicago.

Laudan, Larry (1969), "Theories of Scientific Method from Plato to Mach", *History of Science* 7: 1–63.

— (1980), "Why was the Logic of Discovery Abandoned?", in: Thomas Nickles, ed. *Scientific Discovery: Case Studies.* Dordrecht: Reidel, 173–83.

Nash, Leonard K. (1956), "The Origins of Dalton's Chemical Atomic Theory", *Isis* 47 (2): 101–116.

Newton, Isaac (1687 [1729, 1968]), *Mathematical Principles of Natural Philosophy.* Translated by Andrew Motte. Reprinted with an introduction by I. Bernard Cohen. In two volumes. London: Dawson.

Rashid, Salim (1985), "Dugald Stewart, 'Baconian' Methodology, and Political Economy", *Journal of the History of Ideas* 46: 245–57.

Reid, Thomas (1785), *Essays on the Intellectual Powers of Man.* Edinburgh: John Bell.

Rossi, Paolo (1984), "Ants, Spiders, Epistemologists", in: M. Fattori, ed. *Francis Bacon: Terminologia e Fortuna nel XVII Secolo.* Rome: Edizioni dell'Ateneo, 245–60.

Ruse, Michael (1979), *The Darwinian Revolution: Science Red in Tooth and Claw.* Chicago: University of Chicago Press.

— (1976), "Charles Lyell and the Philosophers of Science", *British Journal for the History of Science* 9: 121–31.

Schweber, S. S., ed. (1981), *Aspects of the Life and Thought of Sir John Frederick Herschel*. New York: Arno Press.

Snyder, Laura J. (1997a), "Discoverer's Induction", *Philosophy of Science* 64: 580–604.

— (1997b), "The Mill-Whewell Debate: Much Ado about Induction", *Perspectives on Science* 5: 159–98.

— (1999), "Renovating the *Novum Organum:* Bacon, Whewell, and Induction", *Studies in History and Philosophy of Science* 30A: 531–57.

— (2006), *Reforming Philosophy: A Victorian Debate on Science and Society*. Chicago: University of Chicago Press.

Stewart, Dugald (1792–1827), *Elements of the Philosophy of the Human Mind*. In three volumes. London: A. Strahan and T. Cadell, Edinburgh: W. Creech.

Thomson, Thomas (1825), *An Attempt to Establish the First Principles of Chemistry by Experiment*. In two volumes. London: Baldwin, Cradock and Joy.

Whewell, William (1831a), "Lyell's *Principles of Geology,* volume 1", *British Critic* 9: 180–206.

— (1830 [1831]), "Modern Science – Inductive Philosophy [Review of J. Herschel, *Preliminary Discourse on the Study of Natural Philosophy* (1830)]", *Quarterly Review* 45 (1831): 374–407.

— (1832), "Lyell's *Principles of Geology,* volume 2", *Quarterly Review* 93: 103–32.

— (1840), *The Philosophy of the Inductive Sciences, Founded Upon Their History*. 1st edition, in two volumes. London: John W. Parker.

— (1847), *The Philosophy of the Inductive Sciences, Founded Upon Their History*. 2nd edition, in two volumes. London: John W. Parker.

— (1837 [1857, 1873]), *History of the Inductive Sciences, from the earliest to the present time*. London: J. W. Parker. Reprint in two volumes, New York: D. Appleton and Co.

— (1858), *Novum Organon Renovatum*. London: John W. Parker.

— (1859), "Prefatory Notice" to W. Whewell, ed. *Literary Remains consisting of lectures and tracts on political economy, by the late Rev. Richard Jones*. London: John W. Murray, ix–xl.

— (1860), *On the Philosophy of Discovery: Chapters Historical and Critical*. London: John W. Parker.

Whewell Papers, Trinity College, Cambridge.

Yeo, Richard (1993), *Defining Science: William Whewell, Natural Knowledge, and Public Debate in Early Victorian Britain*. Cambridge: Cambridge University Press.

From Axioms to Conventions and Hypotheses: The Foundations of Mechanics and the Roots of Carl Neumann's "Principles of the Galilean-Newtonian Theory"

Helmut Pulte

Abstract:

This paper is devoted to the rise of hypothetical thinking in the tradition of 19th century rational mechanics in general, and to the roots of Carl Neumann's paper on the "Principles of the Galilean-Newtonian Theory" within this tradition in particular. While Neumann's analysis of the law of inertia and Newton's concept of space is well known and accepted as an important step towards a better understanding of both, this historical background – which sheds light on Neumann's systematic arguments in different respects – has been widely neglected. It is shown that the rise of "pure mathematics" plays an important role for the rise of hypothetical thinking concerning the foundations of mechanics in general, and that this new understanding of mathematics is of utmost importance for Neumann's hypothetical-deductive concept of science.

1. Introduction

In 1866, the philosopher and psychologist Wilhelm Wundt published *Die physikalischen Axiome und ihre Beziehung zum Kausalprinzip* (Wundt 1866). This book is one of the latest manifests of what may be called "classical mathematical philosophy of nature" (CMN): It expresses the view that natural philosophy can be established on the basis of certain unshakable mathematical "axioms" of mechanics which deal with the movement of "ponderable" masses underlying certain forces and constraints. About four decades later, a second, revised edition of this work appeared under the title *Die Prinzipien der Naturlehre. Ein Kapitel aus einer Philosophie der Naturwissenschaften* (Wundt 1910). Wundt seized this opportunity in order to reflect critically on his former position, and to indicate a dramatic change with respect to the understanding and use of the concept "axiom," both in mechanics and in (pure) mathematics, in the two decades from 1866 onward: "What had been accepted as an axiom in former times was now labelled as "hypothesis," thereby expressing that also alternative systems of premises – perhaps deviating essentially from the established system – can be chosen, as long as they serve the purpose of linking the phenomena which have to be described" (Wundt 1910, 2).

In fact, the two decades which – according to Wundt – undermined the traditional "axiomatic view" and paved the way for a new "hypothetic view" in mechanics, include the first public debate about Bernhard Riemann's *Ueber die Hypothesen, welche der Geometrie zu Grunde liegen* (Riemann 1854 [1892], first publ. 1867) and – sometimes without a clear demarcation from Riemann's approach – on the Non-Euclidean Geometries of Gauss, Lobatschewskij and the Bolyais. They also include the rise of electrodynamics and thermodynamics as fields of mathematical physics in their own rights, and the rise of a broad discussion on the epistemological status and tasks of natural science, highlighted *inter alia* by the "description versus explanation-discussion" provoked by G. R. Kirchhoff's *Mechanik* from 1876 (cf. Kirchhoff 1876 [1897]). Wundt's historical explanation of the decline of CMN is restricted to three aspects: *(1)* the foundational debate in geometry, *(2)* the rise of the concept of energy, which undermined the traditional basis of mechanics and *(3)* the rise of electrodynamics and its radical "descendant," the electron theory of matter (Wundt 1910, 3). Today, a well-informed historian of science will add *at least* one more reason: *(4)* The rise of a new strand of phenomenalism within philosophy and the sciences, which is – with respect to the destruction of mechanical "axioms" – most obvious in the work of E. Mach.

As far as the criticism of Newton's theory of absolute space and the law of inertia is concerned, Mach had to accept – and frankly did accept (Mach 1872 [1909], 47) – one mathematician as his precursor who was obviously neither an adherent of phenomenalism, nor fitted well into Wundt's historical analysis of the decline of CMN: Carl Neumann, a son of the mathematical physicist Franz Ernst Neumann. Neumann the elder founded with Carl G. J. Jacobi the Königsberg seminar for mathematics and physics, which can be seen as the "nucleus" of German theoretical physics in the second half of the 19th century (see Olesko 1991). In 1869, Carl Neumann gave his lecture *On the Principles of the Galilean-Newtonian Theory* (Neumann 1869b) that – in sharp conflict with Wundt's position from 1866 – expressed emphatically a modern, even "Popper-like" hypothetical-deductive understanding of mathematical natural philosophy in general and especially a modern concept of mechanics (MMN: "Modern Mathematical Philosophy of Nature"), thus opening a vivid discussion on its foundations which lasted until Einstein and, in a way, made Einstein's revolution possible.

These cursory remarks point to the two main objectives of this paper: First and in general, I am interested in the way *how* – and the reason *why* – hypothetical thinking at first penetrated the discussion on the principles of classical mechanics in the course of the early 19th century. Wundt's analysis, elaborated and extended later by the history of science and the history of philosophy of science, concentrates on a relatively late period. Even a superficial glimpse reveals the development in question: The basic laws of mechanics, may they be formulated in a synthetic, Newtonian style or in the later, analytical manner, are at the beginning of the 19th century labelled as "axioms" (for example by J. L. Lagrange or J. Herschel), as "necessary truths" or "indubitable principles" (for example by P. S. de Laplace or W. Whewell) or – by a transcendental transformation of these properties – as "synthetic principles a priori" (by I. Kant, W. R. Hamilton and others). The set of basic laws used for the organization of theoretical mechanics was understood as a *unique* one, and each principle was dignified not only as general, but also as *certain* and *evident*, though the epistemological justifications of these features differed considerably among philosophers and scientists. In the second half of the nineteenth century, however, we meet with quite different notions for the *same* laws: After the "turning point" noticed by Wundt, they are labelled as "conventions" (by H. Poincaré, for example, though *not* for the first time), as mere "hypotheses" (by B. Riemann, C. Neumann, L. Boltzmann and others) or as provisional "descriptions" (see G. R. Kirchhoff or E. Mach, for example). This change of "second-order labels" is easily visible, but indicates a profound change of the understanding of rational mechanics as a both *mathematical* and *empirical* science that is less visible and the reasons of which are not completely understood until now. In short, this development can be described as removal of a traditionally mechanical Euclideanism – I am using this "Lakatosian" term deliberately as it is "epistemologically neutral"[1] – by a modern, "hypothetico-deductive" understanding of science

---

1 Euclideanism according to Lakatos expresses the view that the "ideal theory is a deductive system with an indubitable truth-injection at the top (a finite conjunction of axioms) – so that truth, flowing down from the top through the safe truth-preserving channels of valid inferences, inundates the whole system"; its basic aim "is to search for self-evident axioms – Euclidian methodology is puritanical, antispeculative" (Lakatos 1978, II, 28 and 29). Euclideanism in

for which the fallibility and revisability even of its *first* principles is decisive. Diachronic analysis of the writings of many philosopher-scientists show that the image of science in general, and mechanics in particular, underwent such a change during the last decades of the 19th century – Hermann von Helmholtz is perhaps the most impressive and best investigated case study in this respect (see Schiemann 1997). The process in question is not a discontinuous, but a *gradual* one – it is a meta-theoretical *evolution* that, to a certain extent, first paved the way for the later scientific *revolution*. As I have described and examined this evolution with respect to mechanics in some detail elsewhere (Pulte 2005, chs. VI, VII), a short structural analysis of the early history in the second part of this paper will suffice here. This part is restricted to the *early* dissolution of mechanical Euclideanism and focuses on the role of a new understanding of mathematics in order to show that there were reasons within the traditional mechanics of "ponderable" masses independent of and prior to the empirical challenges (i. e. the integration of "new" areas of phenomena like those of electrodynamics and thermodynamics), independent of the rise of the debate about Non-Euclidean geometries and also independent of the rise of modern phenomenalism (or empirocriticism). In other words, the second part of this outline will end *before* these aspects became predominant in the discussion on the principles of mechanics. C. G. J. Jacobi will be the central figure of this part.

The second and more specific objective will be dealt with in the third part of this paper: The literature on Carl Neumann and on his lecture on the Galilean-Newtonian theory takes the new "MMN-position" presented there as something coming "out of the blue." I will try to show, however, that Neumann's turn can only be understood in the context of the earlier development or, to be more concrete, that it is strongly influenced by C. G. J. Jacobi and his new attitude towards mathematics and the mathematically formulated mechanical principles. This result seems to me of some importance for the history of philosophy of science, because it shows that the neo-humanist under-

this sense is *epistemologically neutral* in so far as it is applicable both to traditional rationalism and empiricism: whether the axioms at the top are revealed by the 'light of reason' (see Descartes, for example) or 'deduced from phenomena' (Newton) is not relevant for the above definition of Euclideanism. Moreover, the dichotomy of traditional rationalism and empiricism conceals the common characteristic of infallibility.

standing of mathematics had a direct impact on the foundational discussion on space and the law of inertia in the later 19th century.

2. The Rise of Hypothetical Thinking on the Foundations of Mechanics in the First Half of the 19th Century

2.1 The Scientific and Metatheoretical Background

I. Newton in his *Principia* names his basic laws *axiomata sive leges motus*, thus formulating two different demands for them: they have to describe motion, and they have to organise the science of motion deductively. The latter demand becomes more and more important in the course of the 18th century, as different basic concepts and laws had to be integrated into rational mechanics. The long and complicated development in question is accompanied by a decline of empirical *and* metaphysical justifications of concepts and the laws combining them. This is true especially of the analytical tradition of mechanics, which becomes dominant from the middle of the 18th century onwards. Without doubt, a full understanding of this strand of mechanics can only be reached if the striving of the underlying mechanical Euclideanism for an axiomatic-deductive organization of the whole body of mechanical knowledge is taken into account. It has to be noticed, however, that in the course of this process an important meta-theoretical change takes place: The "first principles" of mechanics become *formal axioms* of science rather than material laws of nature. The principle of virtual velocities, later formulations of the principle of least action or Hamilton's principle clearly show the consequences of this change: The rise of these principles is accompanied by an increase of the deductive demands and, at the same time, a "semantic unloading" of their basic mathematical concepts like moment, action, *vis viva*, potential, or kinetic energy (cf. Pulte 2001, 62, 74–77).

Lagrange's analytical mechanics is most significant in this respect: On the one hand, it continues the efforts of Euler, d'Alembert and others to reach a deductive organization of mechanics, and brings these efforts to an end. On the other hand, however, it marks a *break* with the older tradition, thereby revealing the basic philosophical problems of mechanical Euclideanism: Lagrange wanted to base mechanics on certain and evident *mathematical* principles without any recourse to metaphysical or empirical justification: "Mechanics can be understood as a geometry with four dimensions," and the "analysis of mechanics as an extension of geometrical analysis" (Lagrange 1797 [1813], 337). This kind of mechanics is a logical *consequence* and, at the same time, a *dissolution* of Euclideanism in its older meaning: Axioms become formal principles of organization rather than principles with empirical content, and the whole system is held together by logical coherence rather than by "material" truth. In Lagrange's concept of mechanics, the higher calculus serves as the uniting element in the deductive chains. Insofar as order and unity become the main targets, and the calculus the main means, this mechanics is rightly called *analytical*, thus expressing both the ambitious methodological and the specific mathematical character of this science.

Lagrange shaped the image of analytical mechanics as a model science for more than half a century. His understanding of rational mechanics as a "self-sufficient" and formal mathematical science, however, inevitably leads to a smouldering conflict with the traditional meaning of axiom as a self-evident first proposition, which is neither provable nor in need of a proof. Lagrange wanted to start his mechanics with *one* principle, i. e. the principle of virtual velocities. In the first edition of his *Méchanique Analitique*, he introduced this general principle *verbatim* as a "kind of axiom" (Lagrange 1788, 12). In the second edition, however, he stuck to this title, but had to admit that his principle lacked one decisive characteristic of an *axiom* in the traditional meaning: It is "not sufficiently *evident* to be established as a primordial principle" (Lagrange 1853/55, vol. I, 23, 27). By two different so-called "demonstrations" he tried to *prove* his primordial principle by referring to simple mechanical processes or machines, thus trying to bring back *intuitive* truth to his "axiom."

Lagrange's formulation and his later demonstrations of the principle of virtual velocities posed a challenge for a number of mathematicians, such as Fourier, Laplace, de Prony, Gauss, Carnot, Poisson, Poinsot and Ampère. Their efforts to solve Lagrange's foundational problem show that the *Méchanique Analitique* indeed brought about a "crisis of principles" (Bailhache 1975, 7). All attempts to solve this crisis aimed at *better* demonstrations, giving the principle of virtual velocities a *more secure* foundation and making it *more evident* (cf. Pulte 1998, 158–161). Like Lagrange, the contemporary and following mathematicians applied their refined logical and mathematical methods in order to substantiate the principle of virtual velocities by geometrical and mechanical arguments. Their meta-theoretical position can aptly be described as "a sort of 'Rubber-Euclideanism'," because it "stretches the boundaries of self-evidence."[2] Despite this crisis, of which only the *avant-garde* of the contemporary scientific community was aware, analytical mechanics in the tradition of Lagrange was seen as a model science in influential philosophies of science, for example in A. Comte's *Cours de philosophie positive* (cf. Fraser 1990). Neither the "positivist" Comte nor empiricists like J. Herschel or J. S. Mill were critical about mechanical principles *qua* axioms, nor were semi-Kantians like W. Whewell or W. R. Hamilton. For different philosophical reasons – mainly for their empiricism or apriorism (in the Kantian, synthetic sense) with respect to mathematics – they kept to the traditional CMM-ideal (cf. Pulte 2005, Ch. VI.1).

2.2 C. G. J. Jacobi's Conventional Mechanics

In German speaking countries the image and understanding of mathematics in the early 19th century was strongly influenced by neo-humanism. This movement, then dominant in Germany, strongly emphasised that science and education (*Bildung*) are ends in themselves. Mathematics and the old languages in particular should be regarded as an expression of pure intellectual activity (see Jahnke 1990). Empiricist conceptions of mathematics like those of the French mathematical physics or British empiricism were sharply rejected, both with respect to their philosophical foundations and to their utilitarian consequences. Mathematics was understood as a "pure" and autonomous mental activity, governed only by the rules of logic and destined for the "honour of the human spirit" (cf. Knobloch et. al. 1995, 100–109, esp. 108). Mathematical truth therefore had to be independent from any external experience and also from mediating intuition in the sense of Kant. The neo-humanist ideal of pure mathematics brought about the problem of the *applicability* of mathematics to the empirical sciences in a new and fundamental form insofar as established answers to this problem (traditional metaphysical justifications, empiricist theories of abstraction, Kantian

2 Lakatos 1978, II, 7 and 9. Lakatos himself subsumes Lagrange and other mathematicians of the 18th century under this label. However, he also admits that the history of the decline of Euclideanism (including its degeneration into 'Rubber Euclideanism') in mechanics has still to be written. Pulte 2005 attempts to fulfil this *desideratum*.

synthetic a priori-approaches) lost their plausibility. In this context, the foundational problem of analytical mechanics described above was only *one* aspect – though one of eminent philosophical relevance – and the mathematician C. G. J. Jacobi was the first adherent of this new conception of pure mathematics who addressed this problem.

I will pass in silence over young Jacobi's Platonistic answer to the problem of applicability (cf. Knobloch et. al. 1995, 110–114), and turn immediately to his last *Vorlesungen über Analytische Mechanik* from 1847/48 (Jacobi 1847/48 [1996]; cf. Pulte 1994), the philosophically most interesting part of which was praised by Carl Neumann for the rigour of its criticism of the foundations of mechanics about two decades later (cf. part 3.1). Jacobi's rejection of Lagrange's mechanics is the first and most distinct expression of his criticism. As Jacobi's last lectures from 1847/48 were not published until 1996, his criticism was noticed only by some of his students (like B. Riemann) and other mathematicians (like C. Neumann). It was totally ignored in the histories of mathematics and physics, where Jacobi's contribution to mechanics – under the influence of his published *Vorlesungen über Dynamik* (from 1842/43, publ. 1866; see Jacobi 1884) – is unanimously subsumed into Lagrange's approach. During his time in Berlin, however, Jacobi came to a different estimation; his new attitude towards his old Lagrangian ideal is most lively expressed in a warning to his students at the beginning of his lectures directed against Lagrange's "Rubber Euclideanism," especially his attempts to give a demonstration of his "axiom" of virtual velocities.[3] I will omit the mathematical and physical details of Jacobi's criticism, but should point out the principle difference concerning their understanding of mathematics. He describes Lagrange's approach as follows (Jacobi 1847/48 [1996], 193–194):

> Everything is reduced to mathematical operation … This means the greatest possible simplification which can be achieved for a problem …, and it is in fact the most important idea stated in Lagrange's analytical mechanics. This perfection, however, has also the disadvantage that you don't study

3 "[Lagrange's] Analytical Mechanics is actually a book you have to be rather cautious about, as some of its content is of a more supernatural character than based on strict demonstration. You therefore have to be prudent about it, if you don't want to be deceived or come to the delusive belief that something is proved, which is [actually] not. There are only a few points, which do not imply major difficulties; I had students, who understood the *mécanique analytique* better than I did, but sometimes it is not a good sign, if you understand something" (Jacobi 1847/48 [1996], 26).

> the effects of the forces any longer ... Nature is totally ignored and the constitution of bodies ... is replaced merely by the defined equation of constraint. Analytical mechanics here clearly lacks any justification; it even abandons the idea of justification in order to remain a pure mathematical science.

Jacobi's reproach has two different aspects. First, he rejects Lagrange's purely analytical mechanics for its inability to describe the behaviour of real physical bodies. In this respect, he shares the view of those French mathematicians in the tradition of Laplace, who called for a "mécanique *physique*" instead of a "mécanique *analytique*." This point does not affect the foundations of mechanics itself. The *second* aspect, however, does affect such foundations, because it concerns the *status* of first principles of mechanics. For Lagrange, the principle of virtual velocities was vital to gain an axiomatic-deductive organization of mechanics, and his two proofs were meant to save this Euclidean ideal. In so far as *this* ideal lacks "any justification" and even "abandons the idea of justification," it can rightly be described as "dogmatic" (Grabiner 1990, 4). Jacobi, on the other hand, applies his analytical and algebraic tools critically in order to show that mathematical demonstrations of first principles *cannot* be achieved. At best they can make mechanical principles "intuitive" (*anschaulich*) (Jacobi 1847/48 [1996], 93–96). But intuitive knowledge is no inferential knowledge in his sense; it is *not* based on unquestionable mathematical axioms and strict logical deduction. At this point Jacobi – the exponent of pure mathematics – dismisses Euclideanism as an ideal of *science:* The formal similarity between the mathematical-deductive system of analytical mechanics and a system of pure mathematics (as number theory, for example) *must not* lead to the erroneous belief that both theories meet the same epistemological standards. Indeed, as far as I am aware, Jacobi was the first representative of the analytical tradition who saw and drew this consequence.

Having described the origin and general features of Jacobi's *destructive* criticism of Lagrange's Euclideanism, I should shortly outline his *constructive* view of mechanical principles and the role of mathematics for them. According to Jacobi, mathematics offers a rich supply of *possible* first principles, and *neither* empirical evidence *nor* mathematical or other reasoning can determine any of them as true. Empirical confirmation is necessary, but can never provide certainty. First principles of mechanics, whether analytical *or Newtonian*, are not certain, but only *probably* true. Certainty of such principles, a feature of mechanical Euclideanism, cannot be achieved. Moreover, the search for proper mechanical principles always leaves space for a choice between different alternatives. Jacobi, well educated in classical philology and very conscious of linguistic subtleties, consequently called first principles of mechanics "conventions," exactly 50 years before H. Poincaré did (Jacobi 1847/48 [1996], 3, 5):

> From the point of view of pure mathematics, these laws cannot be demonstrated; [they are] mere *conventions*, yet they are assumed to correspond to nature ... Wherever mathematics is mixed up with anything, which is outside its field, you will however find attempts to demonstrate these merely conventional propositions a priori, and it will be your task to find out the false deduction in each case. ...
>
> There are, properly speaking, no demonstrations of these propositions, they can only be made plausible; all existing demonstrations always presume more or less because mathematics cannot invent how the relations of systems of points depend on each other.

It is important here to take note of Jacobi's "point of view of pure mathematics": He draws a line between mathematics itself and "anything, which is outside its field." Mathematical notions and propositions on the one hand and physical concepts and laws on the other hand must be *sharply* separated. This marks a striking contrast to Lagrange's "physico-mathematician's" point of view and is essential for Jacobi's "conventional turn." This is firstly because his idealistic background prevents him from believing that mechanical principles are grounded in experience. Secondly, he *shares* Lagrange's opinion that no metaphysical justification of such principles is possible. And finally, he *rejects* Lagrange's view that mathematics itself can prove these principles as certain and evident. Mathematics, however, can offer different principles of describing physical reality in an economical way. It is in *mathematics* that the conventional character of these principles has to be located, because mathematics offers more possibilities than nature can realize.

For Jacobi, conventions are neither gained by experience (i. e., they are no inductive generalizations) nor are they a priori-principles (in Kant's sense). Comparable to Poincaré (cf. Pulte 2000), he comes to a "third-way-solution," which makes a choice between different alternatives possible and necessary. Jacobi, too, holds the opinion that this choice is not arbitrary, but restricted by considerations of simplicity and convenience: "... again a convention in form of a general principle will take place. One can demand that the form of these principles is as simple and plausible as possible" (Jacobi 1847/48 [1996], 5). Of course, mechanical conventions, as principles, need to be empirically relevant.

They are assumed in order "to correspond to nature" in a way "that experiments show their correspondence" (Jacobi 1847/48 [1996], 3). Jacobi, however, is not explicit on the question of how conventions are to be handled in case of empirical anomalies. But as he sometimes remarks, that mechanical principles are not certain, but only "probable" (Jacobi 1847/48 [1996], 32–33), he obviously believes that experience is entitled to *falsify* principles. Poincaré, on the other hand, exempts mechanical conventions from empirical falsification.

It is important to note that Jacobi applies his concept not only to analytical principles, where it might be used in the trivial sense that different conventions can be used by merely formal, empirically meaningless operations, but also to Newton's synthetical "axioms," especially to the *law of inertia.* They, too, are first and above all *mathematical* propositions. Here, Jacobi comes close to the semantic aspect of conventions in Poincaré's "hidden definition-interpretation." As is well known, Poincaré regards the law of inertia as a fixation of the meaning of "force-free movement." Other *définitions déguisées* are possible and permissible, for example motions with changing velocity or circular motions. Jacobi's interpretation seems similar (Jacobi 1847/48 [1996], 3–4):

> From the point of view of pure mathematics it is a circular argument to say that rectilinear motion is the proper one, [and that] consequently all others require external action: because you could define [*setzen*] as justly any other movement as law of inertia of a body, if you only add that external action is responsible if it doesn't move accordingly. If we can physically demonstrate external action in any case where the body deviates, we are entitled to call the law of inertia, which is now at the basis [of our argument], a law of nature.

The circular argument presented here suggests that the law of inertia implies a convenient definition: It determines the meaning of "being free of external forces." We are entitled to choose other movements, for example circular movement, if we can trace back any deviation from circular movement to external actions.

To sum up, one can say that Jacobi's "conventional" mechanics marks a sharp break with the older tradition of mechanical Euclideanism, and that his neo-humanist concept of pure mathematics is fundamental for this break. While possible mechanical principles are free inventions of mathematics, a methodologically reflected *decision* is necessary in order to come to empirical laws, which can, however, never gain the certainty of the propositions of pure mathematics. While the older tradition of mathematical physics keeps to an axiomatic-deductive ideal of mechanics (CMN), and strives for making its first principles safe and evident in its scientific practice, Jacobi rejects this ideal as futile and demands the acceptance of the revisability and fallibility of mechanical principles in scientific practice. In this sense, his conventional understanding of mechanics – which is not yet "conventionalistic" in the sense of Poincaré's doctrine elaborated half a century later – is an important early contribution to a modern, hypothetical understanding of mechanics (MMN).

2.3 A Note on the Reception of Jacobi's Lectures

One of the participants at Jacobi's *Vorlesungen über Analytische Mechanik*, delivered in Berlin in 1847/48, was Wilhelm Scheibner, whose notes served as the basis for the later publication (Jacobi 1847/48 [1996]). Scheibner went to Leipzig, where in 1853 he qualified as a university lecturer. The thesis he defended in his *disputatio* was: "The principles leading to the basic equations of mechanics are of a conventional nature, especially the principle of virtual velocities, and the principle named after d'Alembert cannot be demonstrated completely."[4] Other participants likewise passed Jacobi's ideas to colleagues and students (cf. Jacobi 1847/48 [1996], XLIX–LI).

The most eminent mathematician who attended the Berlin *Vorlesungen* was B. Riemann. After his return to Göttingen, he was busy working on the principles of natural philosophy and their epistemological and methodological implications. In this time he wrote the fragment *Neue mathematische Principien der Naturphilosophie* (Riemann 1853 [1892]). The title obviously alludes to Newton's *Principia*. The relation of natural philosophy and physical geometry cannot be discussed here (cf. Pulte 2005, 388–399). It should be noted, however, that the "New Principles of Natural Philosophy" *precede* his famous lecture *Ueber die Hypothesen, welche der Geometrie zu Grunde liegen* (Riemann 1854 [1892]). In his earlier fragment, Riemann picks up Jacobi's rejection of mechanical "axioms" and integrates this point of view into his own more empiricist framework, which likewise has no place for empirical laws which are distinguished by mathematical *certainty*. He does not, however, adopt Ja-

4 See the document 'Diss. Phil. Lip. 1840–1872' in the archives of Universität Halle ('Personalakte Wilhelm Scheibner'); cf. also Jacobi 1847/48 [1996], XLIX–L.

cobi's label "convention," but uses the more traditional notion "hypotheses" to articulate his position (Riemann 1853 [1892], 525):

> Newton's distinction between laws of motion or axioms and hypotheses seems to me untenable. The law of inertia is the hypothesis: If a material point were alone in the world and moved in space with a definite velocity, it would preserve this velocity constantly.

A hypothesis, according to Riemann, is anything which is "added to experience by our thinking" (Riemann 1892, 525). He refuses to accept a mathematical statement as an *axiom* of mechanics, if, as he thinks, it exceeds our experience and, in a certain sense, even contradicts it.[5]

It is important to note that his understanding of mathematics as the driving force of "hypotheticity" both in his mechanics and in physical geometry: Scientific experience needs mathematical conceptualisation, but, vice versa, mathematical concepts can only potentially be applied in present natural philosophy. Hence it follows a genuine *methodological* demarcation between both areas. Physics has to decide by measurement what concepts from the rich "supply" offered by mathematics are suitable for the representation of phenomena. As the mathematical principles can only be deductively checked by empirical facts, it is even possible to have different sets of principles, which are corroborated by the same facts (Riemann 1854 [1892], 273).

Without doubt, Riemann's fragment on "New Mathematical Principles of Natural Philosophy" presents mechanics as a hypothetical-de-

5 *Later* he makes a similar point concerning Euclidean axioms as a basis of physical geometry, when he says, that "we neither perceive whether and how far their connection is necessary, nor a priori, whether it is possible" (Riemann 1854 [1892], 273). When he doubts their *necessity*, he obviously has in mind *other* systems of physical geometry. When he doubts their very *possibility*, he does not only raise the question of logical consistency, but also the question of whether Euclid's axioms are true for physical space or not. A physical realisation of rectilinearity, according to Newton's first law, is a part of this problem, and it seems clear that this part was first understood as *problematic*.

It is a widespread misconception that geometry was understood as an 'independent' basis of mechanics and that, for this reason, hypotheticity of the principles of mechanics was a natural consequence of the hypotheticity of the principles of physical geometry. In Newton's *Principia*, we find an inverse 'foundational relation' between mechanics and geometry, and philosopher-scientists like von Helmholtz adopted this view: Geometry, as an empirical science, depends on mechanics (see Schiemann 1997, 219–234). This aspect is important to overcome the 'standard view' that the rise of MMN is a mere 'epiphenomenon' of the rise of Non-Euclidean geometries.

ductive science in its modern sense. It is as alien to Mechanical Euclideanism as Jacobi's "conventional mechanics." Hence, Riemann's approach is comparable to, but not identical with Jacobi's understanding of mathematics (cf. Pulte 2005, 375–388): It is basically the *autonomy* of mathematics from empirical constraints, which brings about the *hypothetical* in mathematical physics in his framework too. For the scientific community, however, this aspect of his work remained widely unknown, because his fragments on natural philosophy were not published before 1876.

C. Neumann, to whom I will now turn, learned from Jacobi's *Vorlesungen* through the notes taken by W. Scheibner. In 1869, when he gave his lecture *Ueber die Principien der Galilei-Newton'schen Theorie* (Neumann 1869b), he too did not know Riemann's fragments on natural philosophy, nor was he aware of Riemann's *Habilitationsvortrag* on geometry.[6] This is important in order to understand the actual roots and outlook of Neumann's inaugural lecture.

3. The Roots of Carl Neumann's Principles of the Galilean-Newtonian Theory

3.1 The Background until 1869

In order to understand the origin and scope of Carl Neumann's Leipzig "Principles," another inaugural lecture, given four years earlier in Tübingen and published as *Der gegenwärtige Standpunct der mathematischen Physik* (Neumann 1865), is extremely helpful: The first parts of both lectures are nearly identical (cf. Neumann 1865, 1–16 and Neumann 1868b, 1–11), thus making it easy to identify essentially *new* elements in the latter.

The Tübingen "Point of View" deals mainly with the mathematical theory of electricity and magnetism. Mechanics serves as an ideal of scientific theory formation: Its outstanding merit is to bring a great number of phenomena under a small number of "basic ideas" (*Grundvorstellungen*), and these are "inertia" and "attraction" (Neumann 1865, 13–

6 Neumann refers to Riemann's *Hypothesen* in one of the footnotes to his lecture (Neumann 1869b, 31–32, n. 10). These footnotes, however, were added later (see, for example, Neumann 1869b, 24, n. 2). Cf. Pulte 2005, 402–412, for more details.

14). These basic ideas should not be understood as *explananda* in the sense that unknown phenomena are reduced to known phenomena, because the basic ideas themselves are "not more explicable" and even "totally incomprehensible" (Neumann 1865, 14, cf. 34–35). Mechanics forms a model science for other parts of physics for exactly this reason: A perfect science reduces a maximum of phenomena under a minimum of basic ideas, albeit that these ideas themselves may be epistemologically opaque (see Neumann 1865, 17). Neumann dismisses here the traditional CMN-claim for *evidence* of first principles, and anticipates some of Mach's and Kirchhoff's ideas of how scientific theories and phenomena are related. However, the basic ideas of mechanics (inertia, gravity) are neither understood as revisable, arbitrary or matter of choice, nor does Neumann discuss the validity of the mechanical principles related to these ideas (law of inertia, second law of motion, law of gravity) – the Tübingen "Point of View" does not include any critical discussion of the principles of mechanics at all. Finally, the contribution of *mathematics* to the character and status of mathematical physics plays no significant role in this lecture. These features have to be kept in mind with respect to the Leipzig "Principles" from 1869.

In 1868, Neumann moved from Tübingen to Leipzig, where he got the opportunity to study Jacobi's lecture from 1847/48 first hand *via* W. Scheibner (cf. Jacobi 1847/48 [1996], LII, n. 166). One year later, he published a paper on the principle of virtual velocities and discussed Jacobi's mathematical treatment affirmatively. Moreover, he was impressed by Jacobi's philosophical analysis of the principles of mechanics. In comparison to the Königsberg *Dynamik*, he states, Jacobi's Berlin lecture "distinguishes itself by a criticism of the *foundations of mechanics* which – in this rigour – has never been articulated in public until now" (Neumann 1869a, 257). From Neumann's marks and marginal notes in Scheibner's copy we know which of Jacobi's remarks he was most interested in; those discussed above (part 2.2) belong to them (see Jacobi 1847/48 [1996], LXII, 3–4). At the third of November 1869, when he was fully aware of Jacobi's point of view, Neumann gave his inaugural lecture in Leipzig. While former reconstructions of this lecture assumed it as a "given" starting point of the public debate on the foundations of mechanics (see, for example, DiSalle 1993), the background sketched here seems important to me for an understanding of the origin as of the content of this lecture. In what follows, I will leave aside Neumann's analysis of the law of inertia, absolute space and the "body alpha" (cf. Pulte 2005, 421–429), and confine myself to his meta-theoretical point of view.

3.2 "Hypotheticity" in Neumann's Principles of the Galilean-Newtonian Theory

Though Carl Neumann repeats large passages of his Tübingen lecture, his general *objective* in 1869 is quite different from that in 1865. He is now interested in the truth and certainty of the principles of mathematical physics in general, and of mechanics in particular. Mathematics itself becomes important for his argument, and he draws *now* for the first time a sharp demarcation between the principles put first at the deductive structures of physical theories and those of logic or pure mathematics. In full agreement with Jacobi, Neumann sharply defines where the parallel between theories of mathematical physics and pure mathematics ends, i. e. *before* the principles of the theories in question (Neumann 1869b, 12):[7]

> For if we wanted a physical theory that is not based on some incomprehensible and hypothetical fundamental notions, but rather one that proceeds from theorems that bear the stamp of *irrevocable certainty*, that offer in themselves the guarantee of an *unassailable truth*, then we would be forced to take refuge in the theorems of logic or mathematics. But it would prove impossible to deduce a physical theory from such purely formal theorems.

Mathematical physics can not be deduced from propositions of logic and pure mathematics, because these propositions are without empirical content. An empirical theory can profit from the truth and certainty of those propositions only *qua* "deductive certainty," not at the genuine level of principles. This *duality* of certism (with respect to logic and pure mathematics) and fallibilism (with respect to mathematical physics) is not present in Neumann's Tübingen lecture, and it can be traced back to his reception of Jacobi. But there is more to come with respect to the principles of mathematical physics (Neumann 1869b, 12–13):

> We have to admit that for those principles or hypotheses [of physics] – indeed *because* they are incomprehensible, *because* they are arbitrary – one cannot speak of correctness or incorrectness, of probability or improbability, at all. …

7 In the following quotations from Neumann's lecture the English translation from 1993 was used, but modified in several cases (cf. Neumann 1993).

> To be sure we are sometimes able to use the word *probable* as well as the word *true* as an *Epitheton ornans* [i.e. a decorating epithet]. But we shall wish to claim only that until today these principles have best been corroborated, not that they are established forever, and even to a lesser extent that they (like a theorem of logic or mathematics) offer in themselves the guarantee of unassailable stability, the guarantee of irrevocable truth.

Strictly speaking, principles that are starting points of a theory of physics will *never* correctly be called true or probable. Rather, they will always be regarded … as something *arbitrary* and *incomprehensible*.

While Jacobi is willing to accept mechanical principles in the best case as "probably" true, Neumann's *dictum* "neither true nor probable" – reminiscent of A. Osiander's famous preface to Copernicus's *De revolutionibus* (Neumann 1869b, 12, 24–25) – goes further: As mathematical physics is strictly deductive, neither truth nor probability (in the sense of "degree of certainty") can be transferred to the principles at the top (cf. part 3.3). Therefore the first principles are not immune from empirical falsification. Neumann's attitude, that even these principles are at stake when a theory is tested, explicitly includes the basic principles of the Galilei-Newtonian *mechanics:* They, too, can be overthrown; they are "arbitrary" and "moveable," as he repeatedly says (Neumann 1869b, 13–15, for example). These characteristics of any principles are rooted in their *mathematical* character: The area of mathematics is "infinite," and therefore the "latitude for the arbitrary choice of principles is extraordinarily large" (Neumann 1869b, 32, 31, n. 10). This does not mean that Neumann asks for an arbitrary proliferation of principles without methodological guidance, but that any claim for their validity depends on empirical tests at the *end* of a deductive chain, and that the process of testing can never – even in the case of repeated corroboration – justify a dogmatic attitude towards the theory in question and the principles at its top (cf. Neumann 1869b, 23).

Like Riemann, Neumann does not adopt Jacobi's notion *convention*, but uses the traditional *hypothesis* for his characterisation of first principles of mathematical physics. (All these "principles" are, basically, "hypothesis"; Neumann 1869b, 12). Like Jacobi, however, he stresses the possibility of choosing quite different principles, thus indicating that different sets of principles and therefore different theories on one area of phenomena are possible (Neumann 1869b, 23). And like Riemann, he *explicitly* rejects not only evidence and certainty of first principles, but also one last *residuum* of traditional CMM: the *uniqueness* of first principles (and theories) of mathematical physics.

This is the reason why Neumann rejects *young* H. von Helmholtz's claim that principles of mathematical physics are to be understood as elements of "objective reality" (Neumann 1869b, 23). It is the "*essence of mathematical-physical theories,*" he says, to be "subjective constructions, originating from us," and "starting from arbitrarily chosen principles and developed in a rigorous mathematical manner," and determined to "supply us with the most accurate picture possible of phenomena" (Neumann 1869b, 22).

Within this philosophical framework, Neumann's discussion of absolute space and the Newtonian law of inertia as well as the introduction of his well-known "body Alpha" – topics not central for my outline itself – gain a definite methodological meaning: Neumann divides up the traditional law of inertia as an indubitable, dogmatic principle into three different principles (existence of Alpha, rectilinearity, uniformity), which together form the empirical content of this law. This decomposition and the explication of different empirical attributes by Neumann are paradigmatic for a modern understanding of mathematical philosophy of nature (MMN): explication and reflection of premises, criticism of hidden (metaphysical) assumptions, operational formulation of empirical tests and other characteristics of a *modern concept of science*[8] can be found in Neumann's Leipzig inaugural lecture. And the origin and meta-theoretical viewpoint of this hallmark of CMN cannot be understood without the rise of a new understanding of mathematics.

3.3 A Note on Neumann as a Precursor of Popper

K. R. Popper, in his article *A Note on Berkeley as a Precursor of Mach*, acknowledged Berkeley's modern, *quasi-Machian* critique of essentialism in general (Popper 1953). Mach frankly acknowledged at least Carl Neumann's priority with respect to the critique of Newton's absolute space and the law of inertia (Mach 1872 [1909], 47, n. 1). It seems, however, that Popper *nowhere* acknowledged Neumann's merits for the rise of a "critical" concept of science in his sense, including a strict fallibilism.

Admittedly, Neumann is not looking for an epistemological and methodological basis of his understanding of scientific theories. There

8 Cf. Diemer 1968, Diemer and König 1991, Schiemann 1998, Part A, esp. A. IV, and Pulte 2005, ch. II for a detailed analysis of the characteristics of 'classical' and 'modern' concepts of science.

is no criticism of induction, no discussion of a criterion of demarcation or an epistemological investigation into the "empirical basis" to be found in his lecture. However, the mathematical physicist Neumann and the philosopher of science Popper share some convictions and insights concerning scientific theories worth mentioning. Both thinkers are anti-essentialists and anti-instrumentalists (cf. Pulte 2005, 418–421). Both are anti-dogmatists and deductivists with respect to scientific theories and hold the view that they are, by and large, determined by their first principles. Both emphasize that the corroboration of principles can never demonstrate their truth or probability. And both understand theory-building as a *creative* process of inventing and testing principles and share the belief in scientific progress as an outcome of this process, as long as it is controlled by a methodological reflection. Popper certainly would have subscribed to the concluding sentence of Neumann's "Principles": "We must always be aware that these principles are something *arbitrary*, and therefore something *mutable*, in order to survey, wherever possible, what effect a *change* of these principles would have on the whole shape of a theory, and to be able to realise such a change at the right time, and (in a word) to be able to preserve the theory from *petrification*, from an *ossification* that can only be pernicious and an obstacle to the advancement of science" (Neumann 1869b, 23; Neumann's emphases).

4. Conclusion

I would like, with three short remarks, to sum up my outline of the structural development of the rise of hypothetical thinking with respect to the foundations of mechanics. Firstly, modern understanding of mechanics as a genuine *physical* science should not blind us to the fact that in the 18th and in early 19th century it was credited with the evidence and certainty of mathematics, being *de facto* regarded as epistemologically equivalent to Euclidean geometry by nearly all scientists and most philosophers of science. *Euclideanism* in Lakatos's sense was, indeed, the dominant image of rational mechanics as a science up to the middle of the 19th century (CMN). Secondly, I have stressed the "top down-perspective" of the working mathematical physicist, in order to show that the dissolution of mechanical Euclideanism and the rise of hypothetical thinking starts *here*, at the "top." And it *had* to start here, because a "bottom up" dissolution (by empirical falsifiers) could take place only after the existence of true "axioms" of mechanics became dubious. The modern understanding of mechanics (MMN), predominant in the last two decades of the 19th century, is an outcome of *both* processes. Thirdly, I have underlined the importance of a new understanding of mathematics for the development in question. In the course of the 19th century, a "shrinking-process" of mathematical evidence and certainty takes place, and not only physical geometry, but also mathematical physics is affected by this process. The concept of *pure mathematics*, isolating arithmetic, algebra and analysis as the remaining mathematical "paradise" of evidence and certainty from the larger area of the mathematical sciences, plays a crucial role in this process. My outline has stressed the position of the prominent mathematicians C. G. J. Jacobi, B. Riemann and C. Neumann, but minor figures like W. Scheibner, W. Schell, O. Rausenberger and others could be added.

While the application of mathematics in the sciences was, for a long time, understood as the best possible expulsion of the "demon named hypotheticity," the rise of modern mathematics and – in its succession – modern logic taught philosophy of science that this kind of "exorcism" will not work for the empirical sciences. Though W. V. O. Quines "Two Dogmas" promoted a new empiricism in the philosophy of mathematics, the older lesson was not lost. And today, there is hardly any scientist or philosopher of science who believes that hypotheticity of principles of empirical theories and, consequently, of empirical theories themselves, is a demon at all.

Bibliography

Bailhache, Patrice (1975), "Introduction et commentaire", in: Louis Poinsot, *La théorie générale de l'équilibre et du mouvement des systèmes*. Edition critique et commentaire par P. Bailhache. Paris: Vrin, 1–199.

Diemer, Alwin (1968), "Die Begründung des Wissenschaftscharakters der Wissenschaft im 19. Jahrhundert – Die Wissenschaftstheorie zwischen klassischer und moderner Wissenschaftskonzeption", in: Alwin Diemer (ed.), *Beiträge zur Entwicklung der Wissenschaftstheorie im 19. Jahrhundert*. Meisenheim: Hain, 3–62.

Diemer, Alwin, and Gert König (1991), "Was ist Wissenschaft?", in: Armin Hermann and Charlotte Schönbeck (eds.), *Technik und Wissenschaft*. Düsseldorf: VDI-Verlag, 3–28.

DiSalle, Robert J. (1993), "Carl Gottfried Neumann", *Science in Context* 6: 345–353.

Fraser, Craig G. (1990), "Lagrange's Analytical Mathematics, its Cartesian Origins and Reception in Comte's Positive Philosophy", *Studies in History and Philosophy of Science* 21: 243–256.

Grabiner, Judith V. (1990), *The Calculus as Algebra: J. C. Lagrange, 1736–1813*. New York: Garland.

Jacobi, Carl Gustav Jacob (1847/48 [1996]), *Vorlesungen über analytische Mechanik*. Ed. Helmut Pulte. Braunschweig/Wiesbaden: Vieweg.

— (1884), *Vorlesungen über Dynamik*. Ed. A. Clebsch. 2nd ed., Berlin: Reimer. (1st ed. 1866).

Jahnke, Hans Niels (1990), *Mathematik und Bildung in der Humboldtschen Reform*. Göttingen: Vandenhoeck Ruprecht.

Kirchhoff, Gustav R. (1876 [1897]), *Vorlesungen über mathematische Physik, Bd. 1: Mechanik*. Ed. Wilhelm Wien. 4th ed., Leipzig: Teubner. (1st ed. 1876).

Knobloch, Eberhard, Herbert Pieper and Helmut Pulte (1995), "'... das Wesen der reinen Mathematik verherrlichen'. Reine Mathematik und mathematische Naturphilosophie bei C. G. J. Jacobi. Mit seiner Rede zum Eintritt in die philosophische Fakultät der Universität Königsberg", *Mathematische Semesterberichte* 42: 99–132.

Lagrange, Joseph Louis de (1788), *Méchanique Analitique*. Paris: Desaint.

— (1797 [1813]), *Théorie des fonctions analytiques*. Nouvelle édition, revue et augmentée par l'auteur. Paris: Bachelier. (1st ed. 1797).

— (1853/55), *Mécanique analytique*. Revue, corrigée et annotée par J. Bertrand. 2 vols., Paris: Courcier. (Vol. 11 and 12 of the *Oeuvres de Lagrange*, publ. par les soins de M. Joseph-Alfred Serret (tomes I–X et XIII) et de M. Gaston Darboux. Paris: Gauthier-Villars 1867–1892, repr. Hildesheim: Olms 1973).

Lakatos, Imre (1978), *Philosophical Papers*. Ed. by J. Worrall/G. Gurrie. 2 vols., Cambridge: University Press.

Mach, Ernst (1872 [1909]), *Die Geschichte und die Wurzel des Satzes von der Erhaltung der Arbeit*. 2nd ed., Leipzig: Barth. (1st ed. 1872).

Neumann, Carl (1865), *Der gegenwärtige Standpunct der mathematischen Physik. Akademische Antrittsrede gehalten in der Aula der Universität Tübingen am 9. November 1865*. Tübingen: Laupp.

— (1869a), "Ueber den Satz der virtuellen Verrückungen", *Berichte über die Verhandlungen der Königlich-Sächsischen Gesellschaft der Wissenschaften zu Leipzig* 21: 257–280.

— (1869b [1870]), *Ueber die Principien der Galilei-Newton'schen Theorie. Akademische Antrittsrede gehalten in der Aula der Universität Leipzig am 3. November 1869*. Leipzig: Teubner.

— (1869c [1993]), "On the Principles of the Galilean-Newtonian Theory. An Academic Inaugural Lecture Delivered in the Auditorium of the University of Leipzig on 3 November 1869", *Science in Context* 6: 355–368. (Engl. transl. of 1869b by Gideon Freudenthal).

Olesko, Kathryn M. (1991), *Physics as a Calling: Discipline and Practice in the Königsberg Seminar for Physics*. Ithaca: Cornell University Press.

Pulte, Helmut (1994), "C. G. J. Jacobis Vermächtnis einer 'konventionalen' analytischen Mechanik. Vorgeschichte, Nachschriften und Inhalt seiner letzten Mechanik-Vorlesung", *Annals of Science* 51: 487–517.

— (1998), "Jacobi's Criticism of Lagrange: The Changing Role of Mathematics in the Foundations of Classical Mechanics", *Historia Mathematica* 25: 154–184.

— (2000), "Beyond the Edge of Certainty: Reflections on the Rise of Physical Conventionalism in the 19th Century", *Philosophiae Scientiae. Travaux d'histoire et de philosophie des sciences* 4: 87–109.

— (2001), "Order of Nature and Orders of Science. Mathematical Philosophy of Nature from Newton and Euler to Kant and Lagrange", in Wolfgang Lefèvre (ed.), *Between Leibniz, Newton and Kant. Philosophy and Science in the Eighteenth Century*. Dordrecht: Kluwer, 61–92.

— (2005), *Axiomatik und Empirie. Eine wissenschaftstheoriegeschichtliche Untersuchung zur Mathematischen Naturphilosophie von Newton bis Neumann*. Darmstadt: Wissenschaftliche Buchgesellschaft.

Riemann, Georg Friedrich Bernhard (1853 [1892]), "Neue mathematische Principien der Naturphilosophie", in: Riemann (1892), 528–532.

— (1854 [1892]), "Ueber die Hypothesen, welche der Geometrie zu Grunde liegen (Habilitationsvortrag vom 10. Juni 1854)", in: Riemann (1892), 272–286.

— (1892), *Gesammelte mathematische Werke und wissenschaftlicher Nachlaß*. Eds. Heinrich Weber and Richard Dedekind. Leipzig: Teubner. (1st ed. 1876).

Schiemann, Gregor (1997), *Wahrheitsgewissheitsverlust. Hermann von Helmholtz' Mechanismus im Anbruch der Moderne. Eine Studie zum Übergang von klassischer zu moderner Naturphilosophie*. Darmstadt: Wissenschaftliche Buchgesellschaft.

Wundt, Wilhelm (1866), *Die physikalischen Axiome und ihre Beziehung zum Kausalprinzip*. Heidelberg: Enke.

— (1910), *Die Prinzipien der mechanischen Naturlehre. Ein Kapitel aus einer Philosophie der Naturwissenschaften*. Stuttgart: Enke.

Contingent Laws of Nature in Émile Boutroux

Michael Heidelberger[1]

Abstract:

In 1874, the French philosopher Émile Boutroux wrote a dissertation on the Contingency of the Laws of Nature that highly influenced academic philosophy during the French Third Republic and led to a more hypothetical view of the natural sciences and mathematics. Boutroux took over the concept of contingency from the neo-Kantian philosopher Eduard Zeller who had insisted against Hegel on the role of contingency in history, and carried it over to nature. From this he tried to show that the sciences are hierarchically structured such that each layer is irreducible to its more basic predecessor. He also distinguished between two kinds of natural laws in a way very similar to Nancy Cartwright in the 1980s. It is finally shown that Boutroux's view on the nature of mathematics as a hypothetical science had a strong impact on Poincaré.

> I started from Natural Science, which compels recognition as a real fact, and which I do recognise as such. I have tried to show that science offers no contradiction to ideas such as individuality, finality, freedom, and the like, which are the basis of our ethical convictions; the collapse of which ideas would necessarily involve the collapse of these convictions. To this end I was bound to demonstrate that science does not necessitate the rigid dogmatism and determinism that so often shelter under its name (Boutroux 1914, as quoted by Benrubi 1926, 55 and 1914, 932).

In this way, the French philosopher Émile Boutroux (1845–1921) summarized his life-work for a Paris daily newspaper on the occasion of a poll among the members of the *Académie française*. Boutroux served as a key figure for the "rapprochement" of science and French philosophy (including their historiographies) in the late 19th century and thereby for the formation of *épistémologie* – the special mix of philosophy of science, philosophy of nature and history of science that is still typical of French philosophical thought even today.

As the quote above shows, Boutroux's philosophy takes its point of departure from a critique of determinism, which is understood as a mis-

---

1 This article is a reworked and greatly expanded translation of Heidelberger 2006. I am very grateful to Warren Schmaus and especially Ernan McMullin for their critical comments as well as for Ernan's, John Michael's and Brian Bodensteiner's help with improving my English.

conceived idea. He tries to show that the laws of nature are not (or only in special cases) to be regarded as expressing necessary relations, that determinism is thus valid only in approximation, and that nature has a place for genuine contingency, spontaneity and uncertainty (*indétermination*). The main goal is therefore not so much the attempt to "save" human freedom in an otherwise mechanical universe – an endeavor which occupied many other French philosophers during the 19th century (for an overview, see Hacking 1983 and 1990, 157–159) – but primarily a claim concerning the philosophy of nature, that the laws of the universe are not as rigid as modern science seems to imply.

Boutroux's claim led to a drastic change in the discussion of the compatibility of moral freedom with the laws of nature and thereby also to a change in the philosophy of science of the time (cp. Nye 1976, 281). After the publication of his doctoral thesis, many philosophers of the Third Republic were influenced one way or another by Boutroux's ideas about the world's contingency, variability and uncertainty, and the conclusions that can be drawn from this for the compatibility of freedom and lawful necessity.[2] But also philosophizing scientists, like Édouard Le Roy, Pierre Duhem and Boutroux's brother-in-law, Henri Poincaré, and even social scientists like Boutroux's disciples Émile Durkheim and the historian Henri Berr, developed a good deal of Boutroux's ideas further (on Durkheim's relation to Boutroux see Jones 1999, 153–160). Some historians even see a relation between Boutroux's "contingentism" and existentialism as well as other developments after World War I. (see e.g. Chaitin 1999).

His criticism of the traditional concept of determinism and law led to a new conception of science in many different respects. It implied a sophisticated and challenging refusal of the mechanistic worldview years before the revolutions in physics could shape the scientific outlook of the 20th century. It was much more sophisticated than, for example, Émile Meyerson's ultimately apologetic interpretation of the mechanistic worldview that was formulated some thirty years later. It also helped

2 The list of his students includes Henri Bergson, Léon Bunschvicg, Émile Durkheim, Gaston Milhaud, Maurice Blondel, Charles Péguy, Xavier Léon, Frédéric Rauh, Victor Delbos, Abel Rey, Pierre Janet, André Lalande, Lucien Lévy-Bruhl and Jean Jaurès. The novelist Marcel Proust, who also studied some time with Boutroux, answered to a questionnaire in 1891 or 1892 when asked about his "favorite heroes in real life": "M. Darlu, M. Boutroux" (Proust 1971, 337). Alphonse Darlu was Proust's (as well as Brunschvicg's, Xavier Léon's and Élie Halévy's) philosophy professor at the *Lycée Condorcet* in Paris.

to overcome, or at least to transform, the three currents dominating the academic philosophy of the day: (1) the eclecticism of Victor Cousin (1792–1867), which, in its distance from science and everyday sentiment alike, was already in decline, (2) the positivism of Auguste Comte (1798–1857), Hippolyte Taine and Émile Littré, which privileged empirical science over anything that smacked of metaphysical thought, and (3) the 'spiritualism' of Félix Ravaisson and Charles Renouvier (1815–1903), which claimed superiority for personality and finality in nature.[3]

In the traditional historiography of French philosophy, Boutroux is frequently counted among the spiritualists, who are said to comprise, besides Ravaisson and Renouvier, also Maine de Biran and Lachelier, and sometimes even Bergson and Brunschvicg. I find 'spiritualism' an unhelpful label for these thinkers because it evokes some kind of nebulous and biased anti-materialism, and a backward opposition to science.[4] It certainly does not fit Boutroux who, in contrast to most of the others, puts science at center-stage in his views. If one were to search for a label for Boutroux's approach, 'non-reductive materialism' would be much more apt, even if it is not an entirely adequate characterization of his outlook. Boutroux can be counted among the spiritualists only in so far as he advocated the autonomy of psychology in relation to the other empirical disciplines. As his disciple Brunschvicg put it (and I think there is good reason to agree with this diagnosis), Boutroux's doctrine is above all "an investigation of science for science – without prejudice against any metaphysics" (Brunschvicg 1922, 271). Or as Boutroux himself expressed it in the introduction to the English translation of 1916 of his major work: "Philosophy should put itself in direct touch with the realities of nature and of life; more particularly it should be grounded on the sciences, for these are the clearest and most faithful image we have of the aspect presented to us by these realities" (Boutroux 1874 [1916], Vf.). He even went so far as to say that philosophy is *the* "science of the sciences," and not a science apart (Boutroux 1926, 12).

3 For recent overviews of French philosophy at the time, see Brooks 1998, ch. 4, Gutting 2001, chs. 1–2 and Copleston 2003, part II. See also Pfeiffer 2002, § 1.9 for an account of the relevant history and philosophy of mathematics in France.

4 See also Copleston 2003, 155.

Very often Boutroux is also counted as a Kantian or neo-Kantian because he highly esteemed the philosophy of Immanuel Kant, and indeed German philosophy in general, to which he devoted many studies (e.g. Boutroux 1897, 1926, 1926a, 1929). This way of classifying him, however, would be equally misleading, even though some of his contemporaries subscribed to it, probably owing to the fact that Kant's fame in France in the late 19th century was largely due to Boutroux's efforts of mediation. It is true that Boutroux was influenced by the neo-Kantianism of his day: when he says that science "compels recognition as a real fact" in the opening quotation above he is very probably alluding to a well-known remark by the German philosopher Hermann Cohen of the Marburg neo-Kantians, to the effect that philosophy should admit the "fact of science" (*das Faktum der Wissenschaft*).

But there are very few elements in Boutroux's philosophy, if any, from which one could see or infer that he identified himself with the philosophy of Kant (see Pillon 1907, 82 ff., Gutting 2001, 23 ff. and Fedi 2001, 107. See also below ch. 3). One should not forget that for most of Boutroux's German contemporaries, whether they in fact favored super- or anti-empirical elements in the sense of the historical Kant or not, the labels 'neo-Kantianism' and 'criticism' often meant nothing more than a sophisticated alternative to "banal" and "naïve" British empiricism, or to certain aspects of Hegelianism (for which see below) and of German Idealism in general, although they often claimed to be idealists in a new sense. This is also true of Boutroux's French contemporary, Charles Renouvier, who is said to be a representative of "neo-criticism" and to have influenced Boutroux. Hence in this loose sense Boutroux *was* indeed a Kantian or neo-Kantian, but in the more literal sense of the word – which implies a significant substantial affinity with the doctrines of Kant – he was not.

In a short description of Kant's reception in France since the 1850s, Boutroux gives us a hint of what it meant to be a 'Kantian' at that time. The study of Kant, he says there, by philosophers like Janet, Lachelier, Renouvier, Pillon has contributed to the awakening of metaphysics in France since the 1850s. "Since then, to come back to the study of Kant does not only mean to act as erudite, historian, dilettante, it means to draw knowledge and valuable energy in order to tackle the problems that impose themselves on us" (Boutroux 1926, 12).[5] In

5 "Dès lors, revenir à l'étude de Kant, ce n'est pas seulement faire œuvre d'érudit, d'historien, de dilettante, c'est puiser des connaissances et des forces utiles pour

other words, to be a 'Kantian' meant no more than to discuss philosophy with the help of the conceptual resources that Kant's philosophy provides, but it did not necessarily mean to accept any of his solutions. As a consequence, Boutroux does not hesitate to subject Kantian doctrines to considerable criticism and to reject most of them for the present.

This somewhat confusing state of affairs seems to have misled Fabien Capeillères, for example, in an otherwise very commendable article. Instead, perhaps, of calling Boutroux an "odd" kind of Kantian due to his rejection of transcendentalism and of the synthetic a priori, he should have rejected the Kantian label for him entirely, even if Boutroux might have called himself a Kantian (which is improbable) or was called a Kantian by his contemporaries (Capeillères 1998, 436–38). These kinds of assessments are of considerable interest in their own right as historical clues as to Boutroux's reception in his own day, but they hardly help us to better understand his philosophy at present.

In what follows, I would like to begin by saying something about Boutroux's life and work in order to elucidate the context in which his specific criticism of determinism and his 'softening' of the rigidity of the laws of nature arose. In the second part, I shall treat Boutroux's concept of contingency. His view of the disunity of science will be dealt with in the third part of the paper. The fourth part then is devoted to Boutroux's distinction between two kinds of laws of nature, which has definite relevance for and affinity with recent philosophical discussion. In the final part, I shall show that Boutroux's philosophy of contingency led to an emphatically hypothetical view of mathematics that was shared by Henri Poincaré.

1. Boutroux's Life and Work[6]

After attending the prestigious *Lycée Henri-IV* in Paris, Boutroux continued his studies from 1865 onwards at the no less esteemed *École Normale Supérieure* (ENS), where he studied mainly under the decidedly

aborder les problèmes qui s'imposent à nous." For another short and succinct description of Kant's reception in France see Boutroux 1895b, 403 f. / transl. 325.

6 For more on Boutroux and his time, see Espagne (2004, 2001), Girel (2003), Fagot-Largeault (2002), 956–966, Espagne (2001), Fedi (2001), Gutting

anti-positivist philosopher Jules Lachelier (1834–1918) (for a recent characterization of Lachelier, cp. Gutting 2001, 14–20). After three years of study, at the suggestion of his later thesis advisor Félix Ravaisson (1813–1900), Boutroux was ordered by the French Minister of Public Instruction, Victor Duruy (who reintroduced the study of philosophy in the *lycées* in 1863, following its abolition ten years earlier), to study at the University of Heidelberg for two years, in order to get acquainted with the German university system and to write a report on it (see Boutroux 1869, Espagne 2004, 152–157, Rollet 2000, 107–111). In the course of his studies there and upon the recommendation of Ravaisson, Boutroux attended lectures by the physicist and physiologist Hermann von Helmholtz (1821–1894), the historian Heinrich Treitschke (1834–1896), and most importantly by the neo-Kantian philosopher Eduard Zeller (1814–1908) (Rollet 2000, 366, 370).[7] He experienced the lively interdisciplinary exchange that was typical of German universities of the time, and to which he was not exposed in France. He noted that, in Heidelberg, the theologian spoke with the naturalist, the jurist talked to the mathematician and the philologist exchanged his views with the physiologist, whereas at a French university or at the ENS, there was no exchange between the *Littéraires* and the *Scientifiques* at all (Rollet 2000, 366 f., 368 f., Boutroux 1869, 186; 1871, 543).

While still in Heidelberg, Boutroux started translating the first volume of Zeller's massive work, *The Philosophy of the Greeks in its Historical Development* (Zeller 1856–68), supplementing it with a long introduction on Zeller and his philosophy that also appeared separately (Boutroux 1877 [1926a]). As we will see shortly, it is true, as Brunschvicg later wrote, that in Boutroux's introduction "one can find the key to his [Boutroux's] later thought" (Brunschvicg 1922, 263). Boutroux's various reports on his stay at Heidelberg give a vivid impression of the atmosphere that existed at a Southwest German University on the eve of the Franco-Prussian war. He does not spare his words about

(2001), 20–25, Gil (2000), Rollet (2000), Capeillères (1998) and Engel (1988). Espagne and Rollet have opened up new and interesting avenues of information in regard to Boutroux from several archives. From the older literature, see Benrubi (1926), 153–161, Schyns (1924), Brunschvicg (1922), Gunn (1922), chs. III & IV, Aliotta (1914), 116–124, James (1910) and Boelitz (1907).

7 For Zeller see Hartung 2009. Zeller was also a protestant theologian and an early follower of the Tübingen Historical School around Ferdinand Christian Baur, who developed the historical-critical method of studying the bible.

the political situation – he notes among the students "a strong hatred against the French nation" – and criticizes the conservative attitude and nationalist spirit of the German professors "more or less modified by Prussian influence" (see his two *rapports* of 1869 in Rollet 2000, 366–371, and Boutroux 1869 & 1871).

Boutroux returned to France on the eve of the Franco-Prussian war of 1870/71, which meant a deep rupture in the relations between the two countries and their cultures, and brought to a more general awareness the inferiority of French higher education in relation to the German model that Duruy had already noticed after the war between Prussia and Austria of 1866 (Espagne 2004, 148, Paul 1972, 9–11). His first position after his return to Germany was at the *lycée* of Caen[8] where he wrote his doctoral thesis *On the Contingency of the Laws of Nature.*[9] This dissertation, which was dedicated to his doctoral advisor, Ravaisson, earned him the degree of a *Docteur ès lettres* in 1874. It contains *in nuce* almost all the philosophical views that Boutroux was to develop later on, and was hugely successful. It was hailed in France as a torchlight that put philosophy back on the right path, second in importance only to Henri Bergson's *Time and Free Will* (*Essai sur les données immédiates de la conscience*) of 1889. Hardly a year after its publication, the philosopher Paul Janet (1823–1899) praised the thesis of his young colleague as a metaphysics that does not separate itself from the sciences, but which appropriates their results in trying to dominate, explicate and surpass them (Janet 1875, 30). William James (1842–1910), one of the founders of pragmatism, later wrote that "many a convert to 'pragmatism' or to 'Bergsonism' has remained ignorant that the ball was set rolling by his [Boutroux's] first publication, *La Contingence des lois de la nature*, away back in 1874" (James 1910, 168; cp. also Perry 1935, II: 567 f.).

By 1929, Boutroux's dissertation had been re-issued ten times, translated into German by the Geneva philosopher Isaac Benrubi in 1911 – at a time when the reception of French philosophy of science and of pragmatism in Germany was at a height – and into English by

8 It is still true that many French professors teach at a *lycée* for some time before they enter university service.

9 The originally planned title was "Le Déterminisme dans ses rapports avec les sciences physiques et les sciences morales" (Boutroux 1873, 216). To put the concept of contingency into the foreground might have been a bow to his teacher Lachelier who had argued in his book on the *Basis of Induction* that all causality is subordinate to purpose, and therefore to contingency.

Fred Rothwell in 1916 (Boutroux 1874 [1916]).[10] In order to obtain his doctor of philosophy Boutroux had to write a second thesis in Latin, as was customary during this period (Boutroux 1874a [1927]). He dedicated *On Eternal Truths in Descartes* to his first and perhaps most important university teacher, Jules Lachelier (as did, incidentally, Henri Bergson with his dissertation 15 years later). A translation into French was made in 1927 by none other than the young Georges Canguilhem, later one of the main representatives of French *épistémologie* and the advisor of Michel Foucault.

During his time in Caen, Boutroux made friends with the influential mathematician Jules Tannery (1849–1910), who had been one of his classmates at the ENS and was now a colleague at the *lycée*. He also met his brother Paul (1843–1904), who worked as an engineer in the service of the state tobacco workshop, but is best-remembered for the notable contributions to the history and philosophy of science he made in his spare time (Pfeiffer 2002, § 1.9). Boutroux admitted both of them into his confidence and they accompanied his dissertation in word and deed (Boutroux 1873, 215 ff.). Another acquaintance he made at the time was with the philosopher Louis Liard (1846–1917), who was a year ahead of Boutroux in publishing his dissertation. He wrote about the role of definitions in geometry, also under Lachelier's guidance, and one can make out clear parallels to the work of Boutroux.

Immediately after receiving his doctorate Boutroux became *chargé de cours* at the University of Montpellier, before moving to the University of Nancy in 1876. There he met and married Aline, sister of the mathematician, physicist and philosopher Henri Poincaré (1854–1912). A year later Boutroux became a *maître de conférences* at the ENS as successor of Alfred Fouillée (1838–1912).[11] In 1880 Aline gave birth to their son,

10 The pre-war period also saw a decisive transfer of French philosophy of science to Vienna that helped to trigger the rise of the Vienna Circle of logical empiricism of the 20th century (cp. Brenner 1998; 2002; 2003, ch. V).

11 Boutroux's student and successor to his chair, Lucien Lévy-Bruhl (1857–1939), recalled Boutroux's Kant-lectures at the ENS of 1877 fifty years later: "He gave us a course on the philosophy of Kant that overwhelmed us to such a degree that we could not stop talking of the *Critique of Pure Reason*" (Lefèvre 1927, 76). – Xavier Léon gave a very affectionate and poignant description of the fascination and charisma Boutroux exerted on his audience and of the love for German philosophy he instilled in his students (Léon 1902, iii). It was in 1893 that Léon co-founded the leading French philosophy journal at the 20th century, the *Revue de métaphysique et de morale*, with Léon Brunschvicg and Élie Halévy, and the *Société française de philosophie* in 1900. He was a good

Pierre Boutroux (1880–1922), who himself later became a distinguished mathematician and historian of science teaching for some time at Princeton.[12]

Beginning in 1885, Boutroux was assigned to teach a complementary course on the history of German philosophy at the Sorbonne where he finally succeeded Paul Janet in the chair of History of Modern Philosophy in 1888. In the winter of 1892/93, he gave a lecture course on the concept of natural law in science and philosophy, which elaborated on the topics of his dissertation. It was published two years later as *Natural Law in Science and Philosophy*. In it he provided a fuller context for his ideas than he had done in his dissertation, and also presented them in a more accessible form (Boutroux 1895 [1914]).

Boutroux thought of himself mainly as a teacher: The list of his disciples reads like a *who's who* of French intelligentsia of the time (cp. Espagne 2001, 210 ff.; 2004, 161–164; Benrubi 1926, 153 f. and note 2 above). He assembled a circle of colleagues from different disciplines around him to foster the type of interdisciplinary exchange he had experienced in Heidelberg (and then found missing in France). Mary Jo Nye coined the expression "Boutroux Circle" for this important group (Nye 1979), whose members included, among others, Boutroux's brother Léon, who was a respected physicist, his brother-in-law Poincaré, the astronomer Benjamin Baillaud, and the brothers Jules and Paul Tannery already mentioned. This circle played a decisive role in the development of conventionalism in French philosophy of science (cp. Rollet 2000, ch. 2 and Rollet 2001).

In 1893/94 Boutroux spent a long sabbatical at Freiburg in the south-west of Germany with the Austrian-German philosopher Alois Riehl (1844–1924), who advocated an original realist version of neo-Kantianism with positivist overtones. Boutroux examined some of Riehl's claims in lectures on Kant he gave in 1896/97 (Boutroux 1926). In 1898, he became a member of the *Institut de France* in the *Aca-*

friend of Henri Poincaré who contributed an article to the first issue of the journal on the mathematical continuum (Félix Ravaisson's article being first). In fact, Poincaré contributed about twenty articles to the journal in the two decades until his death (cp. Rollet 2001) and was one of the first members of the *Société*.

12 Boutroux also had a daughter, Louise, who was older than Pierre and later married Pierre Villey-Desmeserets – a Montaigne specialist who became Professor of French literature at the University of Caen. One of his children was the law philosopher Michel Villey (1914–1988).

démie des Sciences Morales et Politiques. When he received a call to be director of the Parisian *Thiers* foundation in 1902, he gave up his teaching and devoted himself solely to research (his chair was only filled again in 1908 by Lucien Lévy-Bruhl). He generally refrained from pronounced political action. During the Dreyfus affair, he signed a seemingly neutral "Appel à l'Union," together with his brother-in-law Henri Poincaré as well as his cousin-in-law, Raymond Poincaré, who was Prime Minister at times and president of the republic from 1913 to 1920 (Rollet 2000, 107, 259 ff.). Although this expression of opinion is not very conclusive or revealing, Boutroux is counted among the pro-Dreyfus intellectual group (Sánchez 2004, 62). From early on he joined the *Union pour l'Action Morale* – an influential association of intellectuals, which was founded in 1892 and renamed *Union pour la vérité* in 1905 (Beilecke 2003, 136, 146, 233, 353 ff.). It defended republican and enlightenment values against totalitarian tendencies and radical ideologies.

In 1903–04 and 1904–05 Boutroux was invited to Glasgow to deliver the Gifford lectures, which were published in part posthumously (Boutroux 1926b). He became corresponding fellow of the British Academy in 1907. In 1908 he visited the International Congress of Philosophy at Heidelberg where he gave an important report on the development of philosophy in France since 1867 (Boutroux 1908a). In the fall of the same year he made the acquaintance of William James, whom he had influenced from early on. He visited him in 1910 in Cambridge (Massachusetts) where he delivered the *Hyde Lectures* at the *Cercle Français* at Harvard,[13] and later wrote an account of his philosophy (Boutroux 1911). In the 1910s, his attention shifted towards a philosophy of life and, in consequence, his interest in science dropped a little into the background. His last major work was *Science and Religion in Contemporary Philosophy*, which appeared in 1908 and discussed the relation of many thinkers to religion, e.g. of Comte, Spencer, Haeckel, Ritschl and James among others. It was translated (at least) into English, German, Italian, Turkish, Japanese and Russian. The French edition had reached 27.000 copies by 1947.

At the time, Boutroux's contemporaries saw his philosophy as belonging to a distinctively French brand of pragmatism. As already mentioned, he finally became a member of the *Académie française* in January 1912. Together with Alois Riehl, he was invited to the opening of the graduate school of Princeton University a year later. During May of

13 See James 1910, Perry 1935, II: 561–569 and Girel 2003.

1914, he went for a lecture tour through Germany and visited (among others) the universities of Berlin (with Alois Riehl) and Jena (with Rudolf Eucken). In Berlin, he gave an address on "French and German Thought: The Mutual Advantages Each Can Offer the Other" (Boutroux 1914. Cp. Hanna 1996, 10).[14]

With the beginning of the war, however, his decidedly Germanophile attitude abruptly changed to mortal enmity and a great disgust for anything German. It made him reinterpret the Germany of his first stay forty-five years before as already displaying symptoms of German barbarism.[15] This episode shows again that the First World War was the primordial catastrophe, in philosophy as in so much else (cp. Boutroux 1916; 1926a, 115–136 and 231–257, Brunschvicg 1922, 275–79, Espagne 2001, 212 ff.; 2004, 165–168). Boutroux must also have been deeply hurt by the fact that his German friend Riehl had been instrumental in formulating, and among the first to sign, the notorious and disastrous "Aufruf an die Kulturwelt" ("call to the cultural world") of September 1914 – a manifesto of 93 leading German scholars and intellectuals that supported the war and was received abroad as a terrible document of German chauvinism and militarism. The manifesto was answered in March 1915 by an appeal of 100 French artists, scientists and men of letters that included Boutroux's signature as well as those of Paul Claudel, Henri Matisse, André Gide and Anatole France (cp. Hanna 1996, ch. 3, esp. 84 f.). Together with other members of the *Académie française*, Boutroux resigned from the Prussian Academy of Sciences in May 1916, of which he had been a corresponding member since 1908.

There was, however, one exception to Boutroux's newly acquired dislike of Germany: he kept defending Kant's philosophy, even if he did not share most of its aspects (he criticized, for example, Kant's alleged

14 He also delivered the first lecture in the still existing Henriette Hertz "Philosophical Lecture" series of the British Academy, on "Certitude and Truth" in 1914, and a Herbert Spencer Lecture on "The relation of thought and action from the German and the classical point of view" at Oxford in 1917.

15 Boutroux took up his criticism of Zeller again in 1916. He claimed that Zeller had stripped the theories of Socrates, Plato, and Aristotle "of all they contained which was personal and living, and were reduced to abstract formulæ, subordinate to an immanent and necessary dialectic. Ever since that date [i.e. the 1870s, when he was engaged in translating Zeller] my impression of German science [proceeding like this generally] has become increasingly confirmed" (Boutroux 1916, 2). He again turned Zeller's critique of Hegel against him!

scorn for feeling), at a time when many conservative French intellectuals denounced it as part of the barbarous German cultural tradition (see Hanna 1996, ch. 4). The last sentence of a long article he had written on Kant in 1895 remained valid for him in spite of the Great War: Kant's doctrine, it says, "is not the mirror of a single epoch, nor even the expression of a nation's thought [*de la pensée d'un peuple*]: it belongs to the whole of mankind" (Boutroux 1895b [1912], 411/ transl. 330. Cp. also Bois 1916, 12–14). This attitude might also have been responsible for the fact that Boutroux is often seen as a neo-Kantian until today.

2. Boutroux's Concept of Contingency

The concept of contingency occupies a central place in Boutroux's philosophy. For that reason, his philosophy has sometimes also been called "contingentism." This concept carries with it a whole series of connotations that Boutroux did not always spell out unambiguously or separate clearly from one another (cp. Lalande 1976, 182). The first layer of meaning is to be seen as chance (in an ontic sense), as the uncaused occurrence of an event. A second layer results from the contrast that Boutroux sets up between *contingence* and *hasard*. Contingency involves the intersection of lines of causality in unplanned and thus not necessary ways, without denying the presence of causes or the validity of laws in general. The latter contrast is, in some respects, already implicit in Aristotle's distinction between τυχη and αυτοματον (cp. Aristotle, *Physics* II, 4 ff.; *Metaphysics* V, 30), and the former, possibly in Lucretius, whose atoms "swerve" as they move, thus breaking the determinism that would exclude human freedom.

Contingency also means that the natural laws themselves are not necessary, that they could be different without leading to a contradiction. In still another sense, and perhaps the dominant one, contingency in Boutroux means that the laws of nature do not *completely* or *strictly* determine the cases that are subsumed under them. (This is a restriction of the first meaning just mentioned.) The "laws" leave enough leeway for the single case to depart from the general law to a tiny degree. Contingency is the sequence of unforeseeable genuine novelties and unexpected occurrences, which interrupts the natural movement of life, and which cannot be comprehended by means of our intellectual formulas. The resulting micro-fluctuation becomes more and more important as one goes from the purely physical to the biological order and from there to the psychological and sociological realms.

Yet in Boutroux's view it still makes sense to talk of "laws of nature" because they represent a certain general tendency that is inherent in the cases treated without ever being strictly fulfilled in the individual case.

> Indeed, we must guard against confusing determinism with necessity: necessity expresses the impossibility of a thing being different from what it is; determinism expresses the sum total of the conditions which make it necessary for the phenomenon to be stated just as it is, with all its modes of being (Boutroux 1895 [1914], 90).[16]

The contingency of the laws of nature leads to the world's following of a course that is again and again influenced by contingent events (i.e. events that are not determined in all their aspects by former events). The deviation from the determined case can be so small that its effect cannot be detected directly. Tiny fluctuations in the occurrence of events are, however, not to be neglected, because they make a difference in the world that can be significant, even essential. They are responsible for the different qualities of the objects that go to make up their individual non-reductive character:

> To consider quantity with relation to a homogeneous quality, or to leave quality altogether out of account, is to place oneself outside the conditions of reality itself. Everything that is possesses qualities, and consequently participates in that radical indetermination and variability which belong to the essence of quality. Thus, the principle of the absolute permanence of quantity does not apply exactly to real things: these latter have a substratum of life and change, which never becomes exhausted. The singular certainty presented by mathematics as an abstract science does not authorize us to look upon mathematical abstractions themselves, in their rigid monotonous form, as the exact image of reality (Boutroux 1874 [1916], 68 f.).[17]

16 "Il faut bien se garder, en effet, de confondre déterminisme et nécessité: la nécessité exprime l'impossibilité qu'une chose soit autrement qu'elle n'est; le déterminisme exprime l'ensemble des conditions qui font que le phénomène doit être posé tel qu'il est, avec toutes ses manières d'être." (Boutroux 1895, 58)

It is interesting to note that Boutroux did not use the word *déterminisme* in his dissertation but only expressions like *déterminé* or *détermination*, whereas in the later lecture series on natural law in science and philosophy, *'déterminisme'* abounds. It might be possible that as a beginner, Boutroux did not want to produce a phrase that expressly challenged the determinism that Claude Bernard had strongly advocated in 1865.

17 "C'est se mettre en dehors des conditions mêmes de la réalité, que de considérer la quantité relativement à une qualité homogène, ou abstraction faite de toute

In this view, two different kinds of criticism meet: a criticism of Descartes's mathematical rationalism and a criticism of Hegelian philosophy of history.

Descartes is the author of the view rejected by Boutroux, expressed, for example, in the quotation above, that the quantitative aspects of all objects are based on a homogeneous quality to which mathematics refers. Cartesianism, says Boutroux, "reduces physical laws to mathematical determinations, the heterogeneous to the homogeneous" (Boutroux 1912, 97). Universality and necessity in science thus become the result of the determinism of mathematics (Boutroux 1926, 28; 1895 [1914], 136 / transl. 207 f.). External nature is reduced to matter and material objects are reduced to their extension (*res extensa*). According to Descartes, material objects differ from each other in their properties only because of differences in the extension, form and movement of their constituents. For Descartes, God's perfection presupposes that He is not only immutable in Himself, but that He also acts in the firmest and most invariable manner. God's immutability allows us to infer some rules as laws of nature. These laws indicate what happens if an object is changed in its extension, form or movement by an external cause.

This means that nature is, of itself, for Descartes a purely passive, motionless and indifferent system: it would rest for ever in an unchanging state if there were no external causes to effect a change (Descartes 1644, II, 36 ff.). There is no inherent liveliness in nature that is related to any special capacities of different substances as there would be for Aristotle or later for Leibniz or Schelling. God's immutability, His invariable and His unwavering will, is the guarantee of eternal truths, especially those of mathematics. The laws, which govern the effects of external causes, are created truths. There is no separation between God's will and His intellect. Accordingly, Descartes opposes the view that certain truths would also be valid if there were no God – that they are, so to say 'self-contained'. If God willed it, He could make $2 + 2 = 5$. The defining principle of God's nature is free activity which results in the

---

qualité. Tout ce qui est possède des qualités et participe, à ce titre même, de l'indétermination et de la variabilité radicale qui sont de l'essence de la qualité. Ainsi, le principe de la permanence absolue de la quantité ne s'applique pas exactement aux choses réelles: celles-ci ont un fonds de vie et de changement qui ne s'épuise jamais. La certitude singulière que présentent les mathématiques comme sciences abstraites ne nous autorise pas à regarder les abstractions mathématiques elles-mêmes, sous leur forme rigide et monotone, comme l'image exacte de la réalité" (Boutroux 1874, 59 f.).

creation of the essences of finite beings. Hence these essences do not explain His activity, but the other way around. This is not only valid for God, but also for finite beings. From all this it follows for Descartes that the changes in the world occur with mathematical necessity according to immutable and necessary natural laws. Yet it also follows that this necessity is not absolute, because it depends entirely on God's will (cp. Boutroux 1929, esp. 62, 137 ff. and Cronin 1960).

On this very last point, Boutroux can agree with Descartes wholeheartedly, but in all other respects he opposed him. Against Descartes, Boutroux maintains, one has to assume a sort of inherent liveliness of nature that results from the different qualities of the world. As a result, not all changes in the world are lawfully regulated – or rather, lawfulness is only a general trait that is not always or not precisely valid for the single case. Material objects and their qualities cannot be reduced to their extension, i. e. to their mathematical, quantitative properties. Laws of nature do not refer to constant repetition in the course of nature, but they are the result of the special perspective of an intelligent observer who tries to grasp nature according to some special regulative idea that varies with the relevant discipline. The laws are thus, according to Boutroux, imposed by the subject and do not refer to an objective, if approximate, regularity in the physical world.

> [W]e look upon the fundamental laws of each science as the least defective compromises that the mind has succeeded in finding for bringing together mathematics and experience … That which we call the laws of nature is the sum total of the methods we have discovered for adapting things to the mind, and subjecting them to be moulded by the will (Boutroux 1895 [1914], 213 & 217).[18]

God lets his creation undergo a constant process of transformation and thereby reveals His creativity. In this view, the immutability of God manifests itself only in this kind of constantly novel and free creation: "The contingency shown in the hierarchy of the general laws and forms of the world finds its explanation in this doctrine of divine free-

18 "[L]es lois fondamentales de chaque science nous apparaissent comme les compromis les moins défectueux que l'esprit ait pu trouver pour rapprocher les mathématiques de l'expérience … Ce que nous appelons les lois de la nature est l'ensemble des méthodes que nous avons trouvées pour assimiler les choses à notre intelligence et les plier à l'accomplissement de nos volontés" (Boutroux 1895, 140, 143 f.).

dom" (Boutroux 1874 [1916], 179 f).[19] God has not only made the world in an act of free creation, but conferred on it an intrinsic creativity that increases in its effect in going from lifeless matter through the different stages to mankind and manifests itself in a growing degree of contingency. We can see from this that Boutroux's concept of contingency implies not only the rejection of immutable and rigid laws, but also what would be labeled today "emergence" or "emergent" phenomena (cp. Fagot-Largeault 2002): The development of the world is not a law-guided and necessary reshuffling of a once-given matter with an "amount of movement" remaining always constant, as in Descartes, but an incessant process of novel creation which leads to an ever richer world of individualities.

Another notion that is implied by Boutroux's concept of contingency is the Hegelian concept of *Zufälligkeit.* It is apparent that Boutroux developed his conception in the context of so-called German "Late Idealism" [*Spätidealismus*], which he came to know in Heidelberg through his mentor, Zeller. If we look at Boutroux's introduction to Zeller's *History of Greek Philosophy* of 1877, we see that in describing Zeller's reflections on the history of philosophy he uses the term 'contingency' in translating Georg Wilhelm Friedrich Hegel's concept of *Zufälligkeit:*[20] "Without doubt, a fact in the history of philosophy is not simply just the pure and simple product of general laws or, so to speak, a point of intersection of such laws: it includes a contingent element in itself." It comprises "partly necessity and partly contingency." As history poses the problem, it is none other than the problem of "the essence of human freedom and of the relation of this freedom to chance (*hasard*) and to necessity" (Boutroux 1877 [1926a], 21 f.).[21] One should not, Boutroux

19 "Par cette doctrine de la liberté divine, la contingence que présente la hiérarchie des formes et des lois générales du monde se trouve expliquée" (Boutroux 1874, 157 f.).

20 In an article from 1907 where Boutroux took up again Hegel's concept of necessity and contingency, he quotes §§ 247–50 of Hegel's *Encyclopedia* and puts after the translated term, "contingency," the German term '*Zufälligkeit*' as Hegel used it in the text. Later on, he juxtaposes "le contingent, *das Zufällige*, l'individuel" (Boutroux 1907 [1926a], 100 and 103). For a more elaborate presentation of the philosophy of "Late Idealism" see Heidelberger 2004, chs. 1.6 and 8.2.

21 "Sans doute les faits philosophiques, comme tous les faits humains, sont autre chose que le produit pur et simple, et en quelque sorte les points d'intersection, des lois générales: ils enferment un élément contingent. [... Le problème] entre dans cet objet [de l'histoire de la philosophie] une part de nécessité et une part

writes, take contingency as Hegel does, "as an illusory phenomenon that will eventually be subsumed under the category of necessity anyway, but as the element in the act of volition that is not determined completely, neither through the external circumstances nor through the inner nature or the preceding actions of the acting person himself" (Boutroux 1877 [1926a], 23).[22]

Boutroux does not want this indetermination to undermine the validity of natural laws – he still takes laws of nature as an "insurmountable limit" which cannot be questioned by humans. "What will become of science, we are asked, if chance reigns in the universe? – We answer that science does not deal with particular facts, but only with general laws and that the contingency with which we deal here pertains only to individual facts" (*ibid.*, 23 ff.).[23] A dualist solution establishing only an external bond between freedom and necessity – be it a theological solution dissociating divine providence and human freedom or a philosophical one like Descartes's or Kant's – opposes "two contradictory absolutes" to each other and is consequently doomed to failure. Only in the immanent world of the phenomena themselves, i. e. in a system that does not recourse to a transcendent or a noumenal realm, can liberty and necessity possibly be reconciled.

Representatives of German "Late Idealism," like Christian Hermann Weisse, Immanuel Hermann Fichte (the son of the more famous Johann Gottlieb Fichte) and the same Eduard Zeller, strongly opposed Hegel's concept of necessity and, as a consequence, his doctrine of freedom, without however rejecting other important insights of Hegel's philosophy. They expected a better solution to follow from a concept of the Absolute or of God not as impersonal, general and immutable, but as a free and evolving personality with, however, some idiosyncratic features. Reality is the result of the free actions of God. Concrete individual particulars in history and in nature therefore possess their own

de contingence ... Ce problème n'est autre que celui de l'essence de la liberté humaine, et du rapport de cette liberté avec le hasard et la nécessité."

22 "Nous devons entendre par *contingence*, non, avec Hegel, un phénomène illusoire, rentrant, en définitive, dans la nécessité, mais la propriété inhérente à l'acte volontaire de n'être déterminé entièrement, ni par les circonstances extérieures, ni par la nature interne ou les actes antérieurs de l'agent lui-même."

23 "Que devient la science, nous dit-on, si le hasard règne dans l'univers? – Nous répondons que la science a pour objet, non les faits particuliers, mais seulement les lois générales, et que la contingence dont il s'agit ne porte que sur les faits particuliers."

share of autonomy and independence and cannot be reduced to the general and the abstract, as Hegel wanted. "According to the doctrine of contingency, it is erroneous and chimerical to attempt to reduce history to a simple deduction" (Boutroux 1874 [1916], 166).[24]

If, however, being cannot be deduced from thought, and the particulars in history and nature are irreducible facts in their own right, then the possibility of knowledge must be conceived in a new and different way from Hegel's. This new understanding is called *Erkenntnistheorie* by the late idealists and the early neo-Kantians, for whom Zeller spoke in a famous inaugural lecture at the University of Heidelberg in 1862 (see Zeller 1862 [1877]). Boutroux's emphasis on the immanent vitality and liberty of nature echoes themes from the work of Friedrich Wilhelm Joseph Schelling, whose later philosophy strongly influenced the late idealists, among them Boutroux's teacher, Ravaisson, and the Swiss philosopher, Charles Secrétan (1815–1895), author of a *Philosophie de la liberté* that Boutroux read for his dissertation. (See Janet 1897, 2: 451, 457 emphasizing Secrétan's influence on Boutroux.[25]) Both of these authors studied at some point with Schelling in Munich in the 1830s as Victor Cousin already did in 1818.

Boutroux's main argument for the reality of contingency is the richness of the universe, in particular, its variability and its diversity. Without an irreducible individuality based on qualitative differences, the world would be entirely monotonous and uniform:

> The march of observation increasingly reveals a profusion of properties: variety, individuality, life, where appearances have shown only uniform and undistinguishable masses. Hence, is it not likely that the simple repetition of the same quality, a thing devoid of beauty and interest, exists nowhere in nature, and that homogeneous quantity is but the ideal surface of beings" (Boutroux 1874 [1916], 29)?[26]

---

24 "Selon la doctrine de la contingence, il est chimérique, il est faux de prétendre ramener l'histoire à une déduction pure et simple." (Boutroux 1874, 145)

25 In his own article on Secrétan, Boutroux described the first volume of Secrétan's work as an initiation of the French public to the work of Jacob Boehme, Kant, Fichte, Schelling and Hegel. He noted that after Lachelier had resumed teaching at the ENS in 1864 and after the works of Ravaisson and Etienne Vacherot had appeared, an ardor for metaphysics developed among young philosophers in Paris that was impregnated with reflections from Secrétan's philosophy. This can easily be recognized as a description of part of Boutroux's own intellectual development (Boutroux 1895a, 230–232).

26 "Le progrès de l'observation révèle de plus en plus la richesse de propriétés, la variété, l'individualité, la vie, là où les apparences ne montraient que des masses

If the objection were raised that microphysical fluctuations have not been detected and that assuming their existence is therefore illegitimate, the answer must be that they are too small to be detected by the methods that are as yet available:

> Supposing that phenomena were indeterminate, though only in a certain measure insuperably transcending the range of our rough methods of reckoning, appearances would none the less be exactly as we see them. Thus, we attribute to things a purely hypothetical if not unintelligible determination when we interpret literally the principle by which any particular phenomenon is connected with any other particular phenomenon (*ibid.*, 28).[27]

For Boutroux, the assumption of small deviations from rigid law in particular cases is actually less strong than the supposition of a necessitating determinism that governs all cases alike. So even a critic of metaphysics should prefer contingentism over determinism.

It is interesting to see how similar this line of reasoning is to the arguments used later by Charles Sanders Peirce in order to defend his "tychism" or "tychasticism" (cp. Peirce 1892, 1893, and 1903). Peirce wrote for example:

> Try to verify any law of nature, and you will find that the more precise your observations, the more certain they will be to show irregular departures from the law. We are accustomed to ascribe these, and I do not say wrongly, to errors of observation; yet we cannot usually account for such errors in any antecedently probable way. Trace their causes back far enough and you will be forced to admit they are always due to arbitrary determination, or chance (Peirce, 1892, § 46).

The main difference between Peirce's view and Boutroux's is that Peirce drew more heavily on Darwinism, thermodynamics and other scientific developments of the time in order to justify the assumption

uniformes et indistinctes. Dès lors n'est-il pas vraisemblable que la répétition pure et simple d'une même qualité, cette chose dépourvue de beauté et d'intérêt, n'existe nulle part dans la nature, et que la quantité homogène n'est que la surface idéale des êtres?" (Boutroux 1874, 25)

27 "A supposer que les phénomènes fussent indéterminés, mais dans une certaine mesure seulement, laquelle pourrait dépasser invinciblement la portée de nos grossiers moyens d'évaluation, les apparences n'en seraient pas moins exactement telles que nous les voyons. On prête donc aux choses une détermination purement hypothétique, sinon inintelligible, quand on prend au pied de la lettre le principe suivant lequel tel phénomène est lié à tel autre phénomène." (Boutroux 1874, 24) In the first draft of his dissertation, which he sent to Paul Tannery, Boutroux actually calls determinism a "hypothesis" (Boutroux 1873, 216 f.).

of objective "chance-spontaneity." Determinism, Peirce argued, cannot explain the undeniable phenomena of growth, variety and evolution. With "tychism" he was, like Boutroux, opposing the position of "necessitarianism" and "anancism," by which he meant above all Hegel's doctrine of "evolution by mechanical necessity."

In Peirce's case, as in that of Boutroux, Schelling also played an important role. In a letter to William James, Peirce called his philosophy "Schellingism transformed in the light of modern physics" (Perry 1935, II: 416). Peirce objected to Hegel's philosophy on the grounds that "living freedom is practically omitted from its method," but admitted in the same breath that otherwise "the whole idea of the theory is superb, almost sublime" – reproach and praise that could easily have been formulated by Boutroux or Zeller in almost the same terms (Peirce 1893, § 305).[28]

Although Boutroux obviously took his critical attitude toward Hegel from Zeller, he accused the latter of not having gone far enough in his criticism: of having discarded the form, but not the essence, of Hegel's system and of returning again to the doctrine of necessity in history which is so deeply inimical to freedom. In Zeller he "found the distinctive trait of the *esprit allemand* … which sees in the individual as such only a negation and provisional form of being" while for the "*génie français*," which has found its characteristic expression in the philosophy of Descartes, "individual free will is *an end in itself*," as every (French?) human being discovers for him- or herself. The German mind never admits to free will or to "the power to retrieve from divine influence, from a universal tendency, from the infinite," permission to persist along its own path. If freedom is tolerated at all by Germans, then it is only as a means to an end – to uphold an order that, however, is in itself necessary and immutable. Freedom is not restricted by boundaries; it is what makes human beings an image of the creator (Boutroux 1877 [1926a], 29 ff., with reference to Descartes 1641, IV, 8). In spite of his otherwise critical attitude towards Descartes, Boutroux sees in him the spokesman of freedom, nonconformity and individualism *à la française*.

---

28 Cp. Heidelberger 2004, ch. 8 for a more comprehensive account of the connection between Late Idealism and objective indeterminism in the 19th century and the pivotal role that Gustav Theodor Fechner played in this.

Following his example, Boutroux wants to safeguard freedom in a much more forceful and consistent way than does Zeller.[29]

We can conclude, then, that Boutroux's contingentism is influenced in three ways by the philosophy of German Late Idealism: 1. There is the criticism of Hegel's determinist philosophy of history (cp. Brunschvicg 1922, 261–67). 2. There is Schelling's dynamic view of nature. 3. There is a radical view of freedom, of a personal God and of the individuality of God's creation that cannot be reduced to terms of necessity. In the end, whole sections of Cartesian doctrine have to give way to considerations like these. We can therefore agree with Espagne "that through the medium of Boutroux, the history of science [and philosophy] 'à la française' seems to be anchored in a German import of criticism [from Late Idealism] and its reinterpretation" (Espagne 2001, 207).[30] The consequences of advancing these three philosophical theses eventually led to a new philosophy of science and also, as we will see shortly, to a new conception of the relation of mathematics to reality.

3. The Disunity of Science

If, according to Boutroux, Descartes is wrong and the world does not possess an intrinsically mathematical structure, then there is no reason anymore to assume that a common ontology exists for all the different sciences. Higher sciences with their laws are intrinsically autonomous and not reducible to lower ones or finally to a lowest level. Each science has its own viewpoint – a "postulate," a "guiding idea" (*idée directrice*), a "synthesis" or a "new element," from which perspective it conceptualizes its objects, objects that are not shared with any other discipline.

29 Boutroux's attitude is typical for the rising myth of a national "French philosophy" at the time, with Descartes as its exponent. In his splendid history of Descartes's reception in France, François Azouvi argues that this myth became clearly visible at about 1858 (see Azouvi 2005, 412). The common approach of the period was to downplay the pre-eminent importance of German philosophy for French thought by claiming the Cartesian origin of modern philosophy, including German Idealism (*ibid.*, 434). Boutroux is clearly following here the same strategy.

30 "À travers Boutroux, on peut dire que l'histoire des sciences 'à la française' apparaît ancrée dans une importation allemande du criticisme et dans sa réinterprétation." [Brackets are mine as throughout this article.]

Although one might expect Boutroux, as an alleged "spiritualist," to be highly critical of positivism, he readily appealed to the founder of the positivist movement, Auguste Comte, in support of his position. Here once again, he was prepared to sacrifice part of the Cartesian heritage to achieve this agreement:

> According to Descartes, mathematics is realized as such deep within the sensible world; it constitutes the very substance of material things. After Descartes, this point of view became more and more limited and disputed, and the positivism of Auguste Comte summed up the results of criticism by declaring that the higher is not reducible to the lower, and that, the more we would account for a loftier reality, the more we must introduce new laws which have a specificity of their own and cannot be reduced to the preceding ones (Boutroux 1895 [1914], 41).[31]

Boutroux most probably refers to the second lesson of the *Cours de philosophie positive*, where Comte occupies himself with the hierarchy of the positive sciences (Comte 1830, 57–115). He separates natural science into two kinds: the abstract sciences as the fundamental ones and the concrete, special and descriptive sciences, which are secondary and can be derived from the corresponding fundamental ones. The abstract sciences, which make up the *philosophie positive*, are naturally divided into six disciplines. They are ordered by their degree of specialization, complication, successive dependence, individuality, concreteness, precision and proximity to humanity.

At the lowest level is mathematics (the *calcul* as the abstract part and geometry and mechanics as its concrete components) as the least complicated, the most general, the simplest and the most abstract science. The next higher level is occupied by astronomy, then come physics, chemistry, physiology, and finally social physics or sociology. This "encyclopedic formula is the only one among the many classifications possible with the six sciences which comprises the fundamental sciences in conformity to the natural and invariable hierarchy of the phenomena"

31 "Pour Descartes, les mathématiques sont réalisées telles quelles au fond du monde sensible; elles constituent la substance même des choses matérielles. Après Descartes, ce point de vue a été de plus en plus limité et contesté, et le positivisme d'Auguste Comte a résumé les résultats de la critique en professant que le supérieur ne se ramène pas à l'inférieur, et qu'à mesure qu'on veut rendre compte d'une réalité plus élevée, il faut introduire des lois nouvelles douées d'une spécificité propre et irréductibles aux précédentes." (Boutroux 1895, 25)

(*ibid.*, 115).[32] In addition, this hierarchy reflects the history of science: the later stages were only possible because the former ones prepared the ground for the higher viewpoints of the later stages. A scientific education can only achieve success if the sciences are learned in their natural order.

According to Comte, it might seem at first glance as if astronomy and part of acoustics and optics have been reduced to mechanics and geometry and therefore that all objects are finally describable in mathematical terms. But since these areas and disciplines cannot be precisely determined, one cannot take them to be part of concrete mathematics. In addition, mathematics founders even in mechanics in solving the three-body problem. The higher the variability and the complexity of the phenomena gets, the lower is the possibility of reducing a science to mathematics. Higher sciences depend on lower ones but cannot be reduced to them (*ibid.*, 153–164).

Boutroux took all this over with only slight variation. According to him, Comte's hierarchy has to be supplemented at an even more fundamental level than mathematics by the laws of logic, which come in two kinds. First come the fully analytical laws of pure logic, which are independent of experience. (There is a distinctively Hegelian ring to this as regards the concept of logic!) Second are the laws of the syllogism; these already include a small amount of descriptive content. On these rest the laws of mathematics dealing with numbers and extensions; these also have a synthetic component. The laws of mechanics are located at the next higher level, followed by the laws of physics, where the quality of energy is introduced because of the Second Law of thermodynamics; after that come the laws of chemistry. Next, we have the laws of biology, on which are built the laws of psychology (these do not appear in Comte's hierarchy) and finally the laws of sociology. Through the introduction of new and higher points of view, phenomena that were left unexplained on preceding levels are discussed and explained on higher ones. The guiding principles at each stage are intrinsically novel and creative and cannot be reduced to those of the lower stages. We find with each advance new elements not deducible from the preceding. There is a continuous increase in variability and complexity that gives rise to different laws at each level (cp. also Gutting 2001, 22–24).

---

32 "[T]elle est la formule encyclopédique qui, parmi le très-grand nombre des classifications que comporte les six sciences fondamentales, est seule logiquement conforme à la hiérarchie naturelle et invariable des phénomènes."

More closely considered, it turns out that Boutroux's agreement with Comte is not only a repudiation of Descartes's view of mathematics but also of Kant's and generally of Kant's synthetic a priori judgments. For Kant, mathematics consists of synthetic judgments that are constitutive and necessarily valid of their objects. Boutroux, however, asserts that mathematics is not a priori, and neither constitutive of objects in any real sense, nor possessing merely empirical origins.

> We believe everything to be necessarily determined, because we believe everything, in essence, to be mathematical. This belief is the spring, manifest or unperceived, of scientific investigation. What we have to find out is whether this is a truly constitutive principle, or simply a regulating principle and a guiding idea. Does science prove that the basis of things is exclusively mathematical, or does it only assume this (Boutroux 1895 [1914], 207)?[33]

Boutroux emphatically chooses the second horn of the dilemma: Concrete reality refuses to be exhaustively confined in mathematical form. For this reason, the Cartesian belief that things are particular determinations of mathematical essences must be rejected. First of all, mathematics is not necessarily true because it contains some synthetic element that is not knowable independently of experience (this view will be treated more fully below). Second, the concrete sciences, e.g. mechanics, offer "elements that are irreducible to pure mathematical determinations." Third, it would be incorrect to say that there is a science, e.g. mechanics, that – at least in principle – constitutes the entire science of the real. "For, in the present state of our knowledge, science is not one, it is multiple. Science, regarded as including all the sciences, is but an abstraction" (*ibid.*, 138/ transl. 211). In the end, the concept of a synthesis a priori remains strange (Boutroux 1926, 129. See also 130 and 25–29 for a thorough critique of the synthetic a priori in this vein). So our mathematics hardly scratches the surface of things and cannot reach or govern their nature. The impression of a unity of science that mathematics suggests is only an illusion.

What is true of mathematics also goes for causality. It is not an a priori category as Kant had taught but a principle that is derived from ex-

33 "Nous croyons que tout est déterminé nécessairement, parce que nous croyons que tout, en réalité, est mathématique. Cette croyance est le ressort, manifeste ou inaperçu, de l'investigation scientifique. La question est de savoir si c'est là un principe constitutif ou simplement un principe régulateur et une idée directrice. La science établit-elle, ou se borne-t-elle à supposer que le fond des choses est exclusivement mathématique?" (Boutroux 1895, 136)

perience. Boutroux thus comes very close to the empiricism of John Stuart Mill:

> It cannot, then, be said that the principle of causality governing science is a law imposed on things by the mind … [W]e must not forget that experience itself has introduced to the human mind the scientific idea of natural cause and has gradually clarified this idea. The latter is not the idea of a principle a priori which governs the modes of being; it is the abstract form of the relation existing between these modes. We cannot assert that the nature of things has its derivation in the law of causality. To us, this law is but the most general expression of the relations arising from the observable nature of given things (Boutroux 1874 [1916], 25–27).[34]
>
> For science, however, no other causality is necessary than the connecting link provided by mathematics. Causality in the proper sense is composed of a mathematical linkage and contingent contiguity. This is not anymore Kant's thesis (Boutroux 1926, 29).[35]

Boutroux also rejects Kant's reconciliation of human freedom with necessity as one of the "most obscure parts of Kantianism." Kant's mistake was to have used an absolute conception of freedom and a view of natural necessity that derives from Newtonian celestial mechanics. The result is a radical and unacceptable dualism that splits man into two. We should better argue the other way around, says Boutroux, and take seriously the unity of our being. Neither is freedom independent of the empirical phenomena nor is physical causality impenetrable. Freedom and physical necessity are ultimately only abstractions with which one can apprehend nature and living beings only to a small extent. Kant's abstract and a priori idea of science should therefore be abandoned, and it should be admitted that physical necessity is an artifact of our mind:

> Under these circumstances, the concepts presupposed by science appear only as the expression of the needs of our spirit. They are not inherent

34 "On ne peut donc dire que le principe de causalité qui régit la science soit une loi dictée par l'esprit aux choses … Toutefois, il ne faut pas oublier que c'est l'expérience elle-même qui a introduit dans l'esprit humain et progressivement épuré l'idée scientifique de cause naturelle. Cette idée n'est pas celle d'un principe à priori qui régit les modes de l'être, c'est la forme abstraite du rapport qui existe entre ces modes. Nous ne pouvons pas dire que la nature des choses dérive de la loi de causalité. Cette loi n'est pour nous que l'expression la plus générale des rapports qui dérivent de la nature observable des choses données" (Boutroux 1874, 22 f.).

35 "Mais, pour la science, il n'y a pas d'autre causalité nécessaire que la liaison mathématique. La causalité proprement dite se décompose en liaison mathématique et contiguïté contingente. Ce n'est plus là la thèse de Kant" (cp. also Boutroux 1895 [1914], 54/ transl. 35 f.).

in the objects. Only facts are given; science misses out on the veritable causes. We discover only factual regularities; we do not know whether we are in possession of absolute laws (Boutroux 1926, 226).[36]

This sounds very much like Comte, who rejected the search for causes or substances as metaphysical and advocated phenomena and their relation instead. Neither Comte nor Boutroux regard the ensuing laws as absolute but only as more or less imperfect approximations. As a result, Boutroux argues, it is illusory to presuppose a unity of science. We have to take science's variegated nature seriously. In science "anything goes," as long as it is successful:

Amidst one and the same science different, heterogeneous, perhaps even contradictory, principles are nowadays admitted. It is enough that they succeed. The sciences are an assemblage of concepts constructed to summarize experience and to represent for us the forces of nature. They are full of lacunae, of obscure concepts that are admitted because they are convenient (*ibid.*).[37]

Boutroux's contemporaries thoroughly noticed the proximity of these views to those of Comte and Mill. Paul Janet saw at the core of Boutroux's doctrine a theory of causality in the tradition of Hume and Mill, and a rejection of the Kantian solution as useless for "the scientific conception of the world." He could not accept this, and recommended instead a metaphysical principle of causality to Boutroux, because otherwise the philosophy of contingency would turn into a doctrine of

36 "Dans ces conditions, les concepts que suppose la science n'apparaissent plus que comme l'expression des besoins de notre esprit. Ils ne sont plus inhérents aux choses. Les faits seuls sont donnés; les véritables causes échappent à la science. Nous découvrons des lois-faits, nous ne savons si nous sommes en présence de lois absolues." This comes very close to David Hume's views; cp. his *Enquiry Concerning Human Understanding* (IV, 1, § 12): "[T]he utmost effort of human reason is, to reduce the principles, productive of natural phenomena, to a greater simplicity, and to resolve the many particular effects into a few general causes … But as to the causes of these general causes, we … in vain attempt their discovery".

37 "Au sein d'une même science aujourd'hui l'on admet des principes différents, hétérogènes, peut-être contradictoires. Il suffit qu'ils réussissent. Les sciences sont un ensemble de concepts construits de manière à résumer l'expérience et à représenter pour nous les forces de la nature. Elles sont pleines d'hiatus, de concepts obscurs, qu'on admet parce qu'ils sont commodes." Boutroux's description of the sometimes even contradictory principles of a science reminds one of the way how, according to Pierre Boutroux, Poincaré viewed Maxwell's electrodynamics (Boutroux, P. 1913, 162 f.).

chance and fortuity – a view that Boutroux himself would find wanting (Janet 1897, 461 f., 465 f.).

François Pillon (1830–1914), a follower of Renouvier's neo-criticism, analyzed and affirmed Boutroux's affinity to the views of Mill, Comte and Littré in two articles (Pillon 1897, 58–63; 1907, 85–97). He also made a connection of Boutroux's views with William James's "radical empiricism," as well as with Henri Poincaré's conventionalism, and claimed that Boutroux had reached "theist" and "libertarian" conclusions very similar to those of the neo-criticist followers of Renouvier, although this school had arrived at them by a completely different route. (Pillon 1897, 69; 1907, 125. Séailles 1905, 2 confirms this diagnosis and connects Renouvier with the tradition of Late Idealism. cp. Beilecke 2003, 232–240).

Pillon conceded, however, that Boutroux departs from positivism à la Comte in one important respect: Whereas Comte explained habit as a result of the law of inertia, Boutroux turned the tables and took the law of inertia – and indeed all other natural laws – as a result of habit. In this, Pillon claimed, Boutroux radicalized the doctrine of his teacher Ravaisson, who had drawn an analogy between habit and nature (Pillon 1907, 102). We can see again that, in spite of strong empiricist commitments, Boutroux's views gleam with Schellingian thought.

The strong similarity of Boutroux's philosophy of science to the views of empiricists like Mill does not imply, however, that he accepted induction as the method of natural science as the latter did. It is only logical that he discarded it, because otherwise it could be claimed that our knowledge of nature follows mechanically from observation, without any original and spontaneous effort on the mind's part. In the first issue of the *Revue de synthèse historique*, founded by his disciple Henri Berr, Boutroux vehemently rejected Francis Bacon's inductive procedure as isolating the facts from each other and denying their rapport with general ideas. Instead, science progresses by propounding general hypotheses, by a conceptual synthesis that is enabled by an overall view of the ensemble of facts as a whole. There exists no real intelligent and instructive analysis that does not start with a synthesis:

> By radically distinguishing facts from laws and by condemning the role of hypothesis in the search for the latter, Bacon decreed the fabrication of complete tables of facts before seeking the laws that develop from them … Nevertheless the physical sciences took a completely different route from the one that Bacon has prescribed them. Far from clinging first to observation before going about searching for laws by reasoning from the par-

> ticular to the general, they proceeded from the universal to the particular, from the grand laws to the laws of detail, always making sure never to regard the general as more than a hypothesis. It is by following more and more consciously this hypothetico-deductive method that science has definitely progressed (Boutroux 1900, 8 f.).[38]

If one wants to call this doctrine "neo-Kantianism" in order to mark its divergence from empiricism, then so be it! (One would then also have to call Karl Popper a neo-Kantian for his anti-inductivism!) But one should keep in mind that it has only a faint similarity with the original doctrine of Kant himself (cp. also Boutroux 1926, 29). It is interesting to note that Poincaré had a similar view of things:

> The physicist could never be content with an inane experience alone. His object is not the same as that of the historian and an isolated fact does not have any value for him. From there derives the usefulness of generalization, which demands the employment of mathematics. Any generalization presupposes a certain confidence in the unity and simplicity of nature (Poincaré 1902a, 130).[39]

So again: generalization is not just a conjecture coming from 'nowhere', so to say, but a free, spontaneous, creative act of the mind – a synthesis that is guided by a view of the whole.

---

38 "C'est Bacon qui, distinguant radicalement les faits et les lois, et condamnant l'hypothèse dans la recherche de ces dernières, prescrit de dresser d'abord des tables complètes de faits, avant de chercher les lois qui s'en dégagent … Cependant les sciences physiques, en se développant, ne tardaient pas à entrer dans une voie tout autre que celle que leur avait tracée Bacon. Loin de s'en tenir d'abord à l'observation, pour n'aborder qu'ensuite la recherche des lois en allant exclusivement du particulier au général, elles ont procédé de l'universel et du général au particulier, des grandes lois aux lois de détail, en ayant soin seulement de n'admettre jamais le général qu'à titre d'hypothèse. C'est en suivant de plus en plus consciemment cette méthode hypothético-déductive, que la science a pris définitivement son essor" (cp. also Boutroux 1895 [1914], 57/ transl. 37 f.).

39 "Le physicien ne saurait se contenter de l'expérience toute nul. Son objet n'est pas le même que celui de l'historien et un fait isolé est pour lui sans valeur. De là l'utilité de la généralisation qui exige l'emploi des Mathématiques. Cette généralisation suppose une certaine croyance à l'unité et à la simplicité de la nature … Cette croyance, justifiée ou non, est nécessaire à la science." (Cp. also Poincaré 1902, ch. IX. Poincaré there uses the same example of Carlyle on King John Lackland as Boutroux 1900!)

4. Two Kinds of Laws

In his 1895 work on the laws of nature, Boutroux introduces a distinction between two kinds of laws which cuts across the categories he discussed before and which concerns, so to speak, the "closeness" of laws to reality. On the one hand, there are mathematical, i.e. fundamental, highly abstract and nearly necessary laws, which are based on improvable postulates and which are remote from concrete contingent reality. On the other hand, there are inductive and observational laws, which pay attention to detail and are true to reality, but which lack necessity and are completely empirical, yet are more determinist and determinate in their make-up than the abstract laws:

> In a general way, then, there are two kinds of laws: the first, which are more akin to mathematical conjunction and imply considerable elaboration and purification of concepts; the second, which are nearer to observation and induction, pure and simple. The former express a rigorous, if not absolute, necessity, but they remain abstract and incapable of determining the details and the mode of effective realization of the phenomena. The latter treat of the details and the relations which complex and organized wholes have with one another: consequently they are far more determinative than the former; but as they have no other basis than experience and connect together wholly heterogeneous terms, they cannot be regarded as necessitative [*sic!*]. Possible prediction does not imply necessity, since free acts may admit of it. Thus, necessity and determination are distinct from each other; our science cannot blend them into one" (Boutroux 1895 [1914], 214).[40]
>
> Either necessity without determinism or determinism without necessity; such is the dilemma confronting us (*ibid.*, 91).[41]

40 "Il y a donc, d'une manière générale, deux sortes de lois: les unes, qui tiennent davantage de la liaison mathématique et impliquent une forte élaboration et épuration des concepts; les autres, qui sont plus voisines de l'observation et de l'induction pure et simple. Les premières expriment une nécessité rigoureuse, sinon absolue, mais restent abstraites et incapables de déterminer le détail et le mode de réalisation effective des phénomènes. Les secondes portent sur le détail et sur les relations qu'ont entre eux les ensembles complexes et organisés: elles sont donc beaucoup plus déterminantes que les premières; mais, n'ayant d'autre fondement que l'expérience et reliant entre eux des termes tout à fait hétérogènes elles ne peuvent être tenues pour nécessitantes. Prédiction possible n'implique pas nécessité, puisque les actes libres peuvent la comporter. Ainsi nécessité et détermination sont choses distinctes. Notre science ne parvient pas à les fondre en une unité." (Boutroux 1895, 141) Cp. also Benrubi 1926, 157 f.

41 "Ou nécessité sans déterminisme, ou déterminisme sans nécessité: voilà le dilemme où nous sommes enfermés." (Boutroux 1895, 59)

And in the "Preface by the author" to the English translation, Boutroux writes:

> The theory upheld in the present work is that no absolute coincidence exists between the laws of nature, as science assumes them to be, and the laws of nature as they really are. The former may be compared to laws proclaimed by a legislator and imposed a priori upon reality. The latter are harmonies towards which … the actions of different beings really tend. The former are abstract relationships, the elements of which are themselves relationships; the latter are concrete relationships, the terms of which are real subjects, concrete beings" (*ibid.*, 6 ff.). (Translation modified.)

The more general and abstract a natural law is, the greater is its necessity, but the less can it account for the real phenomena in their richness and variety. "Formulated in an absolute way, they [the laws] represent only vague or false generalities" (Boutroux 1926, 130). And inversely, the more a law adapts to particular cases in their actual and concrete form, the more its determinacy is warranted (in the sense of being determined uniquely), but the lesser is the degree of its necessity. The principle of the conservation of force stands in contrast to laws of change like Clausius's law (entropy principle). The negative form of the latter, writes Boutroux, "actually prevents this [conservation] principle from generating a complete determination" of the phenomena (Boutroux 1895 [1914], 89).[42]

The difference between abstract-necessary and empirical-determinate laws is co-extensive with the difference between science as explanation and science as description, which Boutroux had invoked already in his dissertation. There he gives an account of how humans develop an interest in science (leaving open whether we are to imagine this process as an ontogenetic or phylogenetic one). In the beginning, humans are content with describing what they can grasp with their senses:

> This is the first phase of science, that wherein the mind relies on the senses in the task of establishing universal knowledge. And, indeed, the senses afford a primary conception of the world, which they show to be a mass of

42 "Mais ces lois, ni ne se ramènent à la loi de conservation, ni ne suffisent à déterminer avec précision les phénomènes. Déjà la forme négative du principe de Clausius empêche que ce principe n'engendre une détermination complète." (Boutroux 1895, 58) This remark foreshadows Émile Meyerson's work, where the course of nature is reconstructed as a permanent conflict between determinate conservation processes (expressed in the theory by propositions of identity) and the reign of the Second Law.

> facts, endless in their variety. Man may observe, analyse, and describe them with ever-increasing exactness: it is this description that constitutes science.

As soon as regularities and connections between the phenomena can be perceived in these processes, the need arises for a deeper understanding:

> Henceforth, it regards purely descriptive science as inadequate, and even inexact, in that it perverts the relations of things. The mind would add explanatory knowledge thereto, but this the senses are unable to procure.

In the course of time, however, a "divergence from real things" appears and leads them to ask: "Is this category of necessary relation, inherent in the understanding, actually to be found in things themselves?" Or rather, does the understanding deceive the mind (Boutroux 1874 [1916], 3 and 5)?[43]

Boutroux's division between laws that are nearly necessary but do not cover the particular case and those that do more justice to the concrete phenomena but thereby lose their necessary character is reminiscent of the influential distinction Nancy Cartwright has recently drawn between "fundamental" or "theoretical" and "phenomenological" laws:

> In modern physics, and I think in other exact sciences as well, phenomenological laws are meant to describe, and they often succeed reasonably well. But fundamental equations are meant to explain, and paradoxically enough the cost of explanatory power is descriptive adequacy. Really powerful explanatory laws of the sort found in theoretical physics do not state the truth (Cartwright 1983, 3).

Thus, Boutroux and Cartwright agree with each other that there is a difference between explanatory and descriptive laws and that the fundamental explanatory ones do not describe, as is widely assumed, the underlying reality behind the appearances, but are hypothetical creations of

43 "C'est la première phase de la science, celle où l'esprit se repose sur les *sens* du soin de constituer la connaissance universelle. Et les sens lui fournissent en effet une première conception du monde. Selon leurs données, le monde est un ensemble de faits d'une infinie variété. L'homme peut les observer, les analyser, les décrire avec une exactitude croissante. La science est cette description même … La science purement descriptive lui paraît désormais insuffisante, inexacte même, en ce qu'elle fausse les relations des choses. Il y voudrait joindre la connaissance explicative. Cette connaissance, les sens ne peuvent la procurer … Or cette catégorie de liaison nécessaire, inhérente à l'entendement, se retrouve-t-elle en effet dans les choses elles-mêmes?" (Boutroux 1874, 2 and 4; translation amended)

the human mind that "patently do not get the facts right." They are abstract mathematical propositions that cannot claim necessity for empirical reality (Boutroux); they are idealized and "fictional model[s] of the system[s] under study" (Cartwright 1983, 158).

Cartwright and Boutroux also agree in holding that descriptive phenomenological laws are neither provisional nor illusive in their contingent and context-specific character, as the tradition (from Democritus and Descartes to Kant and von Helmholtz) claimed them to be. The messiness and diversity of reality in its deviation from the fundamental picture is not an illusion brought about by ignorance of initial conditions, but is a true representation. Reality is much more diverse than the explanatory laws lead us to expect (Boutroux); the "phenomenological laws are indeed true of the objects in reality – or might be; but the fundamental laws are true only of objects in the model" (Cartwright 1983, 4).

After this common starting point, the views of Boutroux and Cartwright drift apart, however, albeit without one ever actually contradicting the other. Cartwright concentrates her further considerations on the price one has to pay for the setting up of fundamental laws: they have to be taken as "lies," not as literal descriptions of reality. We do not hear from her anything about the costs involved in the reassessment of the observational laws and of their closeness to reality. This is something, however, that Boutroux worries about: the cost to be paid for observational laws is the renouncing of their necessity. Some passages even suggest that, for Boutroux, the set of necessary laws dwindle down more and more until only the laws of pure logic are left, and therefore their actual effectiveness in the realm of experience reduces to no more than to that of a limiting case (Boutroux 1874 [1916], ch. 1).

As they develop their views further, the similarities between Cartwright and Boutroux continue to diminish. For Cartwright, the causal role of theoretical entities becomes a realist substitute, compensating for the antirealist character of the fundamental laws, an option which seems alien to Boutroux's thought. On the other hand, Cartwright would not make the inherent indetermination and spontaneity of nature responsible for the multifarious and almost chaotic phenomena that are the subject of the observational laws as Boutroux does. It is interesting, however, to note that they both take the spirit of positivism as their starting point, in spite of distancing themselves in later respects from this doctrine. Boutroux could have agreed with Cartwright in the following maxim she put at the top of her first chapter: "The picture of science that I present in these essays lacks the purity of positivism. But it shares one deep positivist conviction: there is no better reality besides the reality we have to hand" (Cartwright 1983, 19).

The historical *tertium comparationis* for Boutroux and Cartwright might be Pierre Duhem, a friend of Boutroux, who is referred to more often than anyone else (except the constructive empiricist, Bas van Fraassen) in Cartwright's book. It would carry us too far afield, however, to follow further this tempting thread.[44]

5. The Hypothetical Nature of Mathematics

We have seen how Boutroux freed himself with the help of German Late Idealism from Descartes's necessitarian tendencies and how he exchanged the inexorable necessity of natural laws for flexible contingency. Descartes's determinism proceeded from two sources: from God's unvarying will as well as from the mathematical essence of the external world, i. e. its extension. But if both of these convictions are abandoned, so that mathematics no longer resides objectively in the objects themselves and that God himself, like His creation, acts in a contingent and variable way, what are the consequences for the nature of mathematics and for our scientific conception of the world? It follows that mathematics is nothing more than a voluntary creation of the human mind, which, prompted by experience, devises for itself a regulative idea and sets it down in the form of convenient principles. As a result, mathematical propositions are neither analytic nor synthetic. They are suggested by experience, but contain an element of unempirical and rationalist idealization:

> [W]hat is the origin of the mathematical laws? If they were wholly known à priori, their intelligibility would be perfect. As it is, they involve elements that cannot be fathomed by thought. We are compelled to acknowledge them: we cannot say we find them clearly springing from the fundamental nature of the intellect. Nor can they be connected with knowledge à posteriori, for they deal only with limits. A limit cannot be grasped empirically, since it is the purely ideal term towards which tends a quantity supposed to increase or decrease indefinitely. The mathematical laws presuppose a very complex elaboration. They are not known exclusively either à priori or à

44 For the development of French philosophy of science from Poincaré and Boutroux to Duhem and beyond up to the Vienna Circle see Brenner 1998, 2002, 2003, 2004.

posteriori, but are a creation of the mind; and this creation is not an arbitrary one, but, owing to the mind's resources, takes place with reference to experience and in view of it. Sometimes the mind starts with intuitions which it freely creates; sometimes, by a process of elimination, it gathers up the axioms it regards as most suitable for producing a harmonious development, one that is both simple and fertile. Thus, mathematics is a voluntary and intelligent adaptation of thought to things, [and] it represents the forms that will allow of qualitative diversity being surmounted, the moulds into which reality must enter in order to become as intelligible as possible (Boutroux 1895 [1914], 40 f.).[45]

In his lectures on the philosophy of Kant, Boutroux makes it clear what he means by the mathematical creation of the mind taking place with reference to experience. Since the time of Kant, he argues, there has been a tendency to reduce and isolate the synthetic element of mathematics as far as possible. Von Helmholtz has shown, he says, that the synthetic element of arithmetic ultimately boils down to the "anteriority" of one fact to another, i. e. to the truth that each number is followed by another one. This truth has to be synthetic because the idea of an object does not contain the idea of another object, let alone the idea of an infinite succession of objects. Arithmetic has to depart from something, and the principle of contradiction alone does not furnish us with such a beginning. Progress in arithmetic does not consist in suppressing all synthetic elements, but in isolating the synthetic from the analytic and in taking it as the starting point of mathematics. This allows one

45 "[Q]uelle est l'origine des lois mathématiques? Si elles étaient connues entièrement a priori, elles présenteraient une parfaite intelligibilité. Or, elles impliquent des éléments impénétrables à la pensée. On est forcé de les admettre; on ne peut pas dire qu'on les voie clairement découler de la nature fondamentale de l'intelligence. Elles ne peuvent non plus être rapportées à la connaissance a posteriori, car elles ne portent que sur des limites. Or, une limite ne peut être saisie empiriquement, puisque c'est le terme purement idéal vers lequel tend une quantité qui est supposée croître ou décroître indéfiniment. Les lois mathématiques supposent une élaboration très complexe. Elles ne sont nues exclusivement ni a priori ni a posteriori: elles sont une création de l'esprit; et cette création n'est pas arbitraire, mais a lieu, grâce aux ressources de l'esprit, à propos et en vue de l'expérience. Tantôt l'esprit part d'intuitions qu'il crée librement, tantôt, procédant par élimination, il recueille les axiomes qui lui ont paru le plus propres à engendrer un développement fécond et exempt de contradictions. Les mathématiques sont ainsi une adaptation volontaire et intelligente de la pensée aux choses; elles représentent les formes qui permettront de surmonter la diversité qualitative, les moules dans lesquels la réalité devra entrer pour devenir aussi intelligible que possible" (Boutroux 1895, 24 f.).

to concentrate on developing the analytic rest. The same goes for geometry, where the synthetic element lies in propositions like "A straight line is the shortest path between two points," but above all in the assumption of the general uniformity of space, i. e. of the endless repetition of something that is given. (The relation, however, of Euclidean to non-Euclidean geometry belongs solely to the analytic part of geometry.)

The irreducible synthetic element of mathematics, however, is far from being necessary and a priori in the sense of Kant:

> It seems therefore indisputable that there is a synthetic element at the basis of mathematics. Are these elements also *a priori?* Certain mathematicians, after having noticed the synthetic character of their principles and being able to understand them only by experience, admit that they are *a priori* in the sense that they are formed by the mind in an arbitrary manner. This is, however, an *a priori* in a sense completely different from that of Kant.
>
> But 'arbitrary', is this not too strong a word? I think that the mathematicians only want to say that their principles do not impose themselves by the objects. One says, for example: 'The mind chooses the simplest combination.' In itself, this is, however, an appeal for a reason and thus not arbitrary. The mind is guided in its construction by its sense of intelligibility, by its nature. In this way, we come closer to the Kantian sense. Mathematical principles suppose actions of the mind that are determined by its very nature – by *actions* and not only by impressions that emanate from the *given* (Boutroux 1926, 28).[46]
>
> [O]ne cannot say that these postulates are founded on a real rational necessity. It is true that they are founded on the understanding, but the understanding cannot prove their necessity. Mathematical postulates are the result of a choice that is made among many equally possible postulates; and if this

46 "Il semble donc que la présence d'éléments synthétiques au fond des mathématiques soit incontestable. Ces éléments sont-ils également *a priori?* Certains mathématiciens, ayant constaté le caractère synthétique de leurs principes, et ne pouvant se les expliquer que par l'expérience, admettent qu'ils sont a priori, en ce sens qu'ils seraient formés par l'esprit d'une manière arbitraire. C'est là un a priori tout autre que celui de Kant. Mais 'arbitraire', n'est-ce pas là un mot trop fort? Je crois que les mathématiciens veulent seulement dire que leurs principes ne sont pas imposés par les choses. On dira, par exemple: 'L'esprit choisit la combinaison la plus simple'. Mais cela même est un appel à une raison, et n'est pas l'arbitraire. Ces constructions, l'esprit les fera, guidé par son sens de l'intelligibilité, par sa nature. Nous nous rapprochons ainsi du sens kantien. Les principes mathématiques supposent des actions de l'esprit déterminées par sa nature même, des *actions*, et non pas seulement des impressions émanantes des choses *données*."

choice is not arbitrary, it is at least determined by reasons of convenience (*ibid.*, 67).[47]

"And he who says convention," wrote Pierre Boutroux in an obituary for his uncle Henri, "also says free decision of the mind (*de l'esprit*) (Boutroux, P. 1913, 159).

It seems that Henri Poincaré is one of the "certain mathematicians" just referred to, and that he shared this same outlook with Boutroux.[48] Geometrical axioms are, as Poincaré famously said,

> *neither synthetic* a priori *intuitions nor experimental facts.* They are *conventions.* Our choice among all possible conventions is *guided* by experimental facts; but it remains *free*, and is only limited by the necessity of avoiding every contradiction. Thus it is that postulates may remain *rigorously* true even when the experimental laws which have determined their adoption are only approximate" (Poincaré 1902 [1905], 50).[49]

Hence one cannot talk of truth or falsehood in mathematics but only of greater or lesser convenience in applying a particular mathematical proposition. It is important to realize that Poincaré's conventionalism implies an empirical element for mathematics (without thereby advocating an empirical foundation) in much the same way as Boutroux has conceived of it in his reworking of Comte. As Boutroux wrote of his brother-in-law: "He opposes the school of Weierstrass, which tends to reduce mathematics to pure logic. He maintains the relation of mathematics to the real, be it sensible or super-sensible, and consequently, to the necessary role intuition has to play in the efforts of discovery" (Bou-

---

47 "Selon eux, il faut des postulats, mais on ne peut pas dire que ces postulats soient fondés sur une véritable nécessité rationnelle. Ils sont bien fondés sur l'entendement; mais l'entendement n'en peut démontrer la nécessité. Les postulats mathématiques sont le résultat d'un choix fait entre beaucoup de postulats également possibles; et si ce choix n'est pas arbitraire, il n'est du moins déterminé que par des raisons de convenance."

48 Compare the quotation just given with the opening of Poincaré's *Science and Hypothesis*; the similarities are just striking (see Poincaré 1902, 31 f. = 1905, 31 f.).

49 "*Les axiomes géométriques ne sont donc ni des jugements synthétiques* a priori *ni des faits expérimentaux.* Ce sont des *conventions*; notre choix, parmi toutes les conventions possibles, est *guidé* par des faits expérimentaux; mais il reste *libre* et n'est limité que par la nécessité d'éviter toute contradiction. C'est ainsi que les postulats peuvent rester *rigoureusement* vrai quand même les lois expérimentales qui ont déterminé leur adoption ne sont qu'approximatives." (Poincaré 1902, 75) One should note that the whole chapter from which this quotation is taken was first published in 1891.

troux 1913, 36 f.; cp. also 48 f. and Boutroux 1926, 28).[50] The conviction that Boutroux and Poincaré share with each other can aptly be summed up in the words of the former: "[M]athematics is necessary only with reference to postulates whose necessity cannot be demonstrated, and so is only hypothetical after all" (Boutroux 1895 [1914], 215).[51]

The adoption of this position has far-reaching consequences for the scientific worldview. Recall Boutroux's assertion about the inability of mathematics to picture the world: "The singular certainty presented by mathematics as an abstract science does not authorise us to look upon mathematical abstractions themselves, in their rigid monotonous form, as the exact image of reality" (quoted above; text for fn. 17). Poincaré shares the same viewpoint:

> The object of mathematical theories is not to reveal to us the real nature of things; that would be an unreasonable claim. Their sole purpose is to coordinate the physical laws that we learn from experience, whose formulation we would be unable to manage without the aid of mathematics. Whether the ether exists or not matters little; let us leave that to the metaphysicians. What is essential for us is that everything happens as though it exists and that this hypothesis lends itself to the explanation of phenomena. After all, have we any other reason for believing in the existence of material objects? That, too, is only a convenient hypothesis; only that it will never cease to be such, while some day, no doubt, the ether will be set aside as useless (Poincaré 1902 [1905], 211 f.).[52]

---

50 "Il s'oppose à l'école de Weierstrass, qui tend à réduire les mathématiques à la pure logique. Il maintient la relation des mathématiques au réel, soit sensible, soit suprasensible, et, par suite, le rôle nécessaire de l'intuition dans le travail de la découverte." (On Poincaré's anti-logicism cp. Detlefsen 1992. I do not, however, agree with Detlefsen's view that this anti-logicism is Kantian.).

51 "[L]es mathématiques ne sont nécessaires que par rapport à des postulats dont la nécessité est indémontrable, et ainsi leur nécessité n'est, en définitive, qu'hypothétique." (Boutroux 1895, 141)

52 "Les théories mathématiques n'ont pas pour objet de nous révéler la véritable nature des choses; ce serait là une prétention déraisonnable. Leur but unique est de coordonner les lois physiques que l'expérience nous fait connaître, mais que sans le secours des mathématiques nous ne pourrions même énoncer. Peu nous importe que l'éther existe réellement, c'est l'affaire des métaphysiciens; l'essentiel pour nous c'est que tout se passe comme s'il existait et que cette hypothèse est commode pour l'explication des phénomènes. Après tout, avons-nous d'autre raison de croire à l'existence des objets matériels? Ce n'est la aussi qu'une hypothèse commode; seulement elle ne cessera jamais de l'être, tandis qu'un jour viendra sans doute où l'éther sera rejeté comme inutile." (Poincaré 1902, 215; translation amended)

As this quotation shows, Poincaré goes further, in fact, than Boutroux. He says, in effect, that we can be even less sure of material objects in our everyday life than in science, where the success of mathematical theory gives us at least a hypothetical or fictitious reason to believe in the existence of a material object.

In another text, Poincaré distinguishes between the content (*fond*) and the form of a physical theory. The content concerns the relations between inaccessible and unknown objects, whereas the form relates to the way how we formulate these relations. The form often changes but the content remains:

> Hypotheses concerning what I just called the form [of theories] cannot be true or false, they can only be convenient or inconvenient. So, for example, the existence of the ether and even the existence of external objects are only convenient hypotheses … It is for this reason … that there are certain classes of facts that can equally well be accounted for by two or more different theories in such a way that no experience can ever decide between them" (Poincaré 1902a, 130).[53]

It has been said that Poincaré's "own language is often inadequate because he relies on the terminology which was in use in the higher education of his time. The influence of his brother-in-law, of the philosopher Boutroux, was often more malign than salutary" (Rougier 1946, 15). In the light of the considerations above, it seems that Boutroux's influence on Poincaré was much more substantial than just a matter of terminology and much more beneficial than harmful. Boutroux has taught us that "instead of being a necessity, [the laws of nature] set us free" (Boutroux 1895 [1914], 218).[54] Mathematics cannot of itself lay bare the essence of the empirical world. It is an intelligent, yet genuinely hypothetical, adaptation of human thoughts to things. Boutroux has thereby laid the foundation for a more moderate and skeptical view of the role of mathematics in the sciences than the Cartesian outlook allowed. And he has also prepared the way for an appreciation of pure mathematics that allows for alternative mathematical systems like

53 "Les hypothèses relatives à ce que je viens d'appeler la forme ne peuvent pas être vraies ou fausses, elles ne peuvent être que commodes ou incommodes. Par exemple, l'existence de l'éther, celle même des objets extérieurs ne sont que des hypothèses commodes … C'est pour cela aussi qu'il y a certaines catégories de faits qui s'expliquent également bien dans deux ou plusieurs théories différentes, sans qu'aucune expérience puisse jamais décider."

54 "Loin d'être une nécessité, elles nous affranchissent." (Boutroux 1895, 143)

non-Euclidean geometry, whether or not they fit neatly over empirical reality.

6. Conclusion

In this paper I have tried to come to terms with Boutroux's concept of contingency of the laws of nature and the consequences he drew from them. These consequences are the denial of determinism for the physical and psychical world, the rejection of mechanics as the reductive and uniform basis of all science and the assertion of a hypothetical view of mathematics. According to Boutroux, mathematics is hypothetical in two respects: 1. The alleged necessity of mathematics cannot be demonstrated and thus not conclusively conferred on physical processes; 2. there is no mathematical idea or structure built into the nature of things. It is the human mind that makes up its own mathematical theories. They are unable to provide knowledge about the world: they can only serve as approximate instruments for action and thus at best as hypotheses a long way off from the real nature of things.

I have also tried to elucidate the historical background from which Boutroux's view of contingency developed into a radically anti-Kantian critique of mathematics and to show how this prepared the way for Poincaré's conventionalism. Boutroux's notion of contingency originated with a critique of Hegel's view of history, and its conveyance on nature and natural science involved an intricate process of cultural transfer from Germany to France. But it has also grown from a critique of Descartes's attempt to make mathematics (geometry in particular) the essence of the physical world. And there is finally the firm grounding of Boutroux's views in Auguste Comte's positivism and its doctrine of the emergent hierarchy of the sciences. As Nancy Cartwright's views on the limited access physical theories provide to a "dappled world" show, Boutroux's philosophical outlook is more topical and relevant for us today than might be apparent at first glance.

Bibliography

Aliotta, Antonio (1914), *The Idealistic Reaction against Science*. Transl. by Agnes McCaskill. London: Macmillan. (Italian original *La reazione idealistica contro la scienza*. Palermo: Optima 1912).

Aristotle, *Physics*.

—, Metaphysics.

Azouvi, François (2005), "Descartes", in *Erinnerungsorte Frankreichs*. Ed. by Pierre Nora. Munich: Beck, 411–451. (French original: *Les lieux de mémoire*. 7 vols. in 3, ed. by Pierre Nora. Paris: Gallimard 1984–1992, vol. 3, tome 3, 1992, 735–783. Engl. ed. of the work: *Realms of Memory: The Construction of the French Past*. 3 vols. ed. by Pierre Nora. Engl. language edition ed. by Lawrence D. Kritzman. Transl. by Arthur Goldhammer. New York: Columbia University Press, 1996–98).

Beilecke, François (2003), *Französische Intellektuelle und die Dritte Republik. Das Beispiel einer Intellektuellenassoziation 1892–1939*. Frankfurt am Main/ New York: Campus.

Benrubi, Isaac (1914), "Émile Boutroux und das philosophische Erwachen der Gegenwart", *Internationale Monatsschrift für Wissenschaft, Kunst und Technik* 8 (8): 929–952.

— (1926), *Contemporary Thought of France*. Transl. by Ernest Dicker. London: Williams & Norgate.

Boelitz, Otto (1907), *Die Lehre vom Zufall bei Émile Boutroux. Ein Beitrag zur Geschichte der neuesten französischen Philosophie*. Leipzig: Quelle & Meyer. (First 1907 as a dissertation in philosophy at the Philosophical Faculty of the University of Erlangen).

Bois, Henri (1916), *Kant et l'Allemagne*. Paris: Librairie Protestante.

Boutroux, Émile (1869), "Lettre de Boutroux [à Ravaisson, 4 février 1869]", in: "Lettres à Félix Ravaisson (1846–1892)". *Revue de métaphysique et de morale* 45(2) (1938): 173–202, on pp. 186–188.

— (1871), "La vie universitaire en Allemagne. – Professeurs et étudiants", *La revue politique et littéraire: revue des cours littéraires*, 2e série, 1re année (23) (Dec. 2, 1871): 542–549.

— (1873), "Émile Boutroux à Paul Tannery" [letter of 26 avril 1873], in: *Mémoires scientifiques*. Ed. by J.-L. Heiberg and H.-G. Zeuthen. Vol. XIII: *Correspondance*, ed. by A. Diès. Toulouse: E. Privat & Paris: Gauthier-Villars 1934, 216–219.

— (1874), *De la contingence des lois de la nature. Thèse de doctorat soutenue devant la Faculté des lettres de Paris*. Paris: Germer Baillère. (Repr. accord. to the 2nd ed. of 1905, Paris: Presses Universitaires de France 1991). (Authorized transl. by Fred Rothwell as *The Contingency of the Laws of Nature*. Chicago & London: Open Court 1916. Quoted according to the reprint of 1920. Translation amended when necessary.)

— (1874a), *De veritatibus aeternis apud Cartesium. Haec apud Facultatem litterarum Parisiensem disputabat Emile Boutroux*. Paris: Germer Baillère. (Thèse Latine, transl. by Georges Canguilhem as *Des vérités éternelles chez Descartes*. Preface

by Léon Brunschvicg [= Brunschvicg 1922]. Paris: F. Alcan 1927, repr. 1985 without preface. Paris: Vrin).
— (1877), "Édouard Zeller et sa théorie de l'histoire de la philosophie", in: Boutroux 1926a, 7–77. (Also in *Revue Philosophique de la France et de l'Etranger* 4 (1877): 1–16, 134–166 and in: Edouard Zeller, *La philosophie des Grecs considérée dans son développement historique*. Transl. by Émile Boutroux. Vol. 1. Paris: Hachette 1877, iii–lxxix).
— (1895), *De l'idée de loi naturelle dans la science et la philosophie contemporaines. Cours professé à la Sorbonne en 1892–1893*. Paris: Lecène, Oudin et Cie. (Authorized transl. by Fred Rothwell as *Natural Law in Science and Philosophy*. New York: Macmillan 1914). (First in the *Revue des cours et conférences,* 1892/93).
— (1895a), "La philosophie de Charles Secrétan", in: Boutroux 1927, 221–245. (First 1895).
— (1895b), "Kant", in: Boutroux 1897 [1912], 317–411/ transl. 255–330. (First appeared in the *Grande Encyclopédie*. 31 vols. Paris: Lamirault 1886–1902. Vol. 21 (1895): 403–420).
— (1897), *Études d'histoire de la philosophie*. Paris: F. Alcan. (Authorized transl. by Fred Rothwell as *Historical Studies in Philosophy*. London: Macmillan 1912).
— (1900), "Histoire et synthèse", *Revue de synthèse historique* 1: 8–12.
— (1907), "Sur la nécessité, la finalité et la liberté chez Hegel. Réponse à la thèse présentée par M. René Berthelot", (Séance du 31 janvier 1907) in: Boutroux 1926a, 93–113. (First in *Bulletin de la Société Française de philosophie* 7 (1907): 141–155).
— (1907a), "Descartes and Cartesianism", in: A. W. Ward, G. W. Prothero and S. Leathes (eds.), *The Cambridge Modern History*. Vol. 4: *The Thirty Years War*. Cambridge. Cambridge University Press, 776–799.
— (1908), *Science et religion dans la philosophie contemporaine*. Paris: E. Flammarion. (Transl. by Jonathan Nield as *Science and Religion in Contemporary Philosophy*. London: Duckworth 1909).
— (1908a), "La philosophie en France depuis 1867", in: Boutroux 1927, 139–193. (First in *Revue de Métaphysique et de Morale* 16 (6): 683–716).
— (1911), *William James*. Paris: A. Colin. (Engl. transl. from the 2nd ed., by Archibald and Barbara Henderson as *William James*. London/New York: Longmans, Green 1912).
— (1913), "Henri Poincaré", *La Revue de Paris* Année 20 (Tome 1 (2)): 673–702, Année 20 (Tome 2 (3)): 77–91. (Also in Boutroux 1927, 32–96).
— (1914), "La pensée allemande et la pensée française", in: Boutroux 1926a, 163–193. (First in *La revue politique internationale*, Sept./Oct. 1914, 193–218).
— (1916), *Philosophy and War*. Authorized transl. by Fred Rothwell. London: Constable.
— (1926), *La philosophie de Kant. Cours de M. Émile Boutroux, professé à la Sorbonne en 1896–1897. (Bibliothèque d'histoire de la philosophie)* Paris: Vrin. (Repr. *ibid.* 1965) (First in the *Revue des cours et conférences,* 27 Dec. 1894–26 Nov. 1896, and 1900–1901).

—— (1926a), *Études d'histoire de la philosophie allemande.* Paris: Vrin.
—— (1926b), *La nature et l'esprit.* Paris: Vrin.
—— (1927), *Nouvelles études d'histoire de la philosophie.* Paris: F. Alcan. (Articles on Jules Lachelier, Henri Poincaré, Félix Ravaisson, Charles Secrétan among others).
—— (1929), *La philosophie allemande au XVIIe siècle. Les prédécesseurs de Leibniz: Bacon, Descartes, Hobbes, Spinoza, Malebranche, Locke et la philosophie de Leibniz. Cours de M. Émile Boutroux professé à la Sorbonne en 1887–88. (Bibliothèque d'histoire de la philosophie)* Paris: Vrin.
Boutroux, Pierre (1913), "Henri Poincaré. L'œuvre philosophique", *La revue du mois* 15(1): 155–183.
Brenner, Anastasios (1998), "Les voies du positivisme en France et en Autriche: Poincaré, Duhem et Mach", *Philosophia scientiae* 3: 31–42.
—— (2002), "The French connection: conventionalism and the Vienna Circle", in: Michael Heidelberger and Friedrich Stadler (eds.), *History of Philosophy of Science: New Trends and Perspectives.* Dordrecht: Kluwer, 277–286.
—— (2003), *Les origines françaises de la philosophie des sciences. (Collection Science, histoire et société)* Paris: Presses Universitaires de France.
—— (2004), "Géométrie et genèse de l'espace selon Poincaré", *Philosophiques* 31 (1): 115–130.
Brooks, John I. (1998), *The Eclectic Legacy: Academic Philosophy and the Human Sciences in Nineteenth-Century France.* Newark, Del. & London: University of Delaware Press & Associated University Presses.
Brunschvicg, Léon (1922), "La philosophie d'Émile Boutroux", *Revue de Métaphysique et de morale* 29 (3): 261–283. (Also in Léon Brunschvicg, *Écrits philosophiques*, 3 vols. Paris: Presses Universitaires de France 1954, 2: 210–231, and in the 1927 translation of Boutroux 1874a, ii–xxxix).
Capeillères, Fabien (1998), "Généalogie d'un néokantisme français. À propos d'Émile Boutroux", *Revue de Métaphysique et de Morale* 103 (3): 405–442.
Cartwright, Nancy (1983), *How the Laws of Physics Lie.* Oxford: Clarendon.
Chaitin, Gilbert D. (1999), "From the Third Republic to Postmodernism: language, freedom, and the politics of the contingent", *MLN* 114 (4): 780–815.
Comte, Auguste (1830), *Cours de philosophie positive.* Vol. 1. Paris: Rouen Frères.
Copleston, Frederick (2003), *A History of Philosophy.* Vol. 9: *19th and 20th Century French Philosophy.* London/New York: Continuum. (First ed. 1974).
Cronin, T.J. (1960), "Eternal truths in the thought of Descartes and of his adversary", *Journal of the History of Ideas* 21 (4): 553–559.
Descartes, René (1641), *Meditationes de prima philosophia.* Paris.
—— (1644), *Principia philosophiae.* Amsterdam.
Detlefsen, Michael (1992), "Poincaré against the logicians", *Synthese* 90: 349–378.
Engel, Pascal (1988), "Plenitude and contingency: modal concepts in nineteenth century French philosophy", in: Simo Knuuttila (ed.), *Modern Modalities: Studies of the History of Modal Theories from Medieval Nominalism to Logical Positivism.* Dordrecht: Kluwer, 179–237.

Espagne, Michel (2001), "L'Allemagne d'Émile Boutroux." *Cahiers d'Études Germaniques* 41: 199–215.

— (2004), *En deçà du Rhin. L'Allemagne des philosophes français au XIXe siècle.* Paris: Les éditions du cerf.

Fagot-Largeault, Anne (2002), "L'émergence", in: Daniel Andler, Anne Fagot-Largeault and Bertrand Saint-Sernin (eds.), *Philosophie des sciences.* Vol. 2. Paris: Gallimard, 939–1048.

Fedi, Laurent (2001), "Bergson et Boutroux, la critique du modèle physicaliste et des lois de conservation en psychologie", *Revue de Métaphysique et de Morale* (2): 97–118.

Gil, Didier (2000), "La philosophie de la nature d'Émile Boutroux", in: Olivier Bloch (ed.), *Philosophies de la nature. Actes du colloque tenu à l'Université de Paris I Panthéon Sorbonne, 20 et 27 mars, 27 novembre et 4 décembre 1994.* Paris: Publications de la Sorbonne, 333–343.

Girel, Mathias (2003), "Varieties of experience in Boutroux and James", *Streams of William James* 5 (2): 2–6.

Gunn, John Alexander (1922), *Modern French Philosophy: A Study of the Development since Comte.* With a foreword by Henri Bergson. London: T. F. Unwin.

Gutting, Gary (2001), *French Philosophy in the Twentieth Century.* Cambridge: Cambridge University Press.

Hacking, Ian (1983), "Nineteenth-century cracks in the concept of determinism", *Journal of the History of Ideas* 44 (3): 455–476.

— (1990), *The Taming of Chance.* Cambridge: Cambridge University Press.

Hanna, Martha (1996), *The Mobilization of Intellect: French Scholars and Writers during the Great War.* Cambridge, MA: Harvard University Press.

Hartung, Gerald (ed.) (2009), *Eduard Zeller.* Berlin: de Gruyter.

Heidelberger, Michael (2004), *Nature from Within: Gustav Theodor Fechner and His Psychophysical Worldview.* Pittsburgh: The University of Pittsburgh Press.

— (2006), "Die Kontingenz der Naturgesetze bei Émile Boutroux", in: Karin Hartbecke and Christian Schütte (eds.), *Naturgesetze. Historisch-systematische Analysen eines wissenschaftlichen Grundbegriffs.* Paderborn: mentis, 269–289.

James, William (1910), "A great French philosopher at Harvard", in: William James, *Essays in Philosophy.* Ed. by Frederick H. Burkhardt, Fredson Th. Bowers and Ignas K. Skrupskelis. (*The Works of William James*, vol. 5) Cambridge, Ma.: Harvard University Press 1978, 166–171; notes 226 f. (First in *Nation* 90 (March 31), (1910): 312–314).

Janet, Paul (1875), "MM. Cournot – Naudin – Boutroux", in: Paul Janet, *La philosophie française contemporaine.* Paris: Calmann Lévy 1879, 18–36.

— (1897), "La philosophie de la contingence", *Principes de métaphysique et de psychologie: leçons professées à la Faculté des lettres de Paris, 1888–1894.* 2 vols., Paris: Delagrave, 2: 451–466.

Jones, Robert Alun (1999), *The Development of Durkheim's Social Realism.* Cambridge: Cambridge University Press.

Lalande, André (ed.) (1976), *Vocabulaire technique et critique de la philosophie.* 12th ed. Paris: Presses Universitaires de France.

Lefèvre, Frédéric (1927), "Une heure avec Lévy-Bruhl", *Les Nouvelles littéraires* VI, n° 226, (12 février 1927): 16–77.

Léon, Xavier (1902), *La philosophie de Fichte. Ses rapports avec la conscience contemporaine.* (Précédé d'une préface de M. Émile Boutroux) Paris: F. Alcan.

Nye, Mary Jo (1976), "The moral freedom of man and the determinism of nature: the Catholic synthesis of science and history in the *Revue des questions scientifiques*", *British Journal of the History of Science* 9: 274–292.

— (1979), "The Boutroux circle and Poincaré's conventionalism", *Journal of the History of Ideas* 40 (1): 107–120.

Paul, Harry W. (1972), *The Sorcerer's Apprentice: The French Scientist's Image of German Science, 1840–1919.* (*University of Florida Social Sciences Monograph*, vol. 44) Gainesville, FL: University of Florida Press.

Peirce, Charles Sanders (1892), "The doctrine of necessity examined", in: Peirce 1960, §§ 35–65.

— (1893), "Evolutionary love", in: Peirce 1960, §§ 287–317.

— (1903), "Variety and uniformity", in: Peirce 1960, §§ 67–85.

— (1960), *Collected Papers of Charles Sanders Peirce*, Charles Hartshorne & Paul Weiss (eds.), 8 vols. Vol. 6: *Scientific Metaphysics.* Cambridge, Ma.: Harvard University Press.

Perry, Ralph Barton (1935), *The Thought and Character of William James*, 2 vols. London: Milford.

Pfeiffer, Jeanne (2002), "France", *Writing the History of Mathematics: Its Historical Development.* Ed. by Joseph Warren Dauben and Christoph J. Scriba. Basel: Birkhäuser, 3–43.

Pillon, François (1897), "Les lois de la nature", *Revue philosophique de la France et de l'étranger* 43: 56–79. (Review of the 2nd ed., 1895, of Boutroux 1874)

— (1907), "Les lois de la nature selon M. E. Boutroux", *L'année philosophique* 18: 79–139.

Poincaré, Henri (1902), *La science et l'hypothèse.* Paris: Flammarion. Repr. *ibid.* 1968, with a preface by Jules Vuillemin. (Transl. by W.[illiam] J.[ohn] G.[reenstreet] as *Science and Hypothesis*, with a preface by Joseph Larmor. London & Newcastle-on-Tyne: Walter Scott 1905).

— (1902a), "Analyse des travaux scientifiques de Henri Poincaré faite par lui-même", *Acta mathematica* 38 (1921): 36–135. (Poincaré drafted this in 1902; see Boutroux, P. 1913, 158).

Proust, Marcel (1971), *Contre Sainte-Beuve. Précédé de Pastiches et mélanges et suivi de Essais et articles.* Ed. by Pierre Clarac in collab. with Yves Sandre. (Bibliothèque de la Pléiade, 229) Paris: Gallimard.

Rollet, Laurent (2000), *Henri Poincaré: Des mathématiques à la philosophie. Étude du parcours intellectuel, social et politique d'un mathématicien au début du siècle.* Villeneuve-d'Ascq: Presses Universitaires du Septentrion. (First as thèse de doctorat en philosophie, Université Nancy 2, 1999).

— (2001), "Henri Poincaré sur la scène philosophique française", *Annales de l'Est* 51 (1): 137–151.

Rougier, Louis (1946), "Introduction. La philosophie scientifique d'Henri Poincaré", in: Henri Poincaré, *La valeur de la science.* (*Classiques français du XXe siècle*) Geneva: Éd. du cheval ailé, 13–55.

Schyns, Mathieu (1924), *La philosophie d'Émile Boutroux*. Paris: Fischbacher. (First as a *thèse de doctorat* at the *Faculté des lettres* of Geneva University 1923).

Sánchez, Gonzalo J. (2004), *Pity in fin-de-siècle French culture: "liberté, égalité, pitié"*. Westport, Conn.: Praeger.

Séailles, Gabriel (1905), *La philosophie de Charles Renouvier: Introduction à l'étude du néo-criticisme*. Paris. F. Alcan.

Zeller, Eduard (1856–68), *Die Philosophie der Griechen in ihrer geschichtlichen Entwicklung*. 2nd, completely reworked ed., 5 vols. Leipzig: Fues. (First ed. Tübingen: Fues 1844–52). (Engl. transl. by Sarah Frances Alleyne as *A History of Greek Philosophy, from the Earliest Period to the Time of Socrates*. 2 vols. London: Longmans 1881).

— (1862), "Ueber Bedeutung und Aufgabe der Erkenntnisstheorie" (with an addition from 1877), in: Eduard Zeller, *Vorträge und Abhandlungen. Zweite Sammlung*. 3 vols. Leipzig: Fues 1877, vol. 2: 479–496. (First as *Ueber Bedeutung und Aufgabe der Erkenntniss-Theorie. Ein akademischer Vortrag*. Heidelberg: Groos 1862).

— (1877–1884), *La philosophie des Grecs considérée dans son développement historique*. Transl. from the 4th German edition of 1876 by Émile Boutroux. (Vol. 3: *Socrate et les Socratiques*, transl. by M. Belot). 3 vols. Paris: Hachette.

Pluralism and the Hypothetical in Heinrich Hertz's Philosophy of Science

Andreas Hüttemann

Abstract:

In this paper I argue against readings of Hertz that overly assimilate him into the thought of late 20th century anti-realists and pluralists.

Firstly, as is well-known, various images of the same objects are possible according to Hertz. However, I will argue that this envisaged pluralism concerns the situation before all the evidence is considered i. e. before we can decide whether the images are correct and appropriate. Hertz believes in final and decisive battles of the kind he participated in while doing experiments in electrodynamics.

Secondly, I will argue that the concept of representation is still quite appropriately applied to important aspects of images, namely when it comes to fundamental physical equations. In this context Hertz explicitly allows that "characteristics of our image, which claim to represent observable relations of things, do really and correctly correspond to them" (Hertz [1894] 1956, 9).

A final consideration is Hertz's consistent appeal to the concept of the hypothesis. I will argue that his use of the concept does not indicate that he contributed to an increasing hypothetization of science, if this trend is understood in a strong sense, i. e. as the belief that the correctness of scientific theories cannot be established for principled reasons. As mentioned, when it comes to experimental evidence Hertz believes in decisive battles.

1. Introduction

In the introduction to his *The Principles of Mechanics* Hertz famously claimed:

> We form ourselves images or symbols of external objects; and the form which we give them is such that the necessary consequents of the images in thought are always the images of the necessary consequents in nature of the things depicted … The images we here speak of are our conceptions of things. With the things themselves they are in conformity in *one* important respect, namely, in satisfying the above-mentioned requirement. For our purpose it is not necessary that they should be in conformity with the things in any other respect whatever. As a matter of fact, we do not know, nor have we any means of knowing, whether our conception of things are in conformity with them in any other than in this *one* fundamen-

> tal respect ... Various images of the same objects are possible, and these images may differ in various respects (Hertz [1994] (1956), 1 ff.).

On the basis of these and similar remarks some writers have made a number of claims about Hertz's philosophy of science that I intend to comment on in this paper.

More particularly, I have in mind the following three claims. Firstly, Ludwig Boltzmann argues that Hertz's picture theory implies that it cannot be our *aim* to look for an absolutely correct theory. Rather, there might be different theories that are equally correct.[1] While Boltzmann maybe right that Hertz's theory has this implication, I want to argue that this was not Hertz's view. He conceived of the pluralism of theories or images in mechanics as a *transitory* stage in the development of science and was convinced that only one of the images can be the correct image in mechanics. Hertz – even in *The Principles of Mechanics* – aims to identify the unique theory or image.

Secondly, Gregor Schiemann has argued:

> To one reality, which Hertz, too, conceives realistically, can now correspond a multiplicity of theories. The world seems remote and the concept of representation inappropriate (Schiemann 1998, 30).

Schiemann describes this as a "loss of truth in theoretical cognition" and as the "the loss of world in the image."

While I think there is some truth in this picture of Hertz's account, I also think that the world has not been completely lost in his images. In fact it was one of Hertz's main objectives to figure out exactly what in a theory represents the world (or nature) and what does not.

Thirdly, I want to deal with Schiemann's claim that Hertz was part of a trend in the second half of the nineteenth century towards "an increasing hypothesization of scientific propositions" (Schiemann 1998, 28). While I will not deny that Hertz may have contributed to this development (due to a reception of his writings like Boltzmann's that misconstrues Hertz's views), I will argue that the notion of a "hypothesis" and *a fortiori* "hypothesization" – if taken in a strong sense – is not a very useful tool for characterizing the distinctive features of Hertz's philosophy of science.

The evidence for my claims relies on two points, which I will introduce in sections 2 and 3. In section 2 I will argue that Hertz's main ob-

---

1 "Daraus folgt, dass es nicht unsere Aufgabe sein kann, eine absolut richtige Theorie ... zu finden" (Boltzmann 1905, 216).

jective in his epistemological writings was the question of separating what in our knowledge is due to nature and what we as knowing subjects have added. This is an important point because Hertz believes that there should be an ultimate theory that separates these features. As it will turn out, Hertz's criteria for individuating theories allow for only one such theory. The second observation concerns Hertz's use of the term "hypothesis." In section 3 I will introduce a weak and a strong reading of "hypothesis." Only a strong reading of Hertz's use of "hypothesis" would provide evidence for his alleged pluralism. On the basis of this distinction, and the determination of his main epistemological objective, I will analyze his writings in electrodynamics and mechanics that touch on the issue of hypothesis and pluralism (sections 4 and 5).

2. The Constitution of Matter

In 1884, while in Kiel as a Privatdozent, Hertz delivered a popular lecture course entitled *The Constitution of Matter.* In the introduction to the lecture course he discusses the relation of physics and philosophy with respect to the question of the nature (or constitution) of matter. According to Hertz this used to be a genuine philosophical question. Presumably he has in mind not only ancient and early modern debates, but in particular the disputes about dynamism and atomism in the first half of the 19th century.[2]

However, according to Hertz, the natural sciences had by his time taken over the question from philosophy.

> Today's philosophy, insofar as it is based on Kant, to an increasing extent removes the question of the constitution of matter from the sphere of its interests and assigns it to the natural sciences, reserving for itself at most a control over final results. There is no longer any doubt that we are here dealing with empirical facts and things that – as in the case of the number of planets and the chemical elements – cannot be dealt with a priori (Hertz 1999, 25).[3,4]

2 See (Carrier 1990).

3 "Die heutige Philosophie, so weit sie sich auf Kant stützt, scheidet immer mehr die Frage nach der Constitution der Materie aus ihrem Interessenkreis aus und weist sie den exacten Naturwissenschaften zu, sich höchstens eine Controlle der letzten Resultate vorbehaltend. Es kann kein Zweifel mehr daran bestehen, daß es sich hier rein um Erfahrungsthatsachen [handelt] sowie um Dinge, die sich so wenig a priori entscheiden lassen, wie die Frage nach der Zahl der Planeten und

Hertz goes on to present an overview over the advances in chemistry and the kinetic theory of gases that provide evidence for an atomic structure of matter. Before discussing these issues in detail (this constituting the rest of the lecture course) he deals with some objections a philosopher might raise.

A philosopher might object that the physicist's account cannot answer the original philosophical question. Since the physicist's atoms are extended, the question of the constitution of matter can again be raised with respect to the atoms themselves. Hertz essentially replies that – while conceding that physics has transformed the original question – physics and philosophy deal with different questions and have different aims. The physicist deals with the facts of nature, while the philosopher deals with the difficulties the understanding has in conceiving nature.[5]

What is interesting in this context is the characterization of the philosopher's project. It is the philosopher's job "to present the facts consistently and to separate which of those are due to the things themselves and what we have added."[6] Several times in his life Hertz takes up the question of what is due to nature or things themselves, and what is due to us.

For instance, in a newspaper-article on von Helmholtz's 70th birthday in 1891, Hertz characterizes von Helmholtz's research in physiology in terms of the following questions:

> How is it possible for vibrations of the ether to be transformed by means of our eyes into purely mental processes which apparently can have nothing in common with the former; and whose relations nevertheless reflect with the greatest accuracy the relations of external things? In the formation of mental conceptions what part is played by the eye itself, by the form of the im-

---

der chemischen Elemente ..." (The translations from *Die Constitution der Materie* are mine. In some cases I have consulted Jesper Lützen's translations of some passages.

4 For an analysis according to which both the *Die Constitution der Materie* as well as *The Principles of Mechanics* fall into a Kantian tradition of a "metaphysics of corporeal nature" see (Hyder 2003). For a critical discussion see (Lützen 2005, 123 ff.).

5 "Ich untersuche die Thatsachen der Natur, und Du untersuchst die Schwierigkeiten, welche der menschliche Verstand findet, sie zu begreifen" (Hertz 1999, 32).

6 "... die Thatsachen begrifflich widerspruchsfrei darzustellen, zu sondern, was von ihnen in den Dingen selbst liegt, und was wir hinzuthun, ..." (Hertz 1999, 32).

> ages which it produces, by the nature of its colour-sensations, accommodation, motion of the eyes, by the fact that we possess two eyes? Is the manifold of these relations sufficient to portray all conceivable manifolds of the external world, to justify all manifolds of the internal world? (Hertz 1896, 336).

These questions concerning visual perception can be asked with respect to all of knowledge. Thus Hertz continues:

> We see how closely these investigations are connected with the possibility and legitimacy of all natural knowledge. The heavens and the earth doubtless exist apart from ourselves, but for us they only exist insofar as we perceive them. Part of what we perceive therefore appertains to ourselves: part only has its origin in the properties of the heavens and the earth. How are we to separate the two? (Hertz 1896, 336/7).

In what follows I will try to show that Hertz's epistemological considerations and even some of his work in theoretical physics is best understood as an answer to this question: how are we to separate what is due to the things (or to nature) from what we have added? It will become evident that what he considers as a philosophical question at first will turn into a question that a (theoretical) physicist has to deal with.

To return to *The Constitution of Matter:* Hertz's reply to the second philosophical objection is already an attempt to come to terms with the issue I have just sketched. The objection concerns the properties we attribute to atoms.[7] It seems that we cannot attribute to atoms any of those properties which we attribute to macroscopic objects. On the one hand, there are sensible (secondary) properties like color, which for all we know atoms do not have. But, on the other hand, even in the case of other macroscopic properties like elasticity we cannot attribute them to atoms because the main motivation for atomism is that we envisage explanations of the macroscopic properties in terms of those of the atoms. To attribute macroscopic properties to the atoms would undermine this explanatory project. So it seems that nothing remains.

> Let us answer in the name of physics as follows: First, there still remains something if we leave out everything we have imagined. There remains a system of conceptually defined magnitudes which are connected among themselves and to the macroscopic properties of matter via strict mathematically formulated relations. Even if it is not allowed to consider these for their own sake, and to attribute conceivable meanings to them,

7 For a discussion of the background of this objection see (Heidelberger 1993, 205 ff.).

> they retain their value as auxiliary magnitudes for the sake of those relations.[8]

This seems to be a first attempt to separate what is due to nature and what we have added ourselves. If we subtract that which we have imagined (alles Gedachte), what remains is a "system of conceptually defined magnitudes which are connected among themselves and with the macroscopic properties of matter *via* strict mathematically formulated relations." Hertz considers the following example:

> If it is not, for example, permitted to talk about the diameter of the atom in the strict sense, but what I call the diameter of the atom for a particular gas retains its meaning: It is that length on the basis of which I can establish a relationship between the heat conductivity of a gas, its internal friction, its dielectric constant and its refractability.[9]

In this case the relationship that remains if we subtract everything imagined is that between the heat conductivity of a gas, its internal friction, its dielectric constant and its refractability, whereas the diameter of the atom cannot be taken to be a literal description of what there is – it is (merely) imagined (gedacht). So there is a contrast of two kinds of features – the relationship between heat conductivity, internal friction, etc. can be read literally or realistically, whereas claims about the diameter of the atom should not be read literally or realistically.

But how are we to understand the non-literal descriptions, e.g. of the atom? Hertz considers two options. First, fictionalism: According to Hertz, many physicists are content to consider atoms and their properties merely as useful fictions (Hülfsfictionen). According to such a conception it is the aim of a theory to give a simple description of ob-

---

8 "Lassen sie uns darauf im Namen der Physik das Folgende antworten: Zunächst bleibt immer noch etwas übrig, wenn wir alles Gedachte fortlassen. Es bleibt übrig ein System von begrifflich definirten Größen, welche unter sich und mit den makroskopischen Eigenschaften der Materie durch streng mathematisch formulierte Beziehungen verbunden sind; ist es nicht erlaubt dieselben um ihrer selbst willen zu betrachten, und ihnen vorstellbare Bedeutungen beizulegen, so behalten sie doch ihren Werth als Hilfsgrößen um jener Beziehungen willen" (Hertz 1999, 35).

9 "Ist es mir also z.B. nicht erlaubt, im eigentlichen Sinne von dem Durchmesser eines Atoms zu reden, so behält doch das, was ich den Durchmesser eines Atoms für ein bestimmtes Gas nenne, seine Bedeutung: es ist eine Länge, mit deren Hülfe ich eine Beziehung zwischen Wärmeleitungsfähigkeit des Gases, seiner inneren Reibung, seiner Dielectrizitätsconstanten und seinem Lichtbrechungsvermögen aufzustellen vermag" (Hertz 1999, 35).

servable phenomena. What transcends sensation is classified as a mere fiction, a fiction that helps to achieve simplicity. Given fictionalism, the properties we attribute to matter have to fulfill two conditions: First, their introduction has to be consistent and, second, the calculations that ensue should be as simple as possible – "that is, they have to be appropriate (zweckdienlich)" (Hertz 1999, 35).

The fictionalist reading, however, is *not* the one Hertz advocates. Hertz does not consider the properties and relations in question to be fictions that serve the purpose of achieving a simple description of observed phenomena. Rather, he considers them to be *necessary conditions for imagination* (Vorstellbarkeit). According to Hertz it is a general and necessary condition of the human mind that we can neither represent things intuitively (anschaulich vorstellen), nor define them conceptually (begrifflich definieren), without adding properties (Hertz 1999, 35).

> What we add are therefore not wrong conceptions, rather they are the conditions for imagination. We cannot simply take them away and replace them with better ones; rather, we either have to add them or to do without conceptions in this realm.[10]

It is in this context that Hertz introduces the notion of a *picture* for the first time to capture the idea that theories contain both features which can be taken to represent reality and others which depend ultimately on the human mind.

> Thus let us guard ourselves from believing that we can investigate the nature of the things themselves by considering the atoms; let us also guard ourselves from confusing the non-essential properties, which we are forced to ascribe to them, with the essential properties, which are merely space and time relations. However, let them not make us believe that all labour is lost if of the things which are real but cannot themselves enter the mind we have made pictures, which coincide with the things in certain respects (Beziehungen) whereas in others they depend on our conceptions.[11]

---

10 "Was wir hinzufügen sind dann nicht falsche Vorstellungen, sondern es sind die Bedingungen der Vorstellbarkeit überhaupt; wir könnten sie nicht fortnehmen und bessere an ihre Stelle setzen, sondern wir müssen sie hinzuthun oder auf alle Vorstellungen in diesem Gebiete verzichten" (Hertz 1999, 36). Hertz later changes his mind about whether these conceptions can be replaced by better ones (replacing conceptions is his main purpose in his theoretical work on electrodynamics).

11 "Hüten wir uns also zu glauben, wir könnten durch Betrachtung der Atome das Wesen der Dinge selbst erforschen, hüten wir uns auch die unwesentlichen Eigenschaften, die wir ihnen nothgedrungen beilegen müssen, mit den wesentlichen zu verwechseln, welches lediglich Zeit- und Raumbeziehungen sind; aber

The main point for our question is that there is a contrast between "conceptually defined magnitudes which are connected among themselves and with the macroscopic properties of matter *via* strict mathematically formulated relations" on the one hand, and conceptions (Vorstellungen) on the other. The strict mathematically formulated relations are the kind of things that can be right or wrong. They represent nature. They are what he refers to as "essential properties." Claims about conceptions, (Vorstellungen) however, such as the properties we attribute to atoms (like having a certain diameter), cannot be read realistically, but are nevertheless not *wrong*. Claims about the diameter of an atom are *neither right nor wrong* according to Hertz. These conceptions are what we add to nature or to the things themselves.

3. Hypotheses

This is a good point to introduce the notion of a hypothesis. I take a hypothesis to be a proposition – usually consistent with what we already know – which is introduced as an assumption to explain known phenomena. A proposition can be represented by a statement and is either true or false. There are two features of this characterization that will be relevant for the following discussion. Hypotheses are

(i) either true or false and
(ii) assumptions, i. e. there is insufficient evidence for them.

The lack of evidence mentioned in (ii) may be either a contingent, transitory feature of the current state of science or it may be a principled matter. The first reading as to why there is a lack of evidence is presumably the common sense reading of what we mean by "hypothesis." If propositions are classified as hypotheses due to lack of evidence in the non-principled sense, I will call them *hypotheses in the weak sense.* The second reading can be backed up by considerations like Popper's according to which we are never able to verify a hypothesis. Propositions that lack evidence for principled reasons will be called *hypotheses in the strong sense.* Since the lack of evidence can only be either a matter of

lassen Sie uns auch nicht glauben, wir hätten unsere Mühe verloren, wenn wir von den Dingen die wirklich sind, aber nicht in unseren Geist eingehen, Bilder geschaffen haben, die mit jenen Dingen in einigen Beziehungen übereinstimmen, während sie in anderen wieder den Stempel unserer Vorstellungen tragen" (Hertz 1999, 36).

principle or not a matter of principle, it is an implication of these definitions that a hypothesis cannot be both a hypothesis in the weak sense and a hypothesis in the strong sense.[12]

Returning to *The Constitution of Matter*, we can classify the claims about the relation between conceptually defined magnitudes on the one hand and those about the diameter of the atom on the other as follows: The latter cannot be hypotheses because they are statements that are neither right nor wrong. This classification makes sense because – given the characterization as necessary conditions for imagination – they are not the kind of thing for which it is reasonable to seek evidence. The former, however, (the mathematical relations) can be true or false – they are the kind of thing for which evidence is sought. There is no evidence in the text that indicates whether such claims should be read as hypotheses in the weak or in the strong sense.

4. Maxwell's Theory

Hertz's papers on electromagnetism are particularly interesting because they introduce a two-fold pluralism – of theories on the one hand and representations of theories on the other. Hertz – as we will see – believes that his work has put an end to both of these pluralisms.

In the introduction to his collection of papers on electromagnetism, *Electric Waves*, Hertz characterizes his own achievements by contrasting the situation before and after his experiments.

> From the outset Maxwell's theory excelled all others in elegance and in the abundance of the relations between the various phenomena which it included. The probability of this theory, and therefore the number of its adherents, increased from year to year. But as long as Maxwell's theory depended solely upon the probability of its results, and not on the certainty of its hypotheses, it could not completely displace the theories which were opposed to it. The fundamental hypotheses of Maxwell's theory contradicted the usual views, and did not rest upon the evidence of decisive experiments (Hertz [1892] 1962, 19).

---

12 For this distinction see Gregor Schiemann, "Werner Heisenberg's position on a hypothetical conception of science," this volume. Hypotheses in the weak sense correspond to Schiemann's 'provisional assumptions'. Hypotheses in the strong sense imply the 'abandonment of claims to truth'.

So before Hertz performed his experiments there was no certainty of the hypotheses and there were no decisive experiments.

> In this connection we can best characterise the object and the result of our experiments by saying: the object of these experiments was to test the fundamental hypotheses of the Faraday-Maxwell theory, and the result of the experiments is to confirm the fundamental hypotheses of the theory (Hertz [1892] 1962, 19/20).

Given the use of the past tense and the contrast of the epistemological situation before and after his experiments, I take it that Hertz wants to say that the decisive evidence sought for in Maxwell's theory has now been found.

Let me briefly comment on Hertz's use of "hypothesis" in this context. He frequently uses the term "hypothesis" in the *Electric Waves.* Thus, for example, he refers to the hypothesis that light-waves are identical to electromagnetic waves (Hertz [1892] 1962, 19, 136). He furthermore classifies Maxwell's equations as "hypotheses" as long as his experiments had not taken place (Hertz [1892] 1962, 19). So the question arises whether Hertz takes hypotheses in the weak or in the strong sense. Given the quotations above it is not very plausible to claim that Hertz believed that Maxwell's equations lacked evidence for principled reasons. On the contrary, it seems plausible to assume that Hertz believes that the question of the fundamental equations of electrodynamics has been settled with his experiments. Maxwell's equations used to be hypotheses in the weak sense.

But what exactly has been so decisively confirmed? What is it that we call the Faraday-Maxwell theory?" Hertz famously answered:

> To the question "What is Maxwell's theory?" I know of no shorter or more definite answer than the following: – Maxwell's theory is Maxwell's system of equations. Every theory which leads to the same system of equations, … I would consider as being a form or special case of Maxwell's theory; every theory which leads to different equations, … is a different theory(Hertz [1892] 1962, 21).

What has been confirmed are the mathematically formulated relations between physical magnitudes, (to use the terminology of his *The Constitution of Matter*) namely Maxwell's equations.

With respect to the issue of pluralism it should be pointed out that before Hertz performed his experiments there used to be a pluralism of theories (fundamental equations for electro-dynamic phenomena). This pluralism has now disappeared. Hertz believes that there is decisive evidence against competing theories such as Weber's. So the pluralism of theories has turned out to be transitory.

Furthermore, it follows from Hertz's views that there *cannot* be a pluralism of theories in the future. The reason is this: The evidence Hertz provides is decisive evidence for Maxwell's equation. *A fortiori* an electromagnetic theory either yields Maxwell's equations, or it is wrong. But by definition every theory that yields these equations is identical with Maxwell's theory. So a pluralism of theories that takes account of the evidence is impossible.

The pluralism of theories is not the only pluralism Hertz discusses. He contrasts Maxwell's theory with its representations. Hertz distinguishes three representations of Maxwell's theory: Maxwell's representation, the representation as a limiting case in von Helmholtz's electrodynamics and his own. All of these are representations of the *same* inner significance or content (Inhalt). What is common to all of these representations is the system of Maxwell's equations. For a representation to be a representation of Maxwell's theory it is both a necessary and sufficient condition to yield these equations. Representations add physical significance to the system of equations by invoking physical conceptions (Vorstellungen) such as "pictures of electrified atoms" or "concrete representations (Vorstellungen) of the various conceptions as to the nature of electric polarisation, the electric current etc." (Hertz [1892] 1962, 19). As Michael Heidelberger points out, the representation (the physical conception) "designates the ultimate unobservable agent which produces the phenomena" (Heidelberger 1998, 18). For example, in the case of one body acting on another at a distance, Hertz distinguishes four fundamental conceptions (standpoints): "the pure conception of direct attraction," the pure conception of indirect (or mediated) attraction, as well as two intermediary conceptions. These are claims about the microphysical causes of the observable phenomena.

We have a plurality of fundamental conceptions (standpoints) about apparent action at a distance in the case electro-dynamic phenomena. The different representations of Maxwell's theory (there are three of them) rely on several of these possible standpoints (Hertz discusses four standpoints). Von Helmholtz's, Maxwell's and Hertz's own representation make use of these fundamental conceptions.

With respect to the first standpoint (the pure conception of direct attraction) Hertz is very explicit that it has been rejected:

> Casting now a glance backwards we see that by the experiments above sketched the propagation in time of a supposed action-at-a-distance is for the first time proved. This fact forms the philosophic results of the experiments; and, indeed, in a certain sense the most important result. The proof includes a recognition of the fact that the electric forces can disentangle themselves from material bodies, and can continue to subsist as conditions or changes in the state of space (Hertz [1892] 1962, 19).

Hertz explicitly points out that this result is independent of the correctness of a particular theory. Thus, his experiments have not only implications for competing theories, but also for competing fundamental conceptions. On the one hand, they provide decisive evidence for Maxwell's equations, on the other hand they demonstrate that the fourth standpoint (action is mediated) is not only possible but actually obtains. Thus claims about fundamental conceptions are not beyond the reach of experimental evidence.

Let me come back to the plurality of representations. Hertz criticizes von Helmholtz's representation, which makes use of one of the intermediary standpoints:

> It assumes that the action of the two separate bodies is not determined solely by forces acting directly at a distance. It rather assumes that the forces induce changes in the space (supposed to be nowhere empty), and that these again give rise to new distance forces (Hertz [1892] 1962, 23).

In the limit of diminishing distance forces von Helmholtz's theory yields Maxwell's equations. It is thus a representation of Maxwell's theory. However, Hertz objects that "it is impossible to deny the existence of distance forces, and at the same time to regard them as the cause of the polarizations" (Hertz [1892] 1962, 25). So Hertz rejects von Helmholtz's representation because the physical conceptions it invokes are inconsistent.

Maxwell's own representation is – according to Hertz – an inconsistent mixture of the fourth standpoint and the second standpoint. (I will not go into the details here.) Hertz own representation is the attempt to disentangle the conceptions that Maxwell used in his representation. Hertz attempts to

> exhibit Maxwell's theory, i.e. Maxwell's equations from this fourth standpoint. I have endeavoured to avoid from the beginning the introduction of any conceptions which are foreign to this standpoint and which might afterwards have to be removed (Hertz [1892] 1962, 27).

Hertz describes some of the conceptions that characterize Maxwell's own representation:

> Maxwell originally developed his theory with the aid of very definite and special conceptions as to the nature of electrical phenomena. He assumed that the pores of the ether and of all bodies were filled with an attenuated fluid, which, however, could not exert forces at a distance (Hertz [1892] 1962, 27).

He characterized his theoretical papers as the attempt to develop a representation of the system of Maxwell's equation that can do without pictorial conceptions (Vorstellungen) of the kind just mentioned:

> I have ... endeavoured in the exposition to limit as far as possible the number of those conceptions which are arbitrarily introduced by us, and only to admit such elements as cannot be removed or altered without at the same time altering possible experimental results (Hertz [1892] 1962, 28).

Hertz wants to do without those conceptions that are part of the second standpoint. Maxwell's theory should be presented entirely from the point of the fourth standpoint.

> It is true that in consequence of these endeavours, the theory acquires a very abstract and colourless appearance ... But scientific accuracy requires of us that we should in no wise confuse the simple and homely figure, as it is presented to us by nature, with the gay garment which we use to clothe it. Of our own free will we can make no change whatever in the form of the one, but the cut and colour of the other we can choose as we please (Hertz [1892] 1962, 28).

Hertz considered the issue of separating what is due to nature on the one hand and, on the other, what we have added on as a *philosophical* question in *The Constitution of Matter*. It was therefore a question the physicist did not have to deal with. In the "Introduction" to his *Electric Waves* this question has now been transformed into a question that a (theoretical) physicist needs to deal with. It was important for Hertz to contrast "the simple and homely figure, as it is presented to us by nature" and "the gay garment which we use to clothe it." Though there is some interpretative controversy about what exactly the "gay garment" and the "simple and homely figure" refer to, it is important to notice that there is such a distinction.[13] On the one hand there are certain fea-

13 (Heidelberger 1998, 21) and (Lützen 2005, 106), see also the discussion in section 6.1.

tures which represent nature, while on the other hand those which have no such function.

What seems to be clear is that Maxwell's equations ought to be read realistically, i. e. that they represent nature. It should be equally clear that the physical conceptions, which belong to the second standpoint and which have been eliminated, belong to what Hertz calls the "gay garment." They are those conceptions that have been arbitrarily added by us (or rather by Maxwell).

Let me summarize those points that are relevant for the issues we discussed at the outset.

First, Hertz uses the notion of "hypothesis" frequently but he uses it in the weak sense only. There is no evidence in the texts we have discussed so far that there are parts of theories that are both, either true or false as well as in principle beyond the reach of empirical evidence.

Second, what Hertz ends up with is a distinction between the simple and homely figure as it is presented by nature on the one hand, and the gay garment on the other. Those parts of theories that are constitutive of the former represent nature. The concept of representation is therefore still an appropriate concept to characterize the relation between theory and nature – at least to some extent.

Finally, both the pluralism with respect to theories as well as the pluralism with respect to representations turned out to be transitory. The plurality among theories was narrowed down to one by Hertz's experiments (and his criterion for individuating theories ensures that there cannot be more than one theory); the plurality of representations by invoking criteria such as consistency (in the case of von Helmholtz) and simplicity or economy (in the case of Maxwell). However, whether or not physical conceptions and thus representations are defensible is also a matter of empirical consideration, as Hertz's remark about the most important result of his experiments (propagation in time of an alleged action at a distance force) indicates.

5. The Principles of Mechanics

Finally, I will discuss the questions outlined at the outset with respect to *The Principles of Mechanics*. The central term Hertz invokes in this context is that of an image (*Bild*, "picture" would have been a better translation). This is significant because in *The Constitution of Matter* Hertz had used two metaphors in the context of characterizing how we talk about nature: the notion of a sign and that of an image. In the context of a discussion of the concept of matter he had introduced the notion of a sign:

> I compare [the concept of] matter with paper-money, which our understanding issues in order to organize its relations to the things. Paper-money is a sign for something else, and precisely on the fact that it is a sign depends its value and meaning. Its own character is irrelevant ...[14]

At other times, as we have seen, he invoked the notion of a picture:

> [O]f the things, which are real but cannot themselves enter the mind, we have made pictures, which coincide with the things in certain respects whereas in others they depend on our conceptions (Hertz 1999, 36). (See above for full quote.)

Even though Hertz uses the notion of sign (Zeichen) or symbol occasionally in the "Introduction" to his *The Principles of Mechanics*, image is the predominant concept he makes use of. This is significant because the two metaphors suggest different claims with respect to the question to what extent the world is represented. Whereas "picture" in Hertz's terminology implies that there is a correspondence (or coincidence) with the things pictured in certain respects (but not in others), Hertz's characterization of a sign is less explicit in this respect.

Choosing picture or image rather than sign as the main conceptual tool can therefore be taken as a first indication that Hertz still believes that certain aspects of nature can be represented.

5.1 Images

In order to come to terms with the questions sketched at the outset of this paper I will now analyze Hertz's concept of an image, as he used it in *The Principles of Mechanics*.

According to Hertz, we introduce images when we try to predict phenomena, i.e. what he calls the "most direct, and in a sense most im-

14 "Ich vergleiche die Materie mit einem Papiergeld, welches unser Verstand ausgiebt, um seine Beziehungen zu den Dingen zu regeln. Das Papiergeld ist ein Zeichen für etwas anderes und gerade in diesem, daß es ein Zeichen ist, liegt sein Wert und seine Bedeutung. Seine eigene Beschaffenheit ist gleichgültig ... (Hertz 1999, 117/8). Hertz was familiar with von Helmholtz's theory of signs (see Schiemann 1998).

portant, problem which our conscious knowledge of nature should enable us to solve" (Hertz [1894] 1956, 1):

> We form for ourselves images or symbols of external objects; and the form which we give them is such that the necessary consequents of the images in thought are always the images of the necessary consequents in nature of the things pictured (Hertz [1894] 1956, 1).

The first thing to be noted is that Hertz makes use of the concept of an image in a narrow and in a broad sense. Images in the narrow sense are parts of theories that refer to particular things in nature. This is the sense in which the concept of a symbol or image is used in the above quotation. When he compares the different images of ordinary mechanics it is rather theories as a whole that he has in mind. The above-quoted requirement for images is valid both for the narrow sense as well as for the broad, as becomes clear in the sequel of the introduction where he exclusively deals with images in the broad sense (see Hüttemann 2001).

But how do we compare consequents of images with consequents of things? What are the constitutive elements of images? Hertz refers to *fundamental ideas* and to *principles* which connect the ideas as the main elements that are characteristic for a particular image. Principles of mechanics are defined as

> [a]ny selection from amongst such and similar propositions, which satisfies the requirement that the whole of mechanics can be developed from it by purely deductive reasoning without any further appeal to experience (Hertz [1894] 1956, 4).

The whole experiential content of a theory is captured by its principles.

The examples of images Hertz discusses in the "Introduction" of his *The Principles of Mechanics* are the customary exposition (Darstellung) of mechanics, which is characterized through the fundamental ideas of space, time, mass and force as well as Newton's laws of mechanics and d'Alembert's Principle. The ideas of space, time, mass and energy plus Hamilton's Principle constitute the "energetical" image. Hertz's own image presupposes just three fundamental ideas – space, time and mass – plus a fundamental law that serves as his principle.

Still, the question remains: How are we to compare whether "the necessary consequents of the images in thought are always the images of the necessary consequents in nature of the things pictured." The first book of *The Principles of Mechanics* does not deal with this problem; it treats the fundamental ideas and introduces definitions without making any reference to experience. "The subject matter of the first book is completely independent of experience" (Hertz [1894] 1956, 45). It is only in the second book that such a connection is established. At the beginning of the second book Hertz introduces three *rules* (Festsetzungen) for his fundamental ideas. The first of these rules concerns time:

> Rule 1. We determine the duration of time by means of a chronometer, from the number of beats of its pendulum. The unit of duration is settled by arbitrary convention (Hertz [1894] 1956, 140).

There are similar rules for space and mass. Hertz seems to think of them as providing definite and determinate values for a determinable.

So what we see is that over and above the fundamental ideas and principles these rules are constitutive for the concept of an image as well:

> Thus only through these rules can the symbols (Zeichen) time, space and mass become parts of our images of external objects. Again, only by these three rules are they subjected to further demands than are necessitated by our thought (Hertz [1894] 1956, 141).

By these rules the images become images of external things.

In the "Introduction" to the *Electric Waves* Hertz distinguished between Maxwell's theory on the one hand and its representations on the other. How does the concept of an image connect to these concepts?

Hertz introduces the notion of an image or picture as follows: "The images which we here speak of are our conceptions (Vorstellungen) of things" (Hertz [1894] 1956, 1). So one might hold – as Michael Heidelberger does (Heidelberger 1998, 21) – that an image is the same thing as a representation in the *Electric Waves*, because – as we have seen – representations were essentially constituted by conceptions. However, if one takes into account how Hertz spells out in detail what belongs to the different images of mechanics (image is here taken in the broad sense), there is a conspicuous absence of what Heidelberger has called "the ultimate unobservable agent which produces the phenomena" (Heidelberger 1998, 18). What constitutes images are the fundamental concepts, the fundamental equations formulated in terms of these concepts plus the rules (Festsetzungen). Jesper Lützen points out that this difference, i. e. the presence of conceptions concerning ultimate agents explaining the phenomena in the *Electric Waves* and their absence in *The Principles of Mechanics*, maybe due to the fact that mechanics as opposed to electrodynamics is the fundamental physical theory. In electrodynamics we can develop conceptions (Vorstellungen) of polarization, etc., in terms of mechanical concepts so as to explain electrodynamical phenomena. In the case of mechanics we cannot develop conceptions in terms of more fundamental concepts (see Lützen 2005, 106).

This observation plus the role fundamental equations play both in Hertz's characterization of Maxwell's theory as well as in the images provides evidence for the claim that "image" is more of a successor of "theory" than of "representation." Nevertheless, besides the fundamental equations, which he took to be constitutive for Maxwell's theory, he adds some further constituents of images such as the fundamental concepts and the rules (Festsetzungen). There is one further element that seems to give some support for Heidelberger's reading, namely the hidden masses that Hertz postulates. These masses have some similarities with the conceptions discussed in the *Electric Waves*. However, Hertz nowhere refers to the hidden masses as an image. They are a part of an image, not an image themselves.

The plurality of images we envisage in *The Principles of Mechanics* is a plurality of fundamental mechanical equations, sets of fundamental concepts plus some auxiliary hypotheses (such as the hypothesis concerning hidden masses). There is no longer a two-folded pluralism (of theories and of representations) as there was in the *Electric Waves*.

5.2 Criteria for the Evaluation of Images

Whether Hertz's pluralism is transitory is an issue that concerns the criteria for evaluating the images. The question is whether Hertz believes that these criteria will ultimately single out a unique image.

Hertz introduces three criteria for the evaluation of images: correctness, permissibility and appropriateness. Let me turn to correctness first. Correctness is the requirement that Hertz mentions in the passage we quoted at the very beginning: "the necessary consequents of the images in thought" have to be "the images of the necessary consequents in nature of the things pictured" (Hertz [1894] 1956, 1).

What are the implications of a theory's correctness? Immediately following the above quotation Hertz continues: "In order that this requirement may be satisfied, there must be a certain conformity (Übereinstimmung) between nature and thought" (Hertz [1894] 1956, 1). But what sort of conformity does Hertz have in mind? A few pages later Hertz discusses the customary image of mechanics and, in particular, the fundamental law's credentials with respect to correctness:

> Upon the correctness of the image under consideration we can pronounce judgment more easily ... No one will deny that within the whole range of experience up to the present the correctness is perfect; that all those characteristics of our image, which claim to represent observable relations of things, do really and correctly correspond to them (Hertz [1894] 1956, 9).

Hertz claims that if an image is correct we are allowed to conclude that systems really correspond to our image so far as the fundamental law is concerned.[15] Hertz reaffirms his realism with respect to the fundamental mathematical equations.[16]

As in his earlier writings Hertz keeps to the view that certain parts of theories or images, if correct, represent nature. *Pace* Schiemann (1999) I hold that the concept of representation is still appropriate for characterizing Hertz's views about the relation of images and the world – at least with respect to fundamental equations.

The question of whether or not an image is correct is essentially a matter of whether or not the fundamental law is correct:

> [The experiential part], in so far as it is not already contained in the fundamental ideas, will be comprised in a single general statement which we shall take for our Fundamental Law. No further appeal is made to experience. The question of the correctness of our statements is thus coincident with the question of correctness or general validity of that single statement (Hertz [1894] 1956, 139, translation augmented).

This fundamental law is a hypothesis in the weak sense. Hertz considers it "provable" – at least by future experience:

> We consider the law to be the probable outcome of most general experience. More strictly, the law is stated as a hypothesis or assumption, which comprises many experiences, which is not contradicted by any experience, but which asserts more than can be proved by definite experience at the present time (Hertz [1894] 1956, 145).

---

15 There seems to be one caveat. Hertz talks about *observable* relations, not about relations in general. There is, however, no evidence that Hertz has in mind the observable/unobservable distinction as it was discussed in 20th-century philosophy of science. Rather, as in the case of "experience," his use seems fairly liberal and seems to comprise everything that is within the reach of experimentation.

16 This kind of realism is sometimes called "structural realism"; see (Ladyman 1998).

To sum up: Images ought to be correct – that is why we construct them in the first place. What this boils down to essentially[17] is the requirement that the fundamental law has to be correct. A fundamental law or principle – if not yet confirmed – is a hypothesis in the weak sense only; it can – at least in principle – be decisively confirmed.[18] Hertz is very explicit that these hypotheses are either right or wrong. Comparing his image of mechanics with the customary image Hertz writes:

> … it is important to observe that only the one or the other of the two images can be correct: they cannot both at the same time be correct … This is the field in which the decisive battle between these different fundamental assumptions of mechanics must be fought out (Hertz [1894] 1956, 40/41).

Hertz does believe that there will be a battle between the various images and he believes that the battle will be decisive. Thus there will be a decisive answer to the question whether one or the other theory is correct.

One thing we learn from this is that Hertz did not believe that images are hypotheses in a strong sense – i. e. hypotheses that we are unable to provide sufficient evidence for some principled reason.

The above quote, however, does not yet imply that Hertz considered the pluralism of images he described as merely transitory. Hertz introduces two further criteria for the evaluation of images. These are the criteria of *permissibility* and *appropriateness*. An image is permissible if it does not contradict the laws of our thought, i. e. if it is logically consistent. Images, which contradict laws of thought, are *inadmissible*. Hertz holds that "two permissible and correct images of the same external objects may yet differ in respect of appropriateness" (Hertz [1894] 1956, 2).

An image can be appropriate in two respects. Firstly, it can be more appropriate than another image because it is more *distinct*. This is the case if it "pictures more of the essential relations of the object" than its competitor (Hertz [1894] 1956, 2). What Hertz has in mind can

17 I write "essentially" because Hertz introduces further empirical hypotheses, e. g. the hypothesis of concealed masses and certain assumptions about the continuity of nature (Hertz [1894] 1956, 146).

18 There is one passage where Hertz claims that "that which can be derived from experience can be annulled through experience" (Hertz [1894] 1956, 9). This, however, should not be taken to indicate that Hertz believes in hypotheses in the strong sense. As the immediately preceding passage makes clear, what Hertz wants to claim is simply that our evaluations of correctness can take into account only past (and present) experiences. So our evaluations might be revised in the light of future experiments.

best be illustrated by his discussion of the customary image of mechanics. We know of essential features of forces which are not an integral part of the customary picture of mechanics. Therefore, it is not as distinct as Hertz's own image.

> Of natural motions, forces and fixed connections we can predicate more than the accepted fundamental laws do. Since the middle of this century we have been firmly convinced that no forces actually exist in nature which would involve a violation of the principle of conservation of energy ... Again these elementary forces are not free. We can assert as a property which they are generally admitted to possess, that they are independent of place and time (Hertz [1894] 1956, 10).

Secondly, an image may be more appropriate than another if it is more simple, i. e. if it contains "in addition to the essential characteristics, the smaller number of superfluous or empty relations" (Hertz [1894] 1956, 2). Again, what Hertz means by this criterion can best be illustrated by his discussion of the role of forces in the customary picture.

> We see a piece of iron resting upon a table, and we accordingly imagine that no causes of motion – no forces – are there present. Physics, which is based on the mechanics considered here and necessarily determined by its basis, teaches us otherwise. Through the force of gravitation every atom of the iron is attracted by every other atom in the universe. But every atom of the iron is magnetic, and is thus connected by fresh forces with every other magnetic atom in the universe. Again, bodies in the universe contain electricity in motion, and this latter exerts further complicated forces which attract every atom of the iron. In so far as the parts of the iron contain themselves electricity, we have fresh forces to take into consideration; and in addition to these again various kinds of molecular forces. Some of these forces are not small: if only a part of these forces were effective, this part would suffice to tear the iron to pieces. But in fact, all the forces are so adjusted among each other that the effect of the whole lot is zero; that in spite of a thousand existing causes of motion, no motion takes place; that the iron remains at rest. Now if we place these conceptions before unprejudiced persons, who will believe us? (Hertz [1894] 1956, 13).

An image of mechanics (such as Hertz's own), which avoids such forces, is a simpler image of mechanics.

Images will picture typically essential as well as inessential relations. As we have seen, Hertz advocates a realistic attitude towards essential relations (mathematical structure). Claims about these essential relations – if not already confirmed – are hypotheses in the weak sense.

With respect to the inessential or empty relations Hertz holds:

> Empty relations cannot be altogether avoided: they enter into the images because they are simply images, – images produced by our mind and necessarily affected by the characteristics of its mode of portrayal (Hertz [1894] 1956, 2).

At this point we have to draw the same conclusion we drew with respect to what Hertz called the "inessential relations" in *The Constitution of Matter.* Claims concerning these ineliminable empty relations are neither right nor wrong. They are not the kind of things for which evidence is possible. They are therefore no hypotheses – neither in the weak nor in the strong sense.

It is exactly with respect to these features of images that Schiemann's claim concerning the limited applicability of the concept of representation is correct. Empty relations do not represent anything.

Let me come back to the issue of pluralism. As I already mentioned, Hertz believes that images do not only differ with respect to correctness and permissibility but also with respect to appropriateness. Whether or not an image is permissible and whether or not it is correct are questions that allow for decisive answers (though his discussions of the details of the images shows that it is not so easy to figure out to what extent an image in fact conforms to these criteria).

The striking thing is that – despite certain ambiguities – Hertz seems to believe that ultimately it will be possible to find a unique image, which does best with respect to appropriateness.

> ... we cannot decide without ambiguity whether an image is appropriate or not; as to this difference of opinion may arise. One image may be more suitable for one purpose, another for another; only by gradually testing can we finally succeed in obtaining the most appropriate (Hertz [1894] 1956, 3).

6. Conclusion

According to Hertz different images of mechanics are possible. But this envisaged pluralism concerns the situation before all the evidence is in, i. e. before we can decide whether the images are correct and appropriate. There is no place at which Hertz suggests that there might be a plurality of images that equally conforms to all three criteria. Hertz believes in final and decisive battles of the kind he participated in while doing experiments in electrodynamics.

Admittedly, Hertz's rhetoric in *The Principles of Mechanics* sounds much more pluralistic than in the *Electric Waves*. This is presumably due to the fact that in *The Principles of Mechanics* he defends a minority view of mechanics. The pluralistic rhetoric is a way of introducing the image of mechanics he favors. At the end of the day, however, according to Hertz there will be a decisive battle and the stage of pluralism will *a fortiori* turn out to be transitory – just as it was the case in electrodynamics. So Boltzmann misrepresents Hertz's views when he claims that Hertz's picture theory implies that it cannot be our *aim* to look for an absolutely correct theory.

We have also seen that the concept of representation is still quite appropriately applied to important aspects of images. Hertz explicitly allows that "characteristics of our image, which claim to represent observable relations of things, do really and correctly correspond to them" (Hertz [1894] 1956, 9). More particularly, he thinks of the fundamental equations at this point.

Finally: Hertz uses the concept of a hypothesis throughout his work. But he uses it in a weak sense. When it comes to experimental evidence Hertz believes in decisive battles, as we have seen. There is no trend in Hertz towards an increasing hypothesization – given this trend is understood in a strong sense, i.e. as leading to a conception according to which the correctness of scientific theories cannot be established for principled reasons.

Bibliography

Boltzmann, Ludwig (1905), *Populäre Schriften.* Leipzig: Johann Ambrosius Barth.

Carrier, Martin (1990), "Kants Theorie der Materie und ihre Wirkung auf die zeitgenössische Chemie", *Kant-Studien* 81: 170–210.

Heidelberger, Michael (1993), *Die innere Seite der Natur: Gustav Theodor Fechners wissenschaftlich-philosophische Weltauffassung*, Frankfurt am Main: Klostermann (Engl. transl. *Nature from Within*, Pittsburgh: University of Pittsburgh Press 2004).

— (1998), "From Helmholtz's Philosophy of Science to Hertz's Picture Theory", in: D. Baird, R. I. G. Hughes, A. Nordmann (eds.) *Heinrich Hertz: Classical Physicist, Modern Philosopher.* Dordrecht: Kluwer, 9–24.

Hertz, Heinrich (1892), *Untersuchungen über die Ausbreitung der elektrischen Kraft.* Leipzig: Johann Ambrosius Barth. Engl. transl. by D. E. Jones: *Electric Waves*, New York: Dover 1962.

— (1894), *Die Prinzipien der Mechanik*. Leipzig: Johann Ambrosius Barth. English translation by D. E. Jones and J. T. Walley *The Principles of Mechanics Presented in a New Form*. New York: Dover 1956.

— (1896), "Hermann von Helmholtz", in: *Miscellaneous Papers*. London, 332–340.

— (1999), *Die Constitution der Materie*, ed. Albrecht Fölsing, Heidelberg: Springer.

Hüttemann, Andreas (2001), "Heinrich Hertz and the Concept of a Symbol", in: *Symbol and Physical Knowledge*, ed. by M. Ferrari und I.-O. Stamatescu. Heidelberg: Springer, 109–121.

Hyder, David (2002), "Kantian Metaphysics and Hertzian Mechanics", in: *Vienna Circle Institute Yearbook 2002*, edited by F. Stadler. Dordrecht: Kluwer. 35–46.

Ladyman, James (1998), "What is Structural Realism", in: *Studies in History and Philosophy of Science*, 409–424.

Lützen, Jesper (2005), *Mechanistic Images in Geometric Form*. Oxford: Oxford University Press.

Schiemann, Gregor (1998), "The Loss of World in the Image: Origin and Development of the concept of Image in the Thought of Hermann von Helmholtz and Heinrich Hertz", in: D. Baird, R. I. G. Hughes, A. Nordmann (eds.) *Heinrich Hertz: Classical Physicist, Modern Philosopher*. Dordrecht: Kluwer, 25–38.

Hypotheses and Conventions in Poincaré

Gerhard Heinzmann[1]

Abstract:

Poincaré considers hypotheses as necessary tools for experimentalists as well as for mathematicians. Indeed, although hypotheses may be only approximate and provisionary, mathematics is not an approximate science. I will argue for the thesis that, if considered in Poincaré's wide semantic field, the term "hypothesis" is not ill chosen in order to show not only contrasts but also methodological analogies concerning the process of comprehension in mathematics and physics. The methodological point is to show how Poincaré, by focusing on the epistemic aspect, traces the uncharted frontier between convention and hypothesis in the sciences from arithmetic and geometry to mechanics and experimental physics by respecting as far as possible the equilibrium between exactness and objectivity, this latter being an attribute of reality. In this way it seems to be possible to recognize methodological analogies and links between a priori principles, conventions and verifiable hypotheses. These are all guided by experience without being (entirely) experimental.

1. Introduction

The title of Poincaré's celebrated opus *Science and Hypothesis* (Poincaré 1902; abbreviated to SH) was probably suggested by Gustave Le Bon, who left it to Poincaré to modify if he so desired.[2] Poincaré accepted the title, assembled a selection of previously published articles, drafted an introduction, and the book was ready to go.

In the introduction to SH, Poincaré considers hypotheses as necessary tools for experimentalists and mathematicians. The latter transgress classical rationalism for which "mathematical truths are derived from a few self-evident propositions" (SH, xxi). But by doing so, he continues, "there is a question whether all these constructions are built on solid foundations" (*ibid.*, xxii), because hypotheses are "merely approximate

1 I am grateful for comments I received from Michael Heidelberger, Gregor Schiemann, David Stump, Paolo Mancosu, Scott Walter and from the anonymous reviewer. I am indebted to Joe Lochte and Scott Walter for their help in correcting my English.

2 Undated letter (1902) sent by Ernest Flammarion to Poincaré, Poincaré Archives, Université de Nancy 2; Engl. translation in: Rollet 2002, 155.

and provisory" (*ibid.*, 89). But surely, Poincaré comments, mathematics – and especially arithmetic – is not an approximate science.

Is not the title of the book then ill chosen? Poincaré distinguishes different senses of the word "hypothesis" and speaks, for example, in geometry only of *apparent hypotheses* that are not experimental hypotheses but conventions. In arithmetic he feigns no hypotheses or conventions. But how can hypothesis then be a necessary tool for mathematics?

In this paper I argue for the thesis that, if considered in Poincaré's wide semantic field, the term "hypothesis" is not ill chosen: it endorses analogies and contrasts in mathematics and physics. The role that plays *experience* in the logical genesis of

a) complete induction in arithmetic,
b) the concept of groups in geometry,
c) verifiable hypotheses in mechanics,

acts as a criterion of the analogies. Every time the experience leads to a process whose results possess, admittedly, a different character: the *a priori principle* of the mathematical induction is compared concerning the indefinite repetition to the *natural hypothesis* of the empirical induction, the *geometrical conventions* appear as *apparent hypotheses* with regard to the *verifiable hypotheses* of mechanics.

Poincaré's methodological approach has roots in what he calls a semi-skeptical attitude,[3] capable of addressing both the absolute doubt of rationalism (Descartes) and the prejudice of naive realism. This insight seems sufficient, not for restoring absolute foundations but for understanding the scientific process. The best way to grasp understanding is by explaining the neglected distinctions and interconnections "between what is experience, mathematical reasoning, convention, and hypothesis" (SH, 89). To reach this aim, Poincaré emphasizes the genetic point of view; like many philosophers at the end of the 19th century "who saw the main aim of philosophy more in the epistemological analysis of science than in the pursuit of metaphysics" (Giedymin 1991, 4). What distinguishes Poincaré from other authors is his systematic – but non-historical – interpretation of the genesis; resembling what the logical empiricists called a (re)construction program.

---

3 "To doubt everything or to believe everything are two equally convenient solutions" (SH, xxii). "[The] semi-scepticism … is for the *savant* the beginning of his wisdom" (Poincaré 1910, 159).

It is true that Poincaré's commentators are quite right to underline his Kantian background:

> Our aim is to find out whether or not [the agreement of a theory with experiment] exists ... To neglect one for the other would be folly. In isolation, theory is empty and experience blind; both are useless and of no interest by themselves (Poincaré [1908] n.d., 275).[4]

Even so, Poincaré resolves the problem of the unity of spontaneity and receptivity totally without introducing a pure *sensibility*. The arithmetical "pure intuition" he introduces is intellectual in character. So it changes the terms of the Kantian mediation between spontaneity and receptivity. On the other hand, Poincaré is very conscious of von Helmholtz's difficulty in attempting to deduce from psycho-physiological observations hypotheses concerning analytic geometry without succumbing to a vicious circle in defining the rigid body (see Heinzmann 2001). Poincaré's geometrical conventions are neither analytic nor synthetic, but they are a kind of bicephalous selection of analytical but non-logical propositions, "guided" at the same time by experience. In this sense, conventionalism is not a merely linguistic thesis concerning the inter-translatability between different geometrical and physical languages, but also an epistemological thesis concerning the vague and conventional borderline between the theoretical language and the observational language of science. While I concede that Poincaré often conceives convention in the linguistic sense, especially in relation with the question of the metrication of geometry, I have a more fundamental aspect in view: the point is to mediate rigor or "artifice" and objectivity, i. e., to mediate these two requirements by determining the conventional aspects in the pre-theoretical genesis of scientific theories. What has to be done is to avoid the following alternative which is a traditional one: either the impressions meet the theoretical requirements, so that the mind has nothing else to do than register the data in an adequate language, or the impressions do not meet the theoretical requirements so that the linguistic construction misses its aim which is to specify the contents of these data.[5]

Poincaré is regarded as the father of conventionalism. If one maintains that his conventionalism includes an aspect that is susceptible of unifying fact with freedom – this is at variance with the opinion of

---

4 Cf. Nye 1979, Folina 1992 et Crocco 2004; concerning Poincaré's historical connections to the French neo-Kantian Circle, see Rollet 2000.

5 Cf. Meyer 1992, 194.

Meyer 1992[6] – the term "convention" seems to be ill chosen. Jules Vuillemin (1970, 12) has understood this correctly and uses the term "occasionalism," suggesting other philosophical connotations. In order not to lay myself open to additional problems of a terminological nature, I shall abide by the term "convention" but connect it to the term "hypothesis."

The problem Poincaré is concerned with is the equilibrium between exactness and objectivity The latter concerns a presupposed universal consensus with respect to natural relations. To understand this equilibrium, we must, according to Poincaré, "pass in the sciences from arithmetic and geometry to mechanics and experimental physics" (SH, xxiv) in order to distinguish *a priori principles* (e. g., complete induction in arithmetic), *natural hypotheses* (e. g. the principle of empirical induction), *apparent hypotheses* as *conventions* (e. g., the axiom of parallels in geometry), *verifiable hypotheses* (e. g., the hypothesis of central forces) and *principles of physics* (e. g., the principle of inertia).

To begin with (*section 1*), it may be useful to recall Poincaré's classifications of the term "hypotheses": The most important distinctions concern *apparent hypotheses* or *conventions* and *verifiable hypotheses* (The former, exemplified by the axioms of geometry, are discussed in *section 3* and the latter, employed only in the empirical sciences, in *section 4*). *Section 2* deals with Poincaré's view on arithmetic.

2. Poincaré's Classification of Hypotheses

In *Science and Hypothesis*, Poincaré gives two classifications of hypotheses.[7] The first (**A**) can be found in the introduction, the second (**B)** in chapter IX, entitled "Hypotheses in Physics." I will quote both *in extenso:*

> **A** We shall also see that there are several kinds of hypotheses; that some are verifiable, and when once confirmed by experiment, become truths of great fertility **[A1]**; that others, without leading us astray, may be useful to us in fixing our ideas **[A2]**; and finally, that others are hypotheses only in

6 Meyer 1992 claims that Poincaré's conventionalism fails in its attempt to go beyond a dualistic realism in which theory only treats experience without being an element of its constitution. We shall see that this opinion seems too severe.

7 I am indebted to my student Igor Ly for the following distinctions (see Ly 2008, PhD thesis).

appearance, and reduce to definitions or to conventions in disguise **[A3]** (SH, xxii [1902]).

B We must also take care to distinguish between the different kinds of hypotheses. First of all, there are those which are quite natural and necessary [*auxquels on ne peut guère se soustraire*]. It is difficult not to suppose that the influence of very distant bodies is quite negligible, that small motions obey a linear law, and that an effect is a continuous function of its cause. I will say as much as for the conditions imposed by symmetry. All these hypotheses affirm, so to speak, the common basis of all the theories of mathematical physics. They are the last that should be abandoned **[B1]**. There is a second category of hypotheses, which I shall qualify as indifferent. In most questions the analyst assumes, at the beginning of his calculations, either that matter is continuous, or the reverse, that it is formed of atoms. In either case, his results would have been the same ... If, then, experiment confirms his conclusions, will he suppose that he has proved, for example, the real existence of atoms? ... These indifferent hypotheses are never dangerous provided their character is not misunderstood ... They need not therefore be rejected **[B2]**. The hypotheses of the third category are real generalizations. They must be confirmed or invalidated by experiment. Whether verified or condemned, they will always be fruitful; but, for reasons I have given, they will only be so if they are not too numerous **[B3]** (SH, 152–153 [first edited in 1900]).

By way of summary, we thus have the following classifications:

Classification A

A1. Verifiable hypotheses of great fertility,

A2. Hypotheses useful to us in fixing our ideas,

A3. Apparent hypotheses that reduce to definitions or to conventions in disguise.

Classification B

B1. Natural hypotheses that are the last to be abandoned,

B2. Indifferent hypotheses,

B3. Hypotheses as real generalizations confirmed or invalidated by experiment.

We see immediately the concordance of **A1** with **B3** (*verifiable hypotheses*) and **A2** with **B2** (*indifferent hypotheses*); both have a descriptive function. By contrast, **A3**, *apparent hypotheses*, and **B1**, *natural hypotheses* have no visible equivalent in the other classification, so that we have to distinguish, from a synchronical point of view, four types of hypotheses:

1. Verifiable hypotheses
2. Indifferent hypotheses
3. Natural hypotheses
4. Apparent hypotheses

According to Poincaré, all generalizations are genuine hypotheses (SH, 150), i. e., *verifiable hypotheses* that have been confirmed by experiment. The analysis of Poincaré's understanding of the confirmation of a hypothesis will be the subject of *section 4* (physics).

Of course, not all hypotheses are generalizations: there are, for example, *indifferent* hypotheses. They concern the *ontological* – but not the structural – determination of elementary phenomena. Take the propagation of heat, where "the law of great numbers will suffice" and we need not inquire how each molecule radiates. Such hypotheses are mere metaphysical and metaphorical crutches useful for thought but "unverifiable" and "useless" as such (SH, 156). They are conventional in the usual sense of the word; that is to say "arbitrary," but compelled by rational agreement. In general, the ontological determination of singular objects is, from a scientific perspective, an overdetermination: scientific objectivity is purely relational, while the *relata* remain inaccessible to human knowledge.[8] This is, as Zahar remarks, a negation of Quine's slogan "To be is to be the value of a first order variable" (Zahar 2001, 37–38). We will return later to this point.

Concerning the status of *natural hypotheses,* they may be considered as experimentally inaccessible conditions for science or as practical rules. But Poincaré does not expand upon this sort of explanatory hypothesis. I shall return to it in *section 2* and *3*.

Apparent hypotheses that reduce to *definitions* or to *conventions in disguise* seem not to be hypotheses at all – they may only be *confused* with genuine hypotheses. Clearly, not all conventions (such as, for instance, nominal definitions) are hypotheses, nor could they be confused as such. Accordingly, in order to understand the possibility of such confusions, we will take up in detail the science most prone to confusion: geometry (*section 3*).

One should stress the fact that the reader meets the mentioned classifications of hypotheses only after an advanced reading of *Science and Hypothesis.* Indeed, the first two chapters of this book take up arithmetic (chapter I) and geometry (chapter II), where no genuine hypotheses are employed. Nevertheless, since Poincaré's methodological priority lies with mathematics, we will respect his recommendation and com-

8 "The aim of science [*ce qu'elle peut atteindre*] is not things themselves, as the dogmatists in their simplicity imagine, but the relations between things; beyond those relations there is no knowable reality" (SH, xxiv; SH, 161).

ment initially upon the first two chapters. The reason is that these chapters provide in fact the understanding of the role of hypotheses in mechanics (chapter III) and physics (chapter IV). Thus in the following section, I show that the understanding of what Poincaré calls a *synthetic a priori principle* prepares the comprehension of the various kinds of hypotheses.

3. Arithmetic: Synthetic *a priori* Principle

The first chapter on arithmetic takes up the principle of recurrence, also known as mathematical or complete induction.[9] Poincaré considers this principle to be a necessary tool for proof of general propositions. It is a synthetic *a priori* principle because it is "imposed upon us with such a force that we could not conceive of the contrary proposition" (SH, 48).[10] One might argue that Poincaré was mistaken to include this chapter in his book in light of the non-hypothetical nature of arithmetic. What is important here, however, is that this principle is understood to be a structural element of empirical investigations:

> We have the faculty of conceiving that a unit may be added to a collection of units. *Thanks to experiment, we have had the occasion to exercise this faculty and are conscious of it* (SH, 24; my emphasis).

The principle of induction

> is only the *affirmation* of the power of the mind which knows it can conceive of the indefinite repetition of the same act, when the act is once possible. *The mind has a direct intuition of this power, and experiment can only be for it an occasion of using it, and thereby of becoming conscious of it* [SH, 13, my emphasis].

In other words, the indefinite repetition is occasioned by experience without being itself empirical and constitutes as such the theoretical part of our knowledge. Experience is the *ratio cognoscendi* of the *affirmation* of the fact that if a domain can be structured through an act of indefinite repetition, a property is valid for all elements if it is valid for the successor of any element. Now it seems clear why the negation of the principle is inconceivable: the principle of induction is a *form of under-*

9 $[E(I) \wedge \forall x(N(x) \wedge E(x) \rightarrow E(x'))] \rightarrow \forall x(N(x) \rightarrow E(x))$ for any property E and the numerical predicate N.

10 Poincaré's understanding of Kant on synthetic a priori judgments is problematic; see Ben-Menahem 2001, 476, especially note 14.

standing, in the sense that "we can form no conception of an experience which would be best interpreted as a violation of it" (cf. Folina 1992, 32). The universal proposition in the consequent of complete induction, as used by Poincaré, is neither used qua conventional nor qua empirical. In arithmetic, we seem to use neither hypotheses nor conventions. Nevertheless, we will see that the elaboration of *hypotheses* will be analogous to the "genesis" of the induction principle. From this perspective, the principle of complete induction as a necessary tool of mathematics is very close to Poincaré's understanding of the principle of empirical induction. The latter is a *natural* hypothesis or a practical rule in the sense of a necessary tool of physics, which "signifies that the consequent is a continuous function of the antecedent" (Poincaré [1905] 1958, 134 and SH, 152–153; quoted above). The "striking analogy" of complete induction with the "usual processes of induction" lies in their function to be *tools* in order to structure different domains. These tools are suggested by experiences but are themselves inaccessible to experience. There are, however, important differences. Physical induction postulates the general order of the external universe, and its predictions are uncertain, whereas complete induction is based on "the affirmation of a property of the mind itself" (SH, 13), such that there is no uncertainty. Poincaré's inclusion of arithmetic under the book's title *Science and Hypothesis* may thus be justified in the following way: the *a priori* status of complete induction contrasts well with hypothesis, but there is a striking analogy between the logical genesis of the principle of induction and a natural hypothesis.

In general, experience is not a sufficient source of knowledge for Poincaré; it makes us aware of the existence of certain mental structures to which we must accommodate our experience, either directly as in the case of complete induction, or by the introduction of conventions. Conventions, according to Poincaré,

> result from the unrestricted activity of the mind". [The laws of the mind are] "imposed on *our* science ..., [but] are not imposed on Nature. Are they then arbitrary? No; for if they were, they would not be fertile. Experience leaves us our freedom of choice, but it guides us by helping us to discern the most convenient path to follow (SH, xxiii).

Against widespread opinion these conventions guided by experience and not conventions as decisions between linguistic alternatives – determine the character of *apparent hypotheses* in geometry.

4. From Sensible Space to Geometry: Natural Hypothesis, Convention and Apparent Hypothesis

In *Science and Hypothesis*, the word "hypothesis" does not appear between the classification in the *introduction* (quoted above as classification **A**) and the chapter on non-Euclidean geometries (SH, 35, chap. III). In the latter chapter, it is used in the sense of **A3**/**B3**, that is, in the sense of an apparent hypothesis or convention. How does Poincaré use the term "convention" in geometry, and why does he speak of conventions as apparent hypotheses?

On the face of it, Poincaré's geometrical conventionalism consists of three theses:

(i) Experience does not relate to space, but to empirical bodies. Geometry deals with ideal bodies, and it can therefore be neither proved nor disproved by experience. Since the propositions of geometry cannot be analytical either, these propositions must then be conventions, neither true nor false (SH, 50).
(ii) The choice between conventions, and in particular between different geometries, is guided by experience (SH, 50).
(iii) Euclidean geometry has nothing to fear from experience, for it is the most advantageous and convenient one (SH, 73).

The first thesis suggests that geometry can be neither true nor false, because it is conventional without being analytical. But Poincaré stresses that one can very well speak of truth once a convention is adopted.[11] What is the correct position to adopt? There is no place for the acknowledgement of trivial inconsistencies, as we are naturally facing different kinds of conventions and hypotheses, the distinction between them being not always easy to discern.

In an analysis of his own scientific work, Poincaré (1921, 127) underlines two main steps of reflection concerning the construction of space: its psycho-physiological genesis and the true nature of Euclid's postulate. We will see that the first step involves a *natural hypothesis* concerning the presupposition of sensible space and an *artificial convention* in the sense of a classification. In the second step, which concerns the group-theoretical nature of displacements, *apparent hypotheses* or *disguised definitions* make an appearance.

11 Linguistic conventions are the condition of the possibility to speak of facts in an objective way (Poincaré [1905] 1958, 118).

Step 1: Natural Hypothesis and Artificial Convention

Poincaré conceives of the localization of objects as a reflection on non-spatial impressions of muscular sequences of motor-processes. In order to classify muscular sensations, Poincaré introduces, in his article *Foundations of Geometry* (Poincaré 1898; for short: FG), a category "of our understanding" which he calls "sensible space" (FG, 3), insisting that the existence of movement is the necessary material condition which alone enables the category to work (2 sq.): "For a completely immobile being, there would be [no] space" (SH, 48). But, in fact, it is not a real "space" and, in particular, it "has nothing in common with geometrical space" (FG, 5). "It involves no idea of measurement" (FG, 4), but "simply enables us to compare sensations of the same kind [...], to perceive that one sensation is greater than another, but not twice as great nor three times as great" (FG, 3). It follows that the category of sensible space is essentially vague (Poincaré 1908 [n. d.], 120) – not only could we not rigorously "speak of axes invariably bound to the body" (Poincaré [1905] 1958, 48), but also, the absence of precise measurement introduces an irreducible error whose limits correlate to the aim of our actions. Finally, sensible space is by no means built up from motor-sensations by classification (Johnson 1981, 90); rather, it is the necessary condition, the conceptual means for classification of the deliberate and conscious reproduction of muscular sensations involved in reaching the object.[12] In short, sensible space is a form of our understanding (not of sensibility), because individual sensations can exist without it (FG, 3). In this sense it is a *natural hypothesis*.

Poincaré is conscious of the fact that sensible reproduction can only be arranged in sensible space, and that we do not represent geometrical figures, but only reason about them (SH, 56). Under these circumstan-

12 "Our representations are only the reproduction of our sensations. They can therefore only be arranged in the same framework – that is to say, representative space ... When it is said, ... that we 'localize' an object in a point of space, what does it mean? It simply means that we represent to ourselves these movements that must take place to reach that object ... When I say that we represent to ourselves these movements, I mean only that we represent to ourselves the muscular sensations which accompany them" (SH 56 sq.; transl. G. H.; there is a serious error in the translation of the second sentence of this passage: the sentence "elles ne peuvent donc se ranger que dans le même cadre qu'elles, c'est à dire dans l'espace représentatif" (SH, 82, French edition) is translated as "They cannot therefore [!!!] be arranged ..."

ces, how do we link the natural hypothesis of sensible space to geometric space?[13]

The solution turns on a very sophisticated psycho-physiological genesis:[14] geometrical space is obtained by choosing the language of groups to serve as the tool of reasoning about representations of muscular sensations. As the main result of this genesis Poincaré finds that certain actions, when accompanied by muscular sensations, define equivalence classes, called *displacements,* and that each set of displacement classes forms a group in the mathematical sense.

Poincaré, however, does not consider the defining group properties to be empirical. If they are true, they cannot also be the result of an "*a priori* reasoning." On the contrary, experience can easily refute the group connection. Nevertheless, Poincaré considers geometry immune to revision (FG, 11). He introduces the notion of convention to solve this paradox. While experience can teach us only that the compensation generating the inverse group element "has approximately been realized," a convention establishes that it has been realized exactly. Experience provides only the "mind's ... occasion to perform" the classification, but it "is not a crude datum of experience" (FG, 9). Poincaré is very precise about the vagueness he has in mind at this level of reconstruction:

> When experience teaches us that a certain phenomenon does not correspond *at all* to these laws, we strike it from the list of displacements. When it teaches us that a certain change obeys them only approximately, we consider the change, *by an artificial convention,* as the result of two other component changes. The first is regarded as a displacement *rigorously* satisfying the laws [...], while the second component, which is small, is regarded as a qualitative alteration. (FG, 11; emphasis in the original)

In order to obtain a rigorous law, the compensation must be read as an order to act in view of an ideal norm. Experience plays a double role: it

13 The essential properties of geometric space are: "1st, it is continuous; 2nd, it is infinite; 3rd, it is of three dimensions; 4th, it is homogeneous – that is to say, all its points are identical one with another, 5th, it is isotropic. Compare this now with the framework of our representations and sensations, which I may call representative space" (SH, 52). In fact, "it is neither homogeneous nor isotropic; we cannot even say that it is three dimensional" (SH, 56).

14 Poincaré's original conventionalism is not related to the general axiomatic method in Hilbert's sense. Influenced by Sophus Lie's theory of group transformation, he considers geometry to be nothing but the study of certain groups and subgroups.

provides an occasion to take notice of the norm (it serves as *ratio cognoscendi*), and it is an occasion to test the norm in its role as an instrument for the conceptualization of reality. The general concept of the group is, like the principle of mathematical induction, a *form* of reason whose existence is suggested by experience. The accommodation of experience to the form, however, is the result of a conventional decision, and there is no proper abstraction or generalization process. In this way, conventions concern not only the relative choice between theories,[15] but have a prior bearing on classifications in relation to the interpretation of the fundamental concepts of every theory.

Step 2: Apparent Hypothesis and Disguised Definition

Let us grant that Poincaré's psycho-physiological reconstruction, outlined above, is an efficient way to get from our muscular sensations to the group of displacements that are isomorphic with the concept of group of motions. Why then does Poincaré claim that to adopt a metric is not to affirm an axiom, but to adopt an "*apparent hypothesis*" or convention that is neither true nor false?

Poincaré studies the "purely formal" properties of the noted group in order to obtain the geometrical space and its elements (points, straight lines, planes, etc.).[16] He observes that the group is continuous and that one can distinguish displacements which conserve certain quantities, such as rotations about a fixed point, rotations about a fixed axis, or the combination of rotations about an axis and translations parallel to the same axis. Groups with these properties (which form sub-groups) correspond to the geometries of Euclid, Lobachevski and Riemann (see FG, 20). Next, Poincaré proposes a criterion of choice between the three possible geometries: "the existence of an invariant subgroup, of which all the displacements are interchangeable and which is formed out of all translations." This criterion "determines our choice in favor of the geometry of Euclid," because it corresponds to the only

15 Poincaré's geometric conventionalism cannot be reduced to Duhem's thesis (see Michael Friedman 1996, 335).

16 Concerning the mathematical aspects of Poincaré's reconstruction of geometrical space and its elements, see Nabonnand 2000.

group containing such an invariant group well founded by immediate experience (FG, 21).[17]

So the conventional choice between different possibilities resulting in the appropriateness and convenience of Euclidian geometry is determined by experience (in contrast to this, no choices between alternatives were given in arithmetic), where "experience" does not imply a return to a crude fact, but is further enriched by theoretical elaboration.[18] Consequently, the so-called *axiom* of Euclidean distance is not a definition by convention in the proper sense, but the *result* of epistemic classifications. The epistemic character of these conventional classifications is the reason why the result could therefore be confused with a hypothesis: the Euclidean convention is not only instrumentally adequate, but also plays an explanatory role. For this reason Poincaré called it an *apparent hypothesis* or *disguised definition.*

In fact, Poincaré uses the term "disguised definition" already before 1899, the year of Hilbert's famous *Foundations of Geometry*, to express his belief that a language used in an apparently descriptive manner is not purely descriptive. The logical genesis of Euclidean distance outlined above constitutes the objectivity of a fact, which is only apparently described by an axiom. A disguised definition determines its object at the end of a very complex procedure of interdependence between theory and experience. Therefore a disguised definition, although opposed to an ordinary explicit definition, is truly explicit. First and foremost, however, it contradicts an intuitive view of facts. In other words, Poincaré's conventions in the sense of apparent hypotheses could be called *disguised* definitions for two reasons: they are neither proper definitions, nor proper descriptions. They are distinguished by a mixed form, as suggested by Roger Pouivet (1995), who finds a community of spirit between Goodman's reflective equilibrium principle and Poincaré's conventions. We are very far from Hilbert's axiomatic schemata, that in view of their non-propositional character were interpreted, from Louis Rougier (1920) on, as conventions in the sense of language rules such that "disguised definitions" turn into *implicit definitions* in Hilbert's sense.

17 According to Poincaré, geometry precedes the notion of space.

18 The distinction between 'crude fact' and 'scientific fact' is one of degree: "There is no precise frontier between the fact in the rough and the scientific fact; it can only be said that such an enunciation of fact is *more crude* or, on the contrary, more *scientific* than another" (SH, 122).

5. Physics: Verifiable Hypotheses, Principles and Physical Conjectures

We so far discussed *a priori* principles, *natural hypotheses*, and *apparent hypotheses*, but we have not yet met with verifiable hypotheses. These are hypotheses characterized by mechanical laws.

Poincaré's understanding of such hypotheses relies on a prior notion: our experience is made up of *complex* phenomena only, which are reduced by the scientist to a number of elementary phenomena. The best means of reaching the elementary phenomena would be "the instinct of simplicity" or, better, the language of "experiment" (SH, 154–157). More precisely, science is based on "good" experiments. "What, then, is a good experiment? It is that which teaches us something more than an isolated fact. It is that which enables us to predict, and to generalize" (SH, 142). Further, scientific experiments conducted in *physics* are not possible without preconceived ideas.

Now, any good experiment in physics can be generalized in an infinite number of ways, depending on our preconceived ideas, and every generalization results in a hypothesis. This hypothesis is accessible to experiment so that it can be called a *verifiable hypothesis* or *simple law*. Reflecting on the involved generalization process shows that it presupposes a belief in the unity and simplicity of Nature (cf. SH, 142 sq.). The supposed unity allows us to make the experimental result rely on the repetition of a phenomenon, and therefore on empirical induction (SH, xxvi) and probability. The criterion of simplicity is relative to the analytical apparatus we employ and, consequently, may be only "apparent."[19] It may even force us to adjust the experiment.[20] The various decisions involved in the generalization process resulting in laws made it obvious that "if from facts we pass to laws … the part of the free activity of the scientist will become much greater" (Poincaré [1905] 1958, 122).

19 SH, 148/149; "the simplicity of Kepler's laws, for instance, is only apparent" (SH, 150).

20 "Experiment only gives us a certain number of isolated points observed. They must be connected by a continuous line, and this is a true generalization. But more is done. The curve thus traced will pass between and near the points observed; it will not pass through the points themselves. Thus we are not restricted to generalizing our experiment, we correct it; and the physicist who would abstain from these corrections, and really content himself with experiment pure and simple, would be compelled to enunciate very extraordinary laws indeed" (SH, 142/143).

At the same time, however, such hypotheses should always be "as soon as possible submitted to verification" (SH, 150).

Poincaré's vocabulary sounds in passing strange to a philosopher's ears. Did he really believe that a hypothesis could be definitively verified? I do not think so.[21] Poincaré is neither an empiricist who has rigidly fixed the meaning of the descriptive terms occurring in empirical laws (see Zahar 2001, 10/11 for another interpretation), nor is he a nominalist in the sense of Le Roy: if the meaning of descriptive terms were fixed arbitrarily (as Le Roy proposed), and the law were seemingly contradicted, it could nevertheless remain valid: it would be sufficient to change the meaning of the descriptive terms. But then, Poincaré argues, science would not enable us to foresee and would preclude "means of knowledge" or "principles of action" (Poincaré [1905] 1958, 122/123). Nevertheless, to maintain the meaning of descriptive terms independently of actual experimental results would halt progress in science. Concerning the specific law, "Phosphorus melts at 44°," it is surely improbable to "discover a body which, possessing otherwise all the properties of phosphorus, did not melt at 44°," so that we would give it a different name. In principle, however, this situation is not excluded.[22] According to Poincaré, scientific activity is a capacity and not the result of a theory about factual data. Standards of truth depend on the scientific context and remain open to revision.

Until now, we have only considered laws directly accessible to experiment. Things are not always so simple: "It is not enough that each elementary phenomenon should obey simple laws: all those we have to combine must obey the same law" (SH, 158/159). Generalization in physical science has to take the mathematical form of a differential equa-

21 In *Science and Method* he wrote: "When we wish to check a hypothesis, what do we do? We cannot verify all its consequences, since they are infinite in number. We content ourselves with verifying a few, and, if we succeed, we declare that the hypothesis is confirmed, for so much success could not be due to chance" (Poincaré [1908] n.d., 89). In Popper's perspective, it may be noteworthy that falsification of hypotheses leads, according to Poincaré, to scientific progress: "it may even be said that it [the rejected hypothesis] has rendered more service than a true hypothesis" (SH, 151).

22 "Doubtless the law may be found to be false. Then we shall read in the treatises of chemistry: 'There are two bodies which chemists long confounded under the name phosphorus; these two bodies differ only by their points of fusion'. That would evidently not be the first time chemists separated two bodies they were at first not able to distinguish … I do not think the chemists fear greatly that a similar happenstance will befall phosphorus" (Poincaré [1905] 1958, 123).

tion in order to grasp complexity that implies the superposition of elementary self-similar phenomena. The law of inertia, for example, obtains only with respect to the pre-existing category of differential equations. These differential equations are, in the final analysis, the physical laws:

> Newton has shown that a law is only a necessary relation between the present state of the world and its immediately subsequent state. All the laws discovered since are nothing else; they are in sum, differential equations (Poincaré [1905] 1958, 87).

In geometry, Poincaré supposed the pre-existent category of a group. In mechanics, the categorical tool is the *natural hypothesis* of empirical induction and differential equations. So, Poincaré's position effectively integrated some Kantian elements by eliminating completely their transcendental component.

Poincaré distinguished empirical laws from conventional principles. How does one move from "simple" empirical laws, understood as *verifiable hypotheses*, to *principles* including explicitly conventional elements? He continues to employ in mechanics the same methodological procedure as in geometry with regard to classes of displacements and groups:

> When a law has received a sufficient confirmation from experiment, we may adopt two attitudes: either we may leave this law in the fray; it will then remain subject to incessant revision, which without any doubt will end by demonstrating that it is only approximate. Or else we may elevate it into a *principle* by adopting conventions such that the proposition may be certainly true [concerning fictions]. For that the procedure is always the same. The primitive law enunciated a relation between two facts in the rough, *A* and *B*; between these two crude facts is introduced an abstract intermediary *C*, more or less fictitious … And then we have a relation between *A* and *C* that we may suppose rigorous and which is the *principle*; and another between *C* and *B* which remains a *law* subject to revision (Poincaré [1905] 1958, 124).

This procedure applies to the normal case. But as we have seen, some of the mechanical laws, such as the laws of inertia or acceleration, have a more complicated structure. The decision to accept the law of inertia as a principle, for example, rests on the fact that every contrary fiction[23] would be very unlikely with respect to the general law, verified in some particular cases but in its generality impossible to confirm or to contra-

23 On the role of fictions in Poincaré and Vaihinger, see the paper of Christophe Bouriau (this volume).

dict by experiment (see SH, 91–97). The origin of this principle is experimental but its general form is unverifiable so that its conventional character in Poincaré's sense is somehow weakened.

The scheme of generalization in mechanics appears as follows: experience provides complex phenomena which we reduce into a number of elementary phenomena. Through physical induction we move from the phenomenon to the experimental fact and, by means of differential equations, to laws and verifiable hypotheses whose number should be kept as small as possible.[24] Finally, laws can be elevated by decree to the status of principles. Of course, this is only a general scheme that can embrace several variations as we have seen. In any case, the basic principles of mechanics are conventions, not arbitrary conventions – "they would be so if we lost sight of the experiments which led the founders of the science to adopt them" (SH, 110).

The analogy between the genesis of physical principles and that of metric axioms of geometry is quite obvious, and gives rise to the following dilemma:

> Upon first glance, the analogy [with the fundamental propositions of geometry] is complete, and the role of experience seems the same. We are then tempted to say that either mechanics must be looked upon as experimental science and then it should be the same with geometry, or, on the contrary, that geometry is a deductive science, and then we can say the same of mechanics (SH, 136).

For Poincaré however, such a conclusion is illegitimate. In fact, the highly- conventional character of geometry, a science in which hypotheses are only *apparent*, is due to the circumstance that convenient conventions (or definitions) are chosen in view of objects outside of geometry:

> Our fundamental experiences are pre-eminently physiological experiments which do not refer to the space that is the object that geometry must study, but to our body – in other words, to the instrument we use for that study (SH, 137).

In mechanics, on the contrary, conventions attached to *verifiable hypotheses* are convenient with respect to "mechanical objects" (SH, 137): in geometry, observation of impression-changes accompanied by muscular sensations motivated Poincaré to suggest a process of muscular compen-

24 Otherwise, "if experiment condemns it, which of the premises must be changed [... and] conversely, if the experiment succeeds, must we suppose that it has verified all these hypotheses at once?" (SH, 151/152).

sation that itself recalls the pre-existing category of the transformation group as a regulative idea. In mechanics, Poincaré argues, we act directly on given experiences that are reduced and recomposed according to the category of differential equations. It follows that, from the ontological point of view, geometry and physics are independent. This fact is underlined by Poincaré's law of relativity, signifying that the empirical information is independent of the structure of geometric space.[25] Nonetheless, experience is a legitimate tool for geometrical knowledge.

The ontological independence of geometry and physics leads us to another important relationship between geometry and physics. Poincaré's argument in favor of the necessary transmission of the conventional character of geometry to physics is based on an analogy. The methodological function of transformation groups with respect to the representative space corresponds to the function of geometry with respect to physics. The analogy runs as follows: just as the law of displacements may correspond only approximately to the law of groups and is consequently considered as the result of two component changes, the first being a displacement, and the second a qualitative alteration, so Poincaré regards the physical "complex" relation between two bodies *A* and *B* as the result of two components. The first is regarded as a "simple" geometrical principle, while the second is itself composed of two "epistemological laws": the bodies A and B are related to figures A' and B' of the geometrical space so that

$$R(A, B) \leftrightarrow R'(A', B') \wedge r_A(A, A') \wedge r_B(B, B')$$

R' being a geometrical proposition, and the r_i expressing the relationships between objects of the representative space and the geometrical space, such as the relation between solid bodies and motion invariants. Poincaré declares that by changing the relations r_i, the geometrical proposition R'(A', B') could even serve to describe the relationship between two different physical bodies (see Poincaré [1905] 1958, 125). Poincaré thereby rules out any application of geometry based on a structural isomorphism with reality (Mette 1986, 75–80).

25 See Michel 2004, 100. In *Science and Hypothesis,* Poincaré sums this up as follows: "The state of the bodies and their mutual distances at any moment will solely depend on the state of the same bodies and on their mutual distances at the initial moment, but will be in no way depend on the absolute initial position of the system and of its absolute initial orientation. This is what we shall call, for the sake of abbreviation, *the law of relativity*" (SH, 76/77).

In sum, for Poincaré the principles of mechanics (principle of inertia, law of acceleration, principle of relative motion, etc.) are not hypotheses, although they are of course of experimental origin. But they are immune to experimental disconfirmation without being a priori in the classical sense (SH, chap. VI and VII).

Now, in the physical sciences proper, i.e., optics and electrodynamics, the scene changes entirely and principles seem no longer to share the conventional character of the geometrical postulates:

> We meet with hypotheses of another kind ... No doubt, at the outset, theories seem unsound, and the history of science shows us how ephemeral they are; but they do not entirely perish, and from each one some traces still remain. It is these traces which we must try to discover, because in them, and in them alone, is the true reality (SH, xxvi).

Surely, as Vuillemin remarks (1968, 10), there are what Poincaré called "indifferent hypotheses" that the analyst assumes at the beginning of his calculations, and that are neither true nor false, but whose roles are to be positions in a *structure*. As such, they are very different from geometrical conventions. In fact, Poincaré's explicit argument in favor of structural realism is, at best, to interpret in Zahar's sense on the case of high-level theories "as indicating that relations common to successive theories reflect objective aspects of the ontological order." For example,

> Though Fresnel's, Maxwell's, and Lorentz' hypotheses seemingly refer to very different entities, namely to a mechanical medium, to the electromagnetic ether, and to some disembodied field respectively, the basic equations are approximately the same in all three cases (Zahar 2001, 39).

The relata as indifferent hypotheses may be responsible for contradictions between theories by usurping references: the postulated entities do not correspond to Poincaré's relational restriction with respect to ontology. The ontological commitment is only expressed by the structure. Hence to conserve such "contradictory" theories without abandoning one as false constitutes a central tenet of scientific practice (SH, 163/164). So the analogy with geometric conventions fails, at most, on one important point. While geometry rests on conventional decree, the structures of physics (mechanics included) do not *systematically* rest on indifferent hypotheses; they have only a psychological and pedagogic function, and are mental crutches.[26] So, *indifferent hypotheses* about the

26 In SH, 152/153, Poincaré gives the following example: "In optical theories two vectors are introduced, one of which we consider as a velocity and the

ontological "atoms" of physical structures do not have the same function as conventions in geometry, but this difference is not the one targeted by Poincaré's remark quoted above.

In what sense does the conventional element of physical hypotheses change with respect to mechanical hypotheses? Poincaré's general position, in which concrete principles are disconnected from confirmation and falsification, unleashed an unexpected challenge when a new physics was announced in 1897. It is this fact that Poincaré alludes to in his introduction to *Science and Hypothesis,* quoted above.[27] Having already taken up the ambivalent status of the law of inertia, I will confine my remarks to the most interesting case, that of the principle of relativity. In mechanics, the principle of relative motion states that the form of the differential equations does not alter with the change of coordinate axes, be they immobile or moving "uniformly in a straight line" (SH, 112; also see (Poincaré [1913] 1963, 101–102). Poincaré, in his famous plenary address in St. Louis (1904), employs the expression "principle of relativity," which does not apply to "directly observed finite equations, but to differential equations" (Poincaré [1913] 1963, 103). He reports that Hendrik A. Lorentz introduced the hypotheses of "local time" and of the "uniform contraction in the direction of motion" in an attempt to save the principle in its application to the electromagnetic domain (see Poincaré [1905] 1958: 98sqq.). The principle was, in fact, shown to lose its universal validity, because there was no further use for it. It was Poincaré (1906a) who made Lorentz's theory fully compatible with the relativity principle.[28] We are thus *naturally* inclined, remarks Poincaré, to admit the postulate of relativity in every domain. In postulating the principle as a "general law of Nature" (Poincaré 1906a, 495), its extended form finds its cause in the Galilean principle,

other as a vortex. This again is an indifferent hypothesis, since we should have arrived at the same conclusions by assuming the former to be a vortex and the latter to be a velocity ... To give it [vector] the concrete appearance that the fallibility of your minds demands, it was necessary to consider it as a velocity or as a vortex. In the same way, it was necessary to represent it by an x or a y." In SH, 215, Poincaré even took the existence of ether as an indifferent (commode) hypothesis.

27 In *The Value of Science,* he asks: "These principles, on which we have built everything, are they about to crumble away in their turn?" (Poincaré [1905] 1958, 96).

28 The difference between Lorentz's principle and Poincaré's principle is emphasized by Scott Walter (this volume).

and its reason in compatibility with a theory explaining *why* no experiment is susceptible to inform us of absolute motion. So, the principle is an element of a work in progress and there is the possibility that it loses its soundness (Poincaré [1905] 1958, 110). In fact, in the same measure as the complexity of holistic considerations increases, the conventional character decreases. In 1900, Poincaré considers the supposition that "the influence of very distant bodies is quite negligible" as a *natural* hypothesis (SH, 152), i. e. not necessarily accessible to experience. In 1912, the fact that "the reciprocal action of two mechanical systems tends to zero as spatial separation increases indefinitely" (Poincaré [1913] 1963, 107) constitutes for Poincaré the "experimental signification" of the principle of relativity. In general, the change in Poincaré's consideration of hypotheses has a taste of a resigned feeling of "fin d'époque," and his call to nevertheless discover the invariants of "true reality" as a final outcome of an equilibrium process of mental categories and experience, like a wail of regret for a lost paradise:

> We shall be submerged by the ever-rising flood of our new riches, and compelled to renounce any idea of classification – to abandon our ideal, and to reduce science to the mere recording of innumerable recipes (SH, 173).

Poincaré's model of a theory, based on a minimum of hypotheses, natural or well-confirmed either directly by experiment or indirectly by fictions, and from which all consequences could be deduced, has vanished with Maxwell's methodological approach (SH, 213/214):

> The English scientist does not try to erect a unique, definitive, and well-arranged building; he seems rather to raise a large number of provisional and independent constructions, between which communication is difficult and sometimes impossible (SH, 215).

Maxwell's theory leading to the explanation of optical phenomena is due to "daring" starting hypotheses, confirmed only "twenty years later" (SH, 239). Although Poincaré was very much engaged with Maxwell's theory, he remained always skeptical with respect to his methodological approach, as he was never satisfied with Lorentz's theory, at least until 1904, since it contained too many *ad hoc* hypotheses,[29] i. e., hypotheses necessary for the consistency of the theory, but not required for empirical predictions (Zahar 2001, 17sq.). Contrary to Einstein,

29 He nevertheless did not lose the hope that "perhaps everything will arrange itself" (VS, 107).

Poincaré's epistemological interest consists in an equilibrium of theory and experience, which are interdependent. This attitude results in hypotheses linking traditional invariants with new predictions, but not in resorting to daring conjectures, and even less in erecting *ad hoc* postulates that are not *natural hypotheses*.

6. Conclusion

Poincaré's categories of Lie groups describing geometries with constant curvatures do not fit with the general theory of relativity. For philosophical purposes, however, this failure is ancillary. The main point of this paper has been to show how Poincaré by focusing on the epistemic aspect of conventions, traces the uncharted frontier between convention and hypothesis in different sciences by respecting as far as possible the equilibrium between exactness and objectivity.

Complete induction as a synthetic *a priori* principle is one of the main tools of mathematical reasoning for Poincaré. Nevertheless, this Kantian terminology hides from view a non-Kantian conception: our experience of finite repetition of an act, of writing strokes for example, provides an opportunity to employ the mind in conceiving an *indefinite* repetition of the same act. There is no convention in the common sense of an arbitrary choice between alternatives and no hypothesis. And it is only *post festum* that the reader recognizes the methodological analogy and link between the principle of induction, conventions in the sense of apparent hypotheses and verifiable hypotheses. These are all guided by experience without being (entirely) experimental, and involve presupposed categories or natural hypotheses: *verifiable hypotheses* employ conventional elements in the generalization process and involve the *natural hypothesis* of physical induction. Geometrical conventions are *apparent hypotheses* guided by the experience of muscular sensations and involve the category of groups. Both these sorts of hypotheses, verifiable and apparent, employed in physics and geometry, are in Poincaré's hierarchy of sciences framed on one hand by *a priori principles* and *natural hypotheses* that are tools as well for mathematicians (complete induction in arithmetic) as for experimenters, and on the other hand by *indifferent hypotheses* and physical *principles*. The indifferent hypotheses are conventional (in the common sense) stipulations of ontological entities that increase our theoretical insight; the physical principles are either verifiable hypotheses elevated by decree beyond the crucible of experiment or well-founded with respect to the network of actual science.

Bibliography

Ben-Menahem, Yemina (2001), "Convention: Poincaré and some of his critics", *British Journal for the Philosophy of Science* 52 (3): 471–513.

Crocco, Gabriella (2004), "Intuition, construction et convention dans la théorie de la connaissance de Poincaré ", *Philosophiques* 31 (1): 151–177.

Folina, Janet (1992), *Poincaré and the Philosophy of Mathematics.* Houndmills/London: MacMillan.

Friedman, Michael (1996), "Poincaré's Conventionalism and the Logical Positivists", in: Greffe J. L./Heinzmann G./Lorenz, K. (1996), 333–344.

Giedymin, Jerzy (1991), "Geometrical and Physical Conventionalism of Henri Poincaré in Epistemological Formulation", *Stud. Hist. Phil. Sci.* 22: 1–22.

Greffe, Jean-Louis, Heinzmann, Gerhard and Lorenz, Kuno (1996), ed., *Henri Poincaré. Science and Philosophy.* Berlin/Paris: Akademie Verlag/ Blanchard.

Heinzmann, Gerhard (1986), ed., *Poincaré, Russell, Zermelo et Peano. Textes de la discussion (1906–1912) sur les fondements des mathématiques: des antinomies à la prédicativité.* Paris: Blanchard.

— (1995), *Zwischen Objektkonstruktion und Strukturanalyse.* Göttingen: Vandenhoeck & Ruprecht.

— (2001), "The Foundations of Geometry and the Concept of Motion: Helmholtz and Poincaré", *Science in Context* 14 (3): 457–470.

Johnson, Dale M. (1981), "The Problem of the Invariance of Dimension in the Growth of Modern Topology, Part II", *Archive for History of Exact Science* 25: 85–267.

Ly, Igor (2008), *Géométrie et physique dans l'œuvre de Henri Poincaré.* Thèse, Université Nancy 2.

Mette, Corinna (1986), *Invariantentheorie als Grundlage des Konventionalismus. Überlegungen zur Wissenschaftstheorie Jules Henri Poincarés.* Essen: Die blaue Eule.

Meyer, Manfred (1992), *Leiblichkeit und Konvention. Struktur und Aporien der Wissenschaftsbegründung bei Hobbes und Poincaré.* Freiburg/München: Alber.

Michel, Alain (2004), "La réflexion de Poincaré sur l'espace, dans l'histoire de la géométrie", *Philosophiques* 31 (1):, 89–114.

Nabonnand, Philippe (2000), "La genèse psychophysiologique de la géométrie selon Poincaré", *Histoires de Géométries. Textes du séminaire de l'année 2000.* Paris: Maison des Sciences de l'Homme.

Nye, Mary Jo (1979), "The Boutroux Circle and Poincaré's Conventionalism", *Journal of the History of Ideas* 40: 107–120.

Poincaré, Henri (1887), "Sur les hypothèses fondamentales de la géométrie", reprinted in: Poincaré 1916–1956, vol. 11, 79–91.

— (1898), "On the Foundations of Geometry", *The Monist* IX, 1–43; French transl. reprinted in: Rollet 2002, 5–31 (abbreviated FG).

—— (1902), *La Science et l'hypothèse*. Paris: Flammarion, 1968; Engl. translation: *Science and Hypothesis*. London: The Walter Scott Publ., 1905 (abbreviated SH).
—— (1904), "L'Etat actuel et l'avenir de la Physique mathématique", *Bulletin des Sciences mathématiques* 28: 302–324 (cf. Poincaré [1905] 1958, Ch. VII, VIII and IX).
—— (1905), *La Valeur de la science*. Paris: Flammarion, 1970; Engl. translation: *The Value of Science*. New York: Dover 1958.
—— (1906), "Les Mathématiques et la logique", *Revue de métaphysique et de morale* 14: 294–317; reprinted in: Heinzmann 1986, 79–104.
—— (1906a), "Sur la Dynamique de l'électron", *Rendiconti del Circolo matematico di Palermo* 21: 129–176; reprinted in: Poincaré 1916–1956, vol. IX: 494–550.
—— (1908), *Science et méthode*. Paris: Kimé, 1999; Engl. translation: *Science and Method*. London/New York: Thomas Neldon & Sons, no date.
—— (1910), *Savants et écrivains*. Paris: Flammarion, 1910.
—— (1913), *Dernières pensées*. Paris: Flammarion, 1963; Engl. translation: *Mathematics and Science: Last Essays*. New York: Dover 1963.
—— (1921), "Analyse des travaux scientifiques de Henri Poincaré faite par lui-même", *Acta mathematica* 38: 36–135.
—— (1916–1956), *Œuvres*, publiées sous les auspices de l'Académie des Sciences. Paris: Gauthier-Villars.
Pouivet, Roger (1995), "Conventionality and Reflective Equilibrium (On Poincaré and Goodman)", in: K. Zamiara (ed.), *O Nauce i Filozofii Nauki*. Poznan: Fondacia "Humaniora", 243–253.
Rollet, Laurent (2000), *Des Mathématiques à la Philosophie. Etude du parcours intellectuel, social et politique d'un mathématicien au début du siècle*. Villeneuve d'Asq: Presses Universitaires du Septentrion.
—— (2002), ed., *Henri Poincaré. Scientific Opportunism/L'opportunisme scientifique. An Anthology. Compiled by Louis Rougier*. Birkhäuser: Basle.
Rougier, Louis (1920), *La Philosophie géométrique de Henri Poincaré*. Paris: Alcan.
Vuillemin, Jules (1968), "Preface", in: Poincaré [1902] 1968. Paris: Flammarion, 7–19.
—— (1970), "Preface", in: Poincaré [1905] 1970. Paris: Flammarion, 7–15.
Zahar, Elie (2001), *Poincaré's Philosophy*. Chicago/La Salle: Open Court.

Hypothesis and Convention in Poincaré's Defense of Galilei Spacetime

Scott Walter[1]

Abstract:

According to the conventionalist doctrine of space elaborated by the French philosopher-scientist Henri Poincaré in the 1890s, the geometry of physical space is a matter of definition, not of fact. Poincaré's Hertz-inspired view of the role of hypothesis in science guided his interpretation of the theory of relativity (1905), which he found to be in violation of the axiom of free mobility of invariable solids. In an effort to save the Euclidean geometry that relied on this axiom, Poincaré extended the purview of his doctrine of space to cover both space and time. The centerpiece of this new doctrine is what he called the "principle of physical relativity," which holds the laws of mechanics to be covariant with respect to a certain group of transformations. For Poincaré, the invariance group of classical mechanics defined physical space and time (Galilei spacetime), but he admitted that one could also define physical space and time in virtue of the invariance group of relativistic mechanics (Minkowski spacetime). Either way, physical space and time are the result of a convention.

Perhaps more than any other figure in contemporary science, Henri Poincaré focused the attention of both philosophers and scientists on the role played by hypothesis in the pursuit of scientific knowledge. The history of twentieth-century philosophy of science is marked by his conventionalist philosophy of geometry, which troubled philosophers from Ernst Cassirer, Moritz Schlick and Hans Reichenbach in the 1910s and 1920s, Philipp Frank, Ernest Nagel and Adolf Grünbaum in the 1950s and 1960s, Lawrence Sklar, Hilary Putnam, David Malament, Michael Friedman and others from the 1970s and 1980s up to the present.[2] Much of this philosophical discussion is concerned with the conventionality of simultaneity in relativity theory, a problem distinct, on one hand, from that of the conventionality of simultaneity in classical mechanics (first discussed by Poincaré in 1898), and on the

1 My thanks go to Yemima Ben-Menahem, Olivier Darrigol, Arthur Fine, Gerhard Heinzmann, Alberto Martìnez, Erhard Scholz, and the editors for discussion and insightful comments.

2 For a comprehensive overview, see Ben-Menahem (2006).

other hand, from that of the Riemann-von Helmholtz-Lie problem of space, which occupied Poincaré and his contemporaries in the late nineteenth century. At the end of his life, Poincaré fused these two problem sets, and realized the overriding necessity of a spacetime convention for the foundations of physics. The following account of Poincaré's progress toward the latter view proceeds chronologically, beginning with the elaboration of his doctrine of physical space (1880–1900), followed by a discussion of Poincaré's understanding of the hypothetical basis of the theory of relativity (1900–1906), and an analysis of the 1912 lecture "L'espace et le temps," in which Poincaré affirmed the central role of hypothesis and convention in the production of scientific knowledge.

1. Poincaré's Doctrine of Physical Space

Poincaré's philosophy of geometry first took form following French debates in the 1870s over the logical coherence and physical meaning of non-Euclidean geometry. While no French mathematician had been directly involved in the reevaluation of the foundations of geometry of the 1820s and 1830s, the ideas advanced by Bernhard Riemann, Eugenio Beltrami, and Hermann von Helmholtz found both partisans and opponents in late nineteenth-century France.

By 1869 at the latest, the French mathematical establishment had recognized the existence of non-Euclidean geometries. In that year the French Academy of Sciences published for the very last time a note purporting to prove the parallel postulate. Its author, Joseph Bertrand, held the chair of general and mathematical physics at the Collège de France and was also Professor of Analysis at the École polytechnique. Bertrand published a demonstration of the parallel postulate over the protests of fellow members of the geometry section, but the ensuing public scandal eventually led him to admit that the proof had met with less than universal approval (Pont 1986, 637).

Debate over the status of non-Euclidean geometry continued well after 1870, particularly among French philosophers. Paul Tannery's empiricism met with opposition from neo-Kantians Charles Renouvier and Louis Couturat, whose pet claim was that only Euclidean geometry could be objective, because it was the only geometry subtended by spatial intuition. Euclidean geometry was thereby an ideal science, and an example of synthetic a priori knowledge in the Kantian scheme.[3]

It was also in the 1870s that France's most brilliant mathematician came of age. Henri Poincaré (1854–1912) was educated at the École polytechnique, under the tutelage of Charles Hermite, Henri Résal, and Alfred Cornu, and at the École des mines, where Henry Le Chatelier taught chemistry. In 1879 he defended a doctoral thesis supervised by Gaston Darboux on the geometric theory of partial differential equations, and after a short stint as a mine inspector in northeastern France, was engaged to teach mathematics at the University of Caen (Rollet 2001).

Less than a year after arriving in Caen, Poincaré entered the competition for the grand prize in mathematical sciences organized by the French Academy of Sciences, which required contestants to "perfect an important element of the theory of linear differential equations in one independent variable." His submission did not win the prize, even though the three supplements to his prize essay established a new class of automorphic functions, that Poincaré called "Fuchsian" functions, in honor of the German mathematician Lazarus Fuchs.

The connections between Fuchsian functions and conventionalist philosophy of geometry are numerous, as Zahar (1997) shows. What Poincaré's supplements reveal is that as early as 1880, Poincaré understood geometry in terms of groups of transformations. Fuchsian functions, Poincaré discovered, are invariant under a certain class of linear transformations that form a group. The study of the group in question reduces to that of the translation group of hyperbolic geometry, prompting the young Poincaré to ask (1997, 35):

> Just what is, in fact, a geometry? It is the study of the group of operations formed by the displacements a figure can go through without deformation. In Euclidean geometry this group reduces to rotations and translations. In Lobachevsky's pseudogeometry it is more complicated.

Geometry is reduced here to group theory, in the spirit of Felix Klein's Erlanger Program, although it appears unlikely that Poincaré had any knowledge of this program at the time. Other printed sources on non-Euclidean geometry were available to him, including French translations of Beltrami and von Helmholtz, and during his student days in

3 On the reception of non-Euclidean geometry among francophone Neokantians, see Panza (1995).

the late 1870s, Poincaré may well have picked up from his teachers (Hermite, Darboux, Jordan) the idea that the motions of a rigid body form a group. There is a strong resemblance between Poincaré's presentation of his model of hyperbolic geometry and that of Beltrami. Both are circle models, differential-geometric and refer to Lobachevsky alone, and although Poincaré does not mention Beltrami by name, the latter's work was surely an inspiration to him (Gray and Walter 1997).

Poincaré's prodigious discovery of Fuchsian functions propelled him into the higher echelons of French mathematics. In 1886 he became a full professor of mathematical physics and probability calculus at the Sorbonne (replacing Gabriel Lippmann), and was elected President of the French Mathematical Society. The next year he was selected to replace Edmond Laguerre in the geometry section of the Academy of Sciences at the *Institut de France.*

Once a member of the *Institut*, Poincaré published his first essay on the foundations of geometry. Strongly influenced by Sophus Lie's writings, the paper concludes with a reflection on the relation between geometry and the physical space of experience (1887, 91):

> [I]n nature there exist remarkable bodies called *solids* and experience teaches us that the diverse motions these bodies can perform are related very closely to the diverse operations of the [Euclid] group. ...
>
> Thus the fundamental hypotheses of geometry are not experimental facts, and yet it is the observation of certain physical phenomena that picks them out from all possible hypotheses. [Original emphasis.]

Geometry is then essentially an abstract science, being the study of groups of transformations. Observation of displaced solids suggests the transformations of one particular geometry, corresponding to the Euclid group. Or as Poincaré put it in a subsequent paper, our experience "played but a single role: it served as an occasion" (1903, 424).

While Poincaré's stance on the formal nature of geometry is unambiguous in his 1887 essay, he does not reflect on the epistemological status of the geometry of physical space. The latter topic is first evoked in Poincaré's next essay on the foundations of geometry, which appeared in Louis Olivier's *Revue générale des sciences pures et appliquées,* and reached a wide readership both in French and English, thanks to a translation published in *Nature.*[4]

4 Poincaré (1891; 1892a). On Poincaré's collaboration with the periodicals edited by Louis Olivier and Xavier Léon see (Rollet 2001).

In the essay "Non-Euclidean geometries," Poincaré presents his conventionalist view of the foundations of geometry and physics. Reiterating his belief in the abstract nature of geometry, Poincaré explains why geometry cannot be an experimental science (1891, 773):

> If geometry were an experimental science, it would not be an exact science, it would be subject to incessant revision. And that is not all: it would even today be shown to be erroneous, since we know that rigorously invariable solids do not exist.

If we were to adopt the empiricist approach to geometry, Poincaré tells his readers in the latter passage, we would eliminate Euclidean geometry straightaway as a candidate geometry for physical space, since there are no perfectly-rigid solids in this space.

When Poincaré points out this conflict between the motion of solids and Euclidean geometry, he bruises our confidence in the truth of the geometric axioms. Worse news is yet to come, however, as Poincaré goes on to tell us that the axioms have no truth-value at all (p. 773):

> *The geometrical axioms are then neither synthetic a priori judgments nor experimental facts.* They are *conventions*; our choice, out of all possible conventions, is *guided* by experimental facts; but it remains *free* and is limited only by the necessity of avoiding all contradiction ... In other words, *the axioms of geometry ... are but definitions in disguise.* This being so, what should one think of the question: "Is Euclidean geometry true?" It is meaningless. [Original emphasis.]

In other words, as Nabonnand (2000) observes, inasmuch as Euclidean geometry is an abstract science, the truth of its theorems may not be ascertained by empirical means.

The same is true for the axioms of non-Euclidean geometry, and to argue the point, Poincaré asks what would happen if the parallax of a given star were observed to have a negative value (corresponding to elliptic space), or if all parallaxes were observed to be greater than a certain positive value (corresponding to hyperbolic space). The answer seems obvious to Poincaré: rather than consider space to be curved, scientists would find it "more advantageous" to suppose that starlight does not always propagate rectilinearly (1891, 774). Poincaré implicitly assumes an alternative, non-Maxwellian optics to be feasible; he later extends the latter assumption to all of physics, by claiming that any experiment at all can be interpreted with respect to either Euclidean or hyperbolic space.

There are two aspects to Poincaré's conjecture I want to underline. First, scientists are free in Poincaré's scheme to choose between the two couples: Euclidean geometry and non-Maxwellian optics, or non-Euclidean geometry and Maxwellian optics. Either way, the geometry of space and the laws of optics result from a convention. In essentials, as Torretti notes (1984, 169), Poincaré's view is equivalent to that of von Helmholtz, to whom Poincaré refers his readers.[5] It also previews Pierre Duhem's holistic view of the structure of physical theory (1906), which led Duhem to reject the possibility of crucial experiments.

Modern commentators like Zahar (2001, 100) understand Poincaré's conjecture to imply a "geometry plus physics" argument ranging over all possible geometries, including those of variable curvature, of the type Einstein employed in his general theory of relativity (1915). For Poincaré and others in the 1890s, however, geometries of *constant* curvature (i. e., Euclidean, hyperbolic, and spherical geometries) were the only plausible candidate geometries for physical space. The motivation for the latter restriction came from physics, which requires a theory of measurement. In the classical physics of von Helmholtz and Poincaré, the act of measurement required free mobility of solids. In turn, free mobility of solids is possible only in Riemannian geometries of constant curvature. As we will see in § 4, the motivation for relaxing the principle of free mobility of solids also came from physics: the physics of relativity.

In the second place, while Poincaré recognized the freedom of scientists to choose a non-Euclidean geometry, he seems convinced that they would never do so. Poincaré's overweening confidence in the convenience of Euclidean geometry for representing natural phenomena, come what may, separates him from most physicists and mathematicians of the late nineteenth century. It is often considered the weak link in Poincaré's philosophy of geometry.[6]

5 Poincaré's notion of phenomenal space as an inseparable couple formed by geometry and physics was foreshadowed by von Helmholtz's appeal to Lipschitz's argument in favor of a dynamics of hyperbolic space, based on the applicability of Hamilton's principle to spaces of nonzero constant curvature (1995, 238). On von Helmholtz's empiricist philosophy of mathematics and geometry, see Volkert (1996) and Schiemann (1997). On Poincaré's reading of von Helmholtz, see Heinzmann (2001).

6 Poincaré's failure to convince scientists and philosophers to adopt his doctrine is noted by Torretti (1984, 256); his failure to convince mathematicians and phys-

Poincaré presented a number of arguments in favor of the convenience of Euclidean geometry, none of which could be considered compelling. One putative advantage of Euclidean geometry is its simplicity, which Poincaré characterized algebraically (Torretti 1984, 335). In the Euclid group, certain "displacements are interchangeable with one another, which is not true of the corresponding displacements of the Lobachevsky group" (1898b, 43). In other words, the Euclid group has a proper normal subgroup corresponding to translations, and according to this definition of simplicity, it is simpler than the hyperbolic group. Alternative criteria for simplicity abound, however, and Poincaré admitted quite freely that we would switch geometries, if confronted with a "considerably different" empirical base (1898b, 42). In sum, Poincaré's stance on the convenience of Euclidean geometry was no dogma. Rather, it reflected his great confidence in the future stability of the established baseline of experimental results, and in the explicative power of the principles of physics then in vigor.

Just what sort of experimental results might have led Poincaré to forgo Euclidean geometry at the close of the nineteenth century? Observations of stellar parallax could not have forced such a change, as we have seen. Poincaré did not elaborate the possibilities; instead, he argued for the possibility of doing physics in hyperbolic space, which was an area of research little explored by nineteenth-century mathematicians (Walter 1999b, 92).[7]

Poincaré's view on the question of the equivalence of Euclidean and hyperbolic geometry is subject to debate. According to Ben-Menahem (2006, 41), Poincaré held all theorems of Euclidean geometry to have counterparts in hyperbolic geometry (and vice-versa). This equivalence in theorems would provide a template of sorts for the elaboration of a physics of hyperbolic space. Yet Poincaré never actually proposed such an equivalence (Torretti 1984, 336). Instead, in an attempt to characterize quadratic geometries, Poincaré wrote that in hyperbolic geometry we have a set of theorems "analogous" to those of Euclidean geometry (1887, 205).

What Poincaré offered in favor of his doctrine of space was a clever thought experiment that builds on a suggestion made by von Helm-

icists is shown by Walter (1997). For the characterization of the doctrine as a weak link, see Vuillemin (1972, 179) and Sklar (1974, 93).

7 Exceptions include Eugenio Beltrami, Wilhelm Killing, and Rudolf Lipshitz, all of whom contributed to the mechanics of non-Euclidean space.

holtz. By wearing glasses fitted with convex lenses, von Helmholtz wrote (1995, 242), we can experience the optical effects of a world in which the natural geometry of space is non-Euclidean. Poincaré (1892b) deftly modified von Helmholtz's example by imagining an apparently non-Euclidean world, that of a heated sphere. In this way, Poincaré refocused his reader's attention away from intuition – a subject dear to von Helmholtz – and toward a subject of his own predilection, the conventionality of laws of physics.

The heated sphere model of hyperbolic space fascinated Poincaré's contemporaries. Imagine a hollow sphere of radius R, heated in such a way that the absolute temperature at a point located a distance r from the sphere's center is proportional to R^2-r^2.[8] All bodies inside the sphere have the same coefficient of thermal dilation, and reach thermal equilibrium instantaneously. The atmosphere is such that the index of refraction is everywhere proportional to the reciprocal of temperature, and the trajectory of a light ray is described by an arc orthogonal to the enclosing sphere. The same arcs are materialized by the shortest distance between two points, as measured by a ruler, and the axis of rotation of a solid body. Poincaré suggests that the natives of such a world would adopt hyperbolic geometry for their measurements.

Highly contrived from a physical standpoint, Poincaré's model conveys quite well the idea that the adoption of Euclidean geometry is conditioned by certain features of our environment (such as the motion of solids). What is more, Poincaré (1895, 646) claimed that if physicists from planet Earth were transported to the heated world, they would continue to use Euclidean geometry, on the grounds that this would be the most convenient option available to them.[9]

The latter thesis is contested by Howard Stein (1987) on the basis of the extreme complexity of doing physics with Euclidean geometry in such a world. A terrestrial, pre-relativist physicist in the heated sphere would be led at first to posit a universal deforming force, only to find that the other laws she contrives conflict with those elaborated by physicists located in other regions of the world. Implicitly, Stein introduces a meta-theoretical commitment to a unified physics, and while Poincaré valued unified, interpenetrating explanations of physical phenomena, he

8 For a rigorous discussion of the heated sphere's two-dimensional counterpart, the heated disk, see Barankin (1942).

9 At first, Poincaré (1892b) maintained only that the sphere's natives would adopt non-Euclidean geometry.

recognized that such values are not imposed on us by the same phenomena. His doctrine of physical space affirms only that Earth-educated physicists *can* use Euclidean geometry inside the heated sphere, not that they *must* do so, or even that it is *in their best interest* to do so. As mentioned above, Poincaré readily admitted that physicists would forgo Euclidean geometry if the circumstances called for it. At the turn of the century, he was also quite sure that such circumstances would never arise.

2. Poincaré's Typology of Hypotheses

Poincaré's mature philosophy of science assigns a leading role to hypotheses. In the previous section, we recalled his belief that the axioms of geometry are conventions (or definitions in disguise), and not hypotheses about the behavior of light rays or solid bodies. In this context, the question naturally arises of the relation between hypothesis and convention.

Hypotheses, according to Poincaré, are not all created equal. Some are more influential than others in determining the course of science, and Poincaré found it useful to categorize the types of hypothesis he encountered according to their truth domain. An important impetus to this theorization of scientific hypotheses was provided by Heinrich Hertz's *Principles of Mechanics* (1894), which made a singular impression on him, in virtue of its epistemic structure, and innovative use of hypothesis (Poincaré 1897, 743):

> While the principles of dynamics have been exposed in many ways, the distinction between definition, experimental truth, and mathematical theorem has never been sufficient. This distinction is still not perfectly clear in the Hertzian system, and what is more, it introduces a fourth element: hypothesis.

This fourth element corresponds to Hertz's assumption of hidden masses, which allowed him to forgo the concept of force. What strikes Poincaré above all is that Hertz's assumption is neither definition, nor experimental fact, nor theorem. It is what Poincaré would later call an "indifferent" hypothesis, in that an alternative hypothesis, or set of hypotheses, leads to the same result.

While Hertz's hypothesis of hidden masses was far too bold for Poincaré's taste, it inspired a new understanding of the role of hypothesis in physics, first unveiled at the international congress of physics held in Paris at the turn of the century. This was the first of two different typologies, which are now addressed in turn.

Poincaré (1900, 1166) identifies three sorts of hypotheses: generalizations, indifferent hypotheses, and natural hypotheses. The first sort is distinguished by the property of susceptibility to experimental corroboration. Once corroborated by experiment, these hypotheses become "fertile truths," without which there can be no increase in knowledge.

The second type of hypothesis, or "indifferent" hypothesis, serves to "fix our thought," but constitutes a non-unique premise in a deductive chain. Examples of indifferent hypotheses include the hidden masses of Hertz's mechanics, and the physical interpretation of the axial and polar vectors in classical optics. It can happen that a theorist prefers one hypothesis to another, to simplify a calculation, for instance, but experimental corroboration of the theory can have no bearing on the truth of the chosen hypothesis.

The third type of hypothesis is the "wholly natural" hypothesis. This vaguely-defined category concerns what might be described as experimental rules of thumb, without which measurement is nigh impossible. Natural hypotheses are accordingly the "last we ought to abandon." They include the law of continuity of cause and effect, and the vanishing force of very remote bodies. The latter hypothesis allows for multiple independent dynamical systems, as we shall see in § 4.

Poincaré's second typology of hypotheses appeared two years after the first, in the introduction to a collection of philosophical essays, *La Science et l'hypothèse* (1902). This revised typology features three categories, including the generalizations and indifferent hypotheses of the first typology, and excluding natural hypotheses. Poincaré may have folded natural hypotheses implicitly into the category of generalizations, since the former are, like the latter, accessible in principle to experiment. It is also possible, however, that neither typology was intended to be exhaustive, as suggested by Poincaré's decision to reprint his first typology in the ninth chapter of *La Science et l'hypothèse*.

A new type of hypothesis appears in Poincaré's 1902 typology: the "apparent" hypothesis. In fact, his third type of hypothesis is not a hypothesis at all, but a definition, or a "convention in disguise." Conventions, or apparent hypotheses, are essential to the activity of theorization in Poincaré's model of science.

For what follows, one final point needs to be underlined concerning the relation of Type I hypotheses (or hypotheses susceptible to experimental confirmation) to Type III hypotheses (or conventions). A physical law, if corroborated by experiment, can become a convention. Such a statutory evolution occurs when the law in question is deemed sufficiently fruitful to warrant protection from new experimental tests. Conventions are of great utility to the pursuit of scientific knowledge, but they are only definitions, and as such, may not be refuted by experiment. As an example, take the law of inertia, which Poincaré considers to be a convention. Any particular observation tending to disconfirm the law of inertia, Poincaré holds, would be dealt with by invoking the effect of invisible bodies, rather than by discarding the law of inertia (1902, 117).

3. Relativity Theory and the Foundations of Geometry

On 5 June 1905, Poincaré presented a note to the French Academy of Sciences that put forth the foundations of the theory of relativity. It is in this short paper that Poincaré expressed the Lorentz transformations in their modern form for the first time, along with the current density transformations (correcting Lorentz). In the 47-page memoir "On the dynamics of the electron" announced by this note, and that appeared in January 1906, Poincaré wrote the velocity transformations, characterized the Lie-algebra of the Lorentz group, and introduced a four-dimensional space in which three coordinate axes are real, and one is imaginary, inaugurating the era of four-dimensional physics. Building on Poincaré's ideas as well as those of Hertz, Lorentz, Einstein and Planck, the Göttingen mathematician Hermann Minkowski elaborated the theory of spacetime, which profoundly marked the philosophy of space and time, and was instrumental to Einstein's discovery of the general theory of relativity (Walter 2007).

Poincaré's contributions to the theory of relativity are well known to historians, but it is not entirely clear how Poincaré understood the theory of relativity to impinge upon his doctrine of space. In part, at least, this is Poincaré's doing, as he did not express himself clearly on this topic. As a result, commentators have offered a wide variety of interpretations of the relation between Poincaré's conventionalist philosophy and his discovery and interpretation of the theory of relativity.[10]

10 For references, and an insightful comparison of Poincaré's and Einstein's philosophical approaches to relativity theory, see Paty (1993).

Although Poincaré does not comment in his 1906 memoir on the eventual influence of the principle of relativity on our conceptions of space and time, he points out an important consequence for the theory of measurement (1906, 132):

> How do we go about measuring? The first response will be: we transport objects considered to be invariable solids, one on top of the other. But that is no longer true in the current theory if we admit the Lorentzian contraction. In this theory, two lengths are equal, by definition, if they are spanned by light in equal times.

Poincaré spies a conflict between the traditional notion of rigidity and the principle of relativity, in that Lorentz-FitzGerald contraction appears to preclude the transport of rigid rulers upon which length measurement depends. Length congruence in Lorentz's theory depends not on the free mobility of invariable solids, but on the light standard.

Does this standard conflict necessarily with Poincaré's doctrine of space? Poincaré will provide an answer to the latter question, but only in the wake of spacetime theory, as discussed in the next section.

4. The Principle of Physical Relativity

From 1905 to the end of his life (on 17 July 1912), Poincaré commented often on the theory of relativity, but only twice on the four-dimensional interpretation he had inaugurated. At first, he compared a possible four-dimensional language for physics to Hertz's mechanics, and observed that working out the corresponding formalism would entail "much pain for little profit" (1907). Poincaré's first and last words on the *philosophical* significance of spacetime were delivered on 4 May 1912, as the second in a series of four lectures at the University of London. His remarks were published posthumously (1912; 1963, 97–109) as "L'espace et le temps," a title recalling that of Minkowski's celebrated 1908 lecture in Cologne, "Raum und Zeit."

Although Poincaré might have derived satisfaction from the fact that Minkowski had based his spacetime theory on essentially the same four-dimensional geometry introduced in Poincaré's 1906 memoir, the Göttingen mathematician had noisily promoted an anti-conventionalist view of physical space and time, which was surely anathema to him (Walter 2009). Like Poincaré, Minkowski recognized that the new mechanics admitted no rigid bodies; unlike Poincaré, Minkowski presented the new intuitions (*Anschauungen*) of space and time not as conventions, but as the result of circumstances both empirical and formal that "forced themselves" upon scientists (1909, 79).

This is just the sort of anti-conventionalist view that Poincaré targets in the opening of his London lecture on space and time (1963, 99):

> Is not the principle of relativity, as Lorentz conceives it, going to impose an entirely new conception of space and time, and force us thereby to abandon conclusions that may have seemed established?

Poincaré sees in Lorentz's principle of relativity a menace to his doctrine of physical space. But what does Poincaré take to be Lorentz's principle of relativity? He defines the latter in terms of group invariance (p. 108):

> The old form of the principle of relativity had to be abandoned; it is replaced by Lorentz's principle of relativity. The transformations of the "Lorentz group" are those that leave unaltered the differential equations of dynamics.

According to Poincaré, Lorentz's principle of relativity is just Lorentz covariance, or what was then understood to be a succinct statement of the content of Einstein's special theory of relativity. Poincaré is poised to reconsider his brief observation of 1905 on the theory of measurement in Lorentz's theory, mentioned above (§ 3), and to show how his doctrine of physical space stands with respect to the theory of relativity.

Poincaré's views on how relativity theory interferes with classical concepts of space and time have significant historical interest, due to his foundational contributions to this theory. His London lecture on space and time represents his final word on this topic, and sets forth a fundamental change in his doctrine of physical space, by extending this doctrine to physical time. This extension is not well known, and is the focus of the following reconstruction of Poincaré's argument.

According to Poincaré, there is a principle of relativity, which he calls the "principle of physical relativity" (PPR), that exists in two forms. One of these is the Lorentz form, which Poincaré refers to as Lorentz's principle of relativity. The other form is what I will refer to as the Galilei form, because it is defined by Galilei group invariance. The PPR holds (p. 102) that the differential equations by which we express physical laws

> are altered neither by a change of fixed rectangular coordinate axes, nor by a change of temporal origin, nor by a substitution of mobile rectangular axes, the motion of which is a uniform, rectilinear translation.

In modern terms, the PPR is equivalent to covariance with respect to a certain group of transformations. Poincaré distinguishes two groups in this context, corresponding to what were later known as the inhomogeneous Galilei transformations of the Galilei group, and the inhomogeneous Lorentz transformations of the Poincaré group.

The PPR has two main features, one of which is its testability. It is what Poincaré describes as an "experimental truth," i. e., a proposition susceptible to experimental disconfirmation. Recalling Poincaré's typology of hypotheses (see above, § 2), the PPR is a Type I hypothesis. He expresses its empirical meaning in terms of two corollaries (p. 106):

> The reciprocal action of two bodies (or mechanical systems) tends to zero as spatial separation increases indefinitely. (1)
>
> Two remote worlds behave as if independent. (2)

Let us examine these corollaries one at a time. The first has a precedent in the 1900 lecture mentioned above, in which Poincaré first sought to characterize the sorts of hypotheses encountered in science. To hold that the "influence of very distant bodies is negligible," Poincaré wrote (1900, 1166), was a "natural" hypothesis. This natural hypothesis was announced some five years before the discovery of relativity theory, and twelve years before Poincaré updated his formula, ostensibly to accommodate the new physics of inertial frames.

The motivation for (1) remained the same over this twelve-year span. This particular natural hypothesis was designed to legislate away the effects of all long-range forces that fall off with increasing separation (such as gravitational and electromagnetic forces), and to create thereby the possibility of separate mechanical systems. As a consequence of what Poincaré calls the "principle of psychological relativity," recognizing the conventional nature of measurements of distance and duration, the existence of distant stars renders inertial frames of reference "purely conventional," obliging us, when we employ the concept of a frame of reference, to forgo "absolute rigor" (p. 103).

The second corollary of the PPR has, like the first, a clear precedent in Poincaré's philosophy. In *La Science et l'hypothèse*, Poincaré analyzed the possibility of generalizing the Galilean principle of relative motion to include rotating frames of reference. If we want to solve a two-body problem based on Newton's Law, Poincaré remarked, we need to know the positions and velocities of the two gravitating bodies, as well as the corresponding initial values, along with "*something else.*" It is this final missing bit of information that worried Poincaré, as Earman observes (1989, 86). This solution element could be either the initial values of acceleration, or the area constant, or the absolute orientation of the universe, or the rate of change of orientation, or the position or velocity of Carl Neumann's Body Alpha. "We have," Poincaré wistfully concluded his earlier analysis, "but a choice of hypotheses" (1902, 137).

Poincaré's position advanced in this case, as well as for (1). His London lecture proposes a new argument in favor of (2), observing in essence, as Kerszberg (1989, 139) notes, that the missing solution element could be objectively determined, if only we disposed of not one universe, but of several universes. Poincaré imagines the situation as follows (p. 105):

> Instead of considering the entire universe, let's imagine small, separate worlds, visible to one another but free from outside mechanical action. If one of these worlds spins, we will see it spin, and recognize that the value to be assigned to the constant just mentioned depends on the spin velocity, and in this way, the convention habitually adopted by dynamicists will be justified.

The PPR, Poincaré realizes here, provides a way out of his earlier dilemma, as it implies the existence of multiple independent mechanical systems in the universe, as expressed by (2).

Along with testability, a second salient feature of the PPR is its capacity to define space and time. The PPR "can serve to define space," by virtue of the fact that we perform measurements (or alternatively, in Poincaré's terminology, "construct space") by displacing solids and defining length congruence as coincidence of figures. The PPR admits the invariance under displacement of the form and dimensions of solids and other sufficiently-isolated mechanical systems, and thereby provides a foundation for length measurement. In Poincaré's words, the PPR provides us with a "new instrument of measurement" (p. 106). Each possible displacement of a solid corresponds to a certain transformation, which leaves the form and dimensions of a given figure invariant. Such transformations, when taken together, form a group: the group of motions of invariable solids. While the principle of free mobility of solids selects any of three motion groups (Euclid, hyperbolic, spherical), the Galilei form of the PPR selects only one of these: the Euclid group. One imagines that Poincaré was uncomfortable with this particular consequence of the PPR, which he did not mention, but could not have ignored. Instead, he stressed the fact that, like the principle of free mobility of solids, the PPR provides a foundation for geometry.[11]

Poincaré justified his replacement of the principle of free mobility of solids by the principle of form-invariance of the differential equations of mechanics in the following way. The motion group of invariable solids on one hand, and the symmetry group of mechanics on the other, give rise to conceptions of space that are not "essentially different" in Poincaré's view, because both groups define space in such a way that solids are unaltered in form when displaced. The role played by solid bodies in the old conception of the foundations of geometry goes over to the more general notion of a mechanical system. In fact, by defining space in terms of the motion group of solids, we affirm that the equations of equilibrium of solids do not vary upon displacement. In other words, we define space in such a way that the equilibrium equations of solids are unaltered by a change of axes. These equations of equilibrium are but a special case of the general equations of dynamics, Poincaré explains, and "according to the principle of physical relativity, they must not be modified by this change of axes." There is, consequently, no essential difference for Poincaré between the old way of defining space (postulating free mobility of solids), and the new one (postulating the symmetry group of mechanics), as far as solids are concerned.

Leaving solids behind, there are two significant differences between these two groups. The symmetry group of mechanics offers greater coverage than the motion group of solids, ranging over both solids and mechanical systems. This juxtaposition of the Euclid group and the symmetry group of mechanics was probably inspired by Minkowski's presentation of the theory of spacetime, which contrasted the Euclid and Galilei groups, although Poincaré neglected to mention Minkowski by name.[12]

The conception of space based on the symmetry group of mechanics differs from the one based on the Euclid group in a second respect, in that it

11 I differ here from Friedman (2008, 214), for whom "Poincaré's conception of space and geometry is entirely based … on the principle of free mobility first formulated by von Helmholtz and then brought to precise mathematical fruition in the Helmholtz-Lie theorem."

12 Minkowski likewise neglected to mention Poincaré in his Cologne lecture, but acknowledged his contributions to relativity, along with those of Lorentz, Einstein, and Planck, in earlier papers (Walter 1999a).

> defines not only space, it defines time. It tells us the meaning of two simultaneous instants, of two equal times, or of a time twice as great as another.

Both space and time are defined by the new view based on the symmetry group of mechanics, and this is significant for students of Poincaré's philosophy, as Paty (1996, 129) underlines, because he had never before admitted that the choice of an invariance group could define space and time. By preferring the symmetry group of mechanics to the motion group of solids, Poincaré considers the laws of mechanics to be more fundamental to our understanding of the world than the axioms of geometry. Put another way, Poincaré finds spacetime to be more fundamental than ordinary Euclidean space.

It may appear at this point that Poincaré has decided to renounce his doctrine of physical space in favor of the PPR. This is not so. The PPR, he explains, is an "experimental fact," and as such, it is "susceptible to incessant revision." The type of revision Poincaré has in mind implies a modification of the geometry of physical space. He was probably thinking of Einstein's program, announced in 1907, to generalize the principle of relativity to uniformly-accelerated frames of reference, and which led him to predict that rays of starlight must bend around the sun. What worries Poincaré is not so much a modification of the PPR as the consequential revision of the geometry of space. For the geometry of space to become immune to revision, the PPR must itself become immune to revision. In Poincaré's philosophical scheme, the only way to render an empirical law immune to revision is to promote it to conventional status. Naturally, this is what Poincaré decides to do, when he writes (p. 107):

> [Geometry] must become a convention again, [and] the principle of relativity must be considered as a convention.

To drive this point home, Poincaré imagines a long-range force that diminishes at first with distance, then increases, producing motion inconsistent with the PPR's first empirical corollary (1). The PPR would then "appear to us as a convention" (*le principe se présente à nous comme une convention*, p. 107), rather than as a Type I hypothesis, ostensibly because we would take measures to save the PPR from any experimental threat, by introducing a hidden mechanism, for example.

What Poincaré asks us to do, in other words, is to reconsider the epistemic status of the PPR, understood as an experimental, Type I hypothesis, and to promote it to a Type III, or apparent hypothesis (i. e., a convention). This promotion means that henceforth, the PPR is immune to empirical disconfirmation. It also means that the geometry of physical space cannot be determined empirically, in complete compliance with Poincaré's doctrine of physical space.

Having explained how the doctrine of space may be salvaged by elevating the PPR to conventional status, Poincaré finally entertains a discussion of the Lorentz form of the PPR. The "recent progress in physics" has brought about a "revolution": Lorentz's principle has replaced the old one. To paraphrase in modern terms, Lorentz covariance has replaced Galilei covariance.

In the same way as the PPR with Galilei covariance can define space and time, the PPR with Lorentz covariance can define space and time (p. 108):

> It is as if time were a fourth dimension of space; and as if the four-dimensional space resulting from the combination of ordinary space and time could rotate not only about an ordinary space axis, in such a way that time is unaltered, but about any axis at all. To get a mathematically exact comparison we would have to assign purely imaginary values to this fourth space coordinate; the four coordinates of a point of our new space would not be x, y, z, and t, but x, y, z, and $t\sqrt{-1}$.

A four-dimensional vector space corresponding to the above description was introduced by Poincaré in the final section of his 1906 memoir on the dynamics of the electron, as a means of identifying Lorentz-invariant quantities to be used in a relativistic law of gravitational attraction. The new space he refers to in his London lecture, however, is not his own, but that of Minkowski, as Paty observes (1996, 132). This much may be inferred from Poincaré's remark that in the new mechanics, and contrary to his earlier (pre-relativistic) analysis of simultaneity relations (1898a), there are events which can be neither the cause nor the effect of other given events. It was Minkowski who first identified such events, situated in a region of spacetime unique to what he called "spacelike" (*raumartigen*) vectors (Fig. 1). This insight was essentially tied to Minkowski's spacetime theory, providing apodictic proof of its fertility.[13]

Poincaré says no more about Minkowski spacetime, which in 1912 was not yet well-known in Great Britain, but had already captured the attention of relativists in Germany and France, including Poincaré's former students Paul Langevin, a physicist at the Collège de France, and

13 Minkowski (1908, § 6; 1909, § III). Poincaré did not employ the Minkowskian term "spacetime" or any of its linguistic variants.

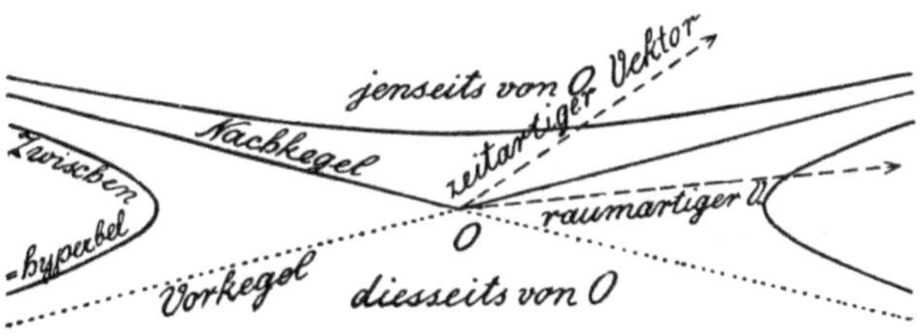

Figure 1: Lightcones in Minkowski spacetime (Minkowski 1909)

Émile Borel, a mathematician at the Sorbonne. He closes his lecture with the following words (p. 109):

> What will our stance be with respect to these new conceptions? Are we going to be forced to modify our conclusions? Of course not: we adopted a convention because it seemed convenient, and we said that nothing could oblige us to abandon it. Nowadays certain physicists want to adopt a new convention. It's not that they are obliged to do so, it's just that they judge the new convention to be more convenient. Those who feel differently may legitimately retain the old convention, so as not to disturb their habits. Between us, I believe this is what they will do for a good while longer.

In the wake of relativity theory, there is, as ever, no fact to the matter of the geometry of physical space, just a principle of (physical) relativity with a choice of invariance group. Poincaré insists that his earlier conclusions need not be modified, and this is true, but in fact he has replaced a convention on space with a convention on spacetime.

The above-cited conclusion of Poincaré's London lecture on space and time raises a number of questions. First of all, does Poincaré's adoption of the PPR with Galilei covariance signal his disavowal of Lorentz covariance? Not at all: Poincaré distinguishes, as shown above, between the PPR as a Type I hypothesis, on one hand, and as a definition of spacetime, on the other. Undoubtedly, the same distinction applies if one associates the PPR with the Galilei group or the Lorentz group. By 1912, after some public hesitation, Poincaré had convinced himself of the experimental soundness of the new mechanics based on Lorentz covariance (1963, 110–111). The PPR with Lorentz covariance was then a viable candidate for elevation to conventional status. But instead of defining space and time in virtue of Lorentz covariance, he prefers to define space and time in virtue of Galilei covariance.

Was such a position coherent at the time? Galilei and Lorentz conventions apply to the same inertial frames, and the quantities measured therein are either real or apparent, depending on the convention. Poincaré's interpretation of such quantities remained what we might call "apparentistic," in that the only true quantities were those of the ether frame. In principle, as Carl Neumann admitted in 1869, any inertial frame at all may be designated as the absolute (or ether) frame (Barbour 1989, 653), although Poincaré does not spell this out. Instead, he maintains (p. 99) that deformation of measuring devices due to motion with respect to absolute space can occur in such a way that this motion can never be detected.[14] The latter proposition is itself a corollary of the PPR, although Poincaré does not present it as such.

From a superficial point of view, Poincaré's position may appear inconsistent, in that he postulates the PPR with Galilei covariance (or Galilei spacetime) while affirming the experimental validity of Lorentz covariance. If we recall his typology of hypotheses, however, the consistency of Poincaré's view is readily apparent. Considered as a Type I hypothesis, Lorentz covariance is an open question for experimenters. Consistency requires only that Lorentz-transformed space and time coordinates be interpreted with respect to the Galilei version of the PPR, the latter principle being understood as a Type III hypothesis. Poincaré satisfies this minimal requirement by referring to quantities measured in inertial frames as "apparent" quantities, with the "true" quantities belonging to the ether frame.

If we grant that Poincaré's defense of Galilei spacetime is both relativist and consistent, it may still appear convoluted, in that one could forgo the indirection of apparentism by adopting the Lorentz version of the PPR (or Minkowski spacetime). By doing so, quantities measured in inertial frames are "true," and the concept of ether is rendered wholly superfluous. I suspect, however, that in 1912 Poincaré's position appeared less convoluted to most physicists than the latter one, due to the conceptual inertia of three centuries of uncritical acceptance of absolute space and time.

The PPR with Lorentz covariance, attributed by Poincaré to "certain physicists," defines space and time in terms of a four-dimensional Minkowski spacetime geometry. The leading proponents of Minkow-

14 For a reevaluation of Poincaré's dynamic approach to relativity, see Brown (2005). Although Brown claims Poincaré never recognized time dilation (p. 147), this effect appears to have been no less real for him than that of length contraction. See, for example, Poincaré's remarks on "The new mechanics," delivered in Lille on 2 August 1909 (1909, 173).

ski's spacetime theory in 1912 included Poincaré's correspondent Arnold Sommerfeld, Minkowski's former students Max Born, Max von Laue, Gunnar Nordström, and Theodor Kaluza, Minkowski's former colleagues Max Abraham and Gustav Herglotz, the Greifswald physicist Gustav Mie, and the mathematical physicist Philipp Frank in Vienna. One has to wonder about Poincaré's characterization of the spacetime theorists understanding of spacetime. Did these theorists consider Minkowski spacetime to be a convention in Poincaré's sense? If this were so, Lorentz covariance could no longer have been a subject of experiment for them. In fact, well before Poincaré's speech, and well after, several physicists and astronomers were engaged in verifying consequences of the theory of relativity. Consequently, Poincaré's attribution of a conventional view of space and time to spacetime theorists is probably best understood as wishful thinking.

Einstein, like most physicists at the time, understood Lorentz covariance as a hypothesis subject to experimental corroboration. In a letter to his friend Friedrich Adler, a Machian anti-relativist imprisoned for the cold-blooded assassination of the Prime Minister of the Austro-Hungarian Empire, Einstein denied that Lorentz covariance was conventional. Referring to the Lorentz transformations in the form

$$x' = \ell\beta(x - ut) \quad y' = \ell y \quad z' = \ell z \quad t' = \ell\beta(t - ux/c^2)$$

where $\beta = \sqrt{1-u^2/c^2}$, Einstein focused his argument on the nature of the constant ℓ:

> It is clear in any event that the choice of ℓ implies no mere formal convention, but a hypothesis characterizing reality ... Thus Bucherer, for example, backed a theory for a while, which comes out of a different choice of ℓ. Nowadays there is no further question of a different choice of ℓ, since the electron's laws of motion have been verified with increased precision.

Einstein does not refer to Poincaré in his letter to Adler, but contemporary correspondence shows that he was familiar with the Frenchman's philosophy of science. The cited passage expresses concern over the epistemological status of Lorentz covariance, which Einstein considers to be a well-verified physical hypothesis, and not a mere definition or convention.[15]

15 Einstein to F. Adler, 29.08.1918 (Einstein 1998, Doc. 628). A month after writing to Adler, Einstein agreed with the German mathematician Eduard Study's criticism of Poincaré's doctrine of space; see Einstein to E. Study, 25.09.1918 (Einstein 1998, Doc. 624). But a few months later, Einstein re-

5. Why Galilei Spacetime?

The two principal approaches to special relativity in 1912, associated, on one hand, with Lorentz and Poincaré, and on the other hand, with Einstein and Minkowski, may be distinguished by their respective ontologies, but also by their performance. We have seen that Poincaré considered two forms of the principle of physical relativity: a Galilean form, defining Galilei spacetime, and a Lorentzian form, defining Minkowski spacetime. We have also seen that Poincaré preferred one form to the other. Since his reasons for preferring Galilei spacetime to Minkowski spacetime are not readily discernible in his London lecture on space and time, we are prompted to look elsewhere in order to understand his view.

The first historian to hazard an explanation for Poincaré's choice was Gerald Holton, who sought to explain Poincaré's attachment to the concept of ether as a consequence of his conservative outlook on science, and described the French mathematician as the "most brilliant conservator of his day" (1973, 189). Holton seems not to have known that Poincaré was in his day one of the best-known critics of the foundations of physics, alongside Ernst Mach and Heinrich Hertz. Poincaré's fellow scientists considered him to be the most lucid of theoretical physicists, on the leading edge of the latest discoveries (Walter et al. 2007). In Émile Borel's opinion, for example, Poincaré "contributed more than anyone to the creation of what may be called the spirit of twentieth-century theories of physics, as opposed to those of the nineteenth century" (1924). In light of such views, it seems impossible to explain Poincaré's attachment to the ether – or Galilei spacetime – as the result of a conservative tendency.

Recent studies of Poincaré's scientific activity suggest a quite different way of understanding his preference for Galilei spacetime. A top graduate of Paris's elite, state-run engineering schools, Poincaré was more skilled than most in the practical arts of civil and mechanical engineering. From the 1890s on he contributed to engineering journals, taught electrical engineering and urged fellow scientists to "increase the output of the scientific machine," as Galison aptly notes (2003,

---

marked to the philosopher Hans Vaihinger that Poincaré's view of the role of Euclidean geometry in science was "wesentlich tiefer" than that of Study (Saß 1979, 319). On Vaihinger and Poincaré, see the contribution by Christophe Bouriau in the present volume.

201). In almost identical terms, Poincaré described the scientific role of theoretical physics, whose duty it was to "guide generalization in such a way as to increase the output ... of science" (1900, 1164).

Polytechnicians like Poincaré were trained to identify the options most likely to enhance productivity, and while Poincaré certainly had several evaluative criteria at hand when comparing Galilei and Minkowski spacetime, that of scientific productivity was likely to have been a leading candidate. From any reasonable standpoint, looking back in time from 1912, the Galilei convention had as much to recommend it as the Lorentz convention, since both approaches could claim relativistic theories of the electron, mechanics, electrodynamics of moving media, and gravitation.

Certain facts of life, however, would have argued against the Galilei convention. For instance, the publishing trend in theoretical relativity favored Minkowski spacetime (Walter 1999b), and the brightest young German and French theorists were either convinced Minkowskians, or were, like Einstein, soon to adopt a Minkowskian spacetime ontology. Poincaré was undoubtedly aware of these facts, and his lecture in London can be read as an effort to stem the tide of Minkowskian relativity. It turned out to be his final effort, as his life ended ten weeks later. He was a skilled and perceptive critic on scientific matters, and one cannot help but speculate upon how he would have reacted to two signal developments that took place the year after his death. Of the two events I have in mind, one would appear to underline the cogency of Poincaré's preference for Galilei spacetime, and the other, its drawbacks.

The limitations of the Galilei form of the PPR are most obvious when Einstein's general theory of relativity is taken into account, the first elaborate expression of which was Einstein and Großmann's *Entwurf* theory (1913). Like almost all of his contemporaries, Poincaré was unprepared for a theory of spacetime as a four-dimensional, pseudo-Riemannian manifold with curvature determined by the distribution of matter and energy. In a sense, the ultimate success of Einstein's theory was also that of the Lorentz form of the PPR, in that Einstein expected Lorentz covariance to hold in the limiting case of weak gravitational fields. At the time of Poincaré's London lecture, however, the secular advance of Mercury's perihelion was an anomaly under both the Galilei and Lorentz forms of the PPR, as Poincaré was the first to point out (Walter 2007, 208).

The second development is that of Niels Bohr's first model of the atom, which discarded a central tenet of classical physics by confining the electron in a hydrogen atom to one of a discrete set of circular orbits, in ordinary space and time. Arnold Sommerfeld altered the model in 1915–1916 to allow for elliptical orbits precessing relativistically, and found an explanation for the fine structure of the spectral lines of hydrogen. In effect, Sommerfeld extended the reach of relativistic dynamics to the inner regions of the hydrogen atom. As a bonus of sorts, the sophisticated analytical methods developed by Poincaré for celestial mechanics in the 1890s soon found application in a modified version of the Bohr atom (Darrigol 1992, chap. 6).

There is ample reason to believe that Poincaré was prepared for a theory such as Bohr's. In the fall of 1911, he participated in discussions of the theory of quanta as a member of the First Solvay Council, along with Einstein, Planck, Sommerfeld, Nernst and others. As Staley (2005) observes, Poincaré was struck by the fact that these physicists already referred to relativistic mechanics as the "old mechanics," the new mechanics being those of energy quanta. Shortly after the Solvay Council, Poincaré showed (as did Paul Ehrenfest) the quantum hypothesis to be necessary and sufficient for the establishment of Planck's law (Prentis 1995). Once Poincaré recognized the mechanics of quanta to be incompatible with *both* ordinary and relativistic mechanics (1963, 111, 125), he had all the more reason to emphasize the conventional nature of both Galilei and Minkowski spacetime.

The principle of physical relativity, which encodes a spacetime view of physical phenomena, was largely ignored in Poincaré's time, while his basic insight later informed a great number of investigations in theoretical and mathematical physics. Poincaré's early study of the role of hypothesis in the physical sciences, inspired in part by his reading of Hertz's mechanics (as shown in § 2), served as an essential resource for his elaboration of the principle of physical relativity, and exemplifies the close intertwining of philosophical reflection and physical understanding at the forefront of research in the natural sciences.

Bibliography

Barankin, E. W. (1942), "Heat flow and non-Euclidean geometry", *American Mathematical Monthly* 49: 4–14.

Barbour, Julian B. (1989), *Absolute or Relative Motion? Vol. 1: The Discovery of Dynamics*. Cambridge: Cambridge University Press.

Ben-Menahem, Yemima (2006), *Conventionalism: From Poincaré to Quine*. Cambridge: Cambridge University Press.

Borel, Émile (1924), "Henri Poincaré", in *Cinquantenaire de la Société Mathématique de France*. Paris: Gauthier-Villars, pp. 49–54.

Brown, Harvey R. (2005), *Physical Relativity: Space-time Structure from a Dynamical Perspective*. Oxford: Oxford University Press.

Darrigol, Olivier (1992), *From c-Numbers to q-Numbers: The Classical Analogy in the History of Quantum Theory*. Berkeley: University of California Press.

Duhem, Pierre (1906), *La Théorie physique: son objet et sa structure*. Paris: Chevalier & Rivière.

Earman, John (1989), *World Enough and Space-Time: Absolute vs. Relational Theories of Space and Time*. Cambridge MA: MIT Press.

Einstein, Albert (1998), *The Berlin Years: Correspondence, 1914–1918*, Schulmann, Robert, Kox, Anne J., Janssen, Michel and Illy, József (eds.), *The Collected Papers of Albert Einstein,* vol. 8. Princeton: Princeton University Press.

Einstein, Albert and Großmann, Marcel (1913), *Entwurf einer verallgemeinerten Relativitätstheorie und einer Theorie der Gravitation*. Leipzig: Teubner.

Friedman, Michael L. (2008), "Space, time, and geometry: Einstein and Logical Empiricism", in: Galison, Peter L., Holton, Gerald and Schweber, Silvan S. (eds.), *Einstein for the 21st Century*. Princeton: Princeton University Press, 205–216.

Galison, Peter (2003), *Einstein's Clocks and Poincaré's Maps: Empires of Time*. New York: Norton.

Gray, Jeremy and Walter, Scott (1997), "Introduction", in: Poincaré (1997), 1–25.

Heinzmann, Gerhard (2001), "The foundations of geometry and the concept of motion: Helmholtz and Poincaré", *Science in Context* 14: 457–470.

Helmholtz, Hermann von (1995), *Science and Culture: Popular and Philosophical Essays,* Cahan, David (ed.). Chicago: University of Chicago Press.

Hertz, Heinrich (1894), *Die Prinzipien der Mechanik in neuem Zusammenhange dargestellt*. Leipzig: J. A. Barth.

Holton, Gerald (1973), *Thematic Origins of Scientific Thought: Kepler to Einstein*. Cambridge MA: Harvard University Press.

Kerszberg, Pierre (1989), *The Invented Universe: The Einstein-De Sitter Controversy (1916–17) and the Rise of Relativistic Cosmology*. Oxford: Oxford University Press.

Minkowski, Hermann (1908), "Die Grundgleichungen für die electromagnetischen Vorgänge in bewegten Körpern", *Nachrichten von der Königlichen Gesellschaft der Wissenschaften zu Göttingen:* 53–111.

—— (1909), "Raum und Zeit", *Jahresbericht der deutschen Mathematiker-Vereinigung* 18: 75–88.

Nabonnand, Philippe (2000), "La polémique entre Poincaré et Russell au sujet du statut des axiomes de la géométrie", *Revue d'histoire des mathématiques* 6: 219–269.

Panza, Marco (1995), "L'intuition et l'évidence; la philosophie kantienne et les géométries non euclidiennes: relecture d'une discussion", in Pont, Jean-Claude and Panza, Marco (eds.), *Les savants et l'épistémologie vers la fin du XIXe siècle.* Paris: Blanchard, 39–87.

Paty, Michel (1993), *Einstein philosophe : la physique comme pratique philosophique*, Paris: Presses Universitaires de France.

—— (1996), "Poincaré et le principe de relativité", in Greffe, Jean-Louis, Heinzmann, Gerhard and Lorenz, Kuno (eds.), *Henri Poincaré: Science et philosophie.* Berlin: Akademie-Verlag, 101–143.

Poincaré, Henri (1887), "Sur les hypothèses fondamentales de la géométrie", *Bulletin de la Société mathématique de France* 15: 203–216.

—— (1891), "Les géométries non euclidiennes", *Revue générale des sciences pures et appliquées* 2: 769–774.

—— (1892a), "Non-Euclidian geometry", *Nature* 45: 404–407.

—— (1892b), "Sur les géométries non euclidiennes", *Revue générale des sciences pures et appliquées* 3: 74–75.

—— (1895), "L'espace et la géométrie", *Revue de métaphysique et de morale* 3: 631–646.

—— (1897), "Les idées de Hertz sur la mécanique", *Revue générale des sciences pures et appliquées* 8: 734–743.

—— (1898a), "La mesure du temps", *Revue de métaphysique et de morale* 6: 1–13.

—— (1898b), "On the foundations of geometry", *Monist* 9: 1–43.

—— (1900), "Les relations entre la physique expérimentale et la physique mathématique", *Revue générale des sciences pures et appliquées* 11: 1163–1175.

—— (1902), *La Science et l'hypothèse.* Paris: Flammarion. Reed. 1968.

—— (1903), "L'espace et ses trois dimensions (II)", *Revue de métaphysique et de morale* 11: 407–429.

—— (1906), "Sur la dynamique de l'électron", *Rendiconti del circolo matematico di Palermo* 21: 129–176.

—— (1907), "La relativité de l'espace", *Année psychologique* 13: 1–17.

—— (1909), "La mécanique nouvelle", *Revue scientifique* 12: 170–177.

—— (1912), "L'espace et le temps", *Scientia* 12: 159–170.

—— (1963), *Dernières pensées.* Paris: Flammarion, 2d edn.

—— (1997), *Trois suppléments sur la découverte des fonctions fuchsiennes*, Gray, Jeremy and Walter, Scott (eds.). Berlin: Akademie-Verlag.

Pont, Jean-Claude (1986), *L'aventure des parallèles: histoire de la géométrie non euclidienne: précurseurs et attardés.* Bern: Peter Lang.

Prentis, Jeffrey J. (1995), "Poincaré's proof of the quantum discontinuity of nature", *American Journal of Physics* 63: 339–350.

Rollet, Laurent (2001), *Henri Poincaré: des mathématiques à la philosophie.* Lille: Éditions du Septentrion.

Saß, Hans-Martin (1979), "Einstein über 'wahre Kultur' und die Stellung der Geometrie im Wissenschaftssystem", *Zeitschrift für allgemeine Wissenschaftstheorie* 10: 316–319.

Schiemann, Gregor (1997), *Wahrheitsgewissheitsverlust: Hermann von Helmholtz' Mechanismus im Anbruch der Moderne*. Darmstadt: Wissenschaftliche Buchgesellschaft.

Sklar, Lawrence (1974), *Space, Time and Spacetime*. Berkeley: University of California Press.

Staley, Richard (2005), "On the co-creation of classical and modern physics", *Isis* 96: 530–558.

Stein, Howard (1987), "After the Baltimore Lectures: some philosophical reflections on the subsequent development of physics", in Kargon, Robert and Achinstein, Peter (eds.), *Kelvin's Baltimore Lectures and Modern Theoretical Physics: Historical and Philosophical Perspectives*. Cambridge, MA: MIT Press, 375–398.

Torretti, Roberto (1984), *Philosophy of Geometry from Riemann to Poincaré*. Dordrecht: Reidel, 2nd ed.

Volkert, Klaus (1996), "Hermann von Helmholtz und die Grundlagen der Geometrie", in: Eckart, Wolfgang U. and Volkert, Klaus (eds.), *Hermann von Helmholtz: Vorträge eines Heidelberger Symposiums anläßlich des einhundertsten Todestages*. Pfaffenweiler: Centaurus, 177–205.

Vuillemin, Jules (1972), "Poincaré's philosophy of space", *Synthese* 24: 161–179.

Walter, Scott (1997), "La vérité en géométrie: sur le rejet mathématique de la doctrine conventionnaliste", *Philosophia Scientiæ* 2: 103–135.

— (1999a), "Minkowski, mathematicians, and the mathematical theory of relativity", in: Goenner, Hubert, Renn, Jürgen, Sauer, Tilman, and Ritter, Jim (eds.), *The Expanding Worlds of General Relativity*, Einstein Studies, vol. 7. Boston/Basel: Birkhäuser, 45–86.

— (1999b), "The non-Euclidean style of Minkowskian relativity", in: Gray, Jeremy (ed.), *The Symbolic Universe: Geometry and Physics, 1890–1930*. Oxford: Oxford University Press, 91–127.

— (2007), "Breaking in the 4-vectors: the four-dimensional movement in gravitation, 1905–1910", in: Renn, Jürgen (ed.), *The Genesis of General Relativity*, 4 vols. Berlin: Springer, vol. 3, 193–252.

— (2009), "Minkowski's modern world", in: Petkov, Vesselin (ed.), *Minkowski Spacetime: A Hundred Years Later*. Berlin: Springer.

Walter, Scott, Bolmont, Étienne, and Coret, André (2007), "Introduction", in: Walter, Scott et al. (eds.), *La Correspondance entre Henri Poincaré et les physiciens, chimistes et ingénieurs. La Correspondance de Henri Poincaré,* vol. 2. Basel: Birkhäuser, ix–xvi.

Zahar, Elie (1997), "Poincaré's philosophy of geometry, or does geometric conventionalism deserve its name?", *Studies in History and Philosophy of Modern Physics* 28: 183–218.

— (2001), *Poincaré's Philosophy from Conventionalism to Phenomenology*. Chicago: Open Court.

Vaihinger and Poincaré: An Original Pragmatism?

Christophe Bouriau[1]

Abstract:

Hans Vaihinger and Henri Poincaré developed a singular form of pragmatism, distinct from that of both James and Peirce. A study of Vaihinger's and Poincaré's conceptions of the hypothetical in scientific practice shows that they share a perspective which seeks to justify in terms of opportunity the employment of propositions and concepts that have no proper truth value. Their pragmatism may be understood as the original recovery of a Kantian position developed in the two sections of the appendix to the Transcendental Dialectic. According to Kant, ideas or propositions devoid of attestable objective value (for example, absolute determinism, or the unity of nature) may nonetheless assume a regulatory function essential to the scientist's task. It is just this operational or practical value of certain ideas that Poincaré and Vaihinger emphasize, albeit in a new perspective.

Introduction

Was Henri Poincaré a pragmatist? Answers to this question have varied with the times. At the dawn of the twentieth century, several writers, particularly in Great Britain and France, viewed Poincaré as a pragmatist, in the sense of William James. In our time, Gerhard Heinzmann also sees in Poincaré a pragmatist, but of the Peirceian, and not the Jamesian variety. The thesis I will defend here is that neither the Peirceian nor the Jamesian variety of pragmatism is the most adequate to describe the pragmatic dimension of Poincaré's thought. Rather, the conception of pragmatism advanced by Hans Vaihinger in his celebrated opus *Die Philosophie des Als Ob* (1911) is best suited to this task. Throughout the discussion of these issues, the different meanings of "hypothesis," of its variants and of related concepts play a decisive role.

The interest of Vaihinger's and Poincaré's approach to hypothesis is twofold. First of all, it shows the "pragmatist" posterity of a certain aspect of Kantian doctrine, developed in the appendix to the Transcendental Dialectic: certain ideas and propositions lacking in attestable ob-

1 Many thanks to Fabien Schang and Scott Walter for their contribution to the English version of my text.

jective value may nonetheless be given a practical or operational value in science under the auspices of conditions of possibility of scientific activity. Secondly, following Kant, Poincaré and Vaihinger insist on the importance of "acting as if" in scientific activity, thereby underlining the essential role of a certain function of the imagination in science. To imagine is, in this instance, to provisionally feint the objectivity of an idea, in order to carry out a given scientific operation with success.

I begin with a brief summary of pragmatic readings of Poincaré's philosophy, and a description of the points of intersection between Poincaré's and Vaihinger's writings. There follows an account of how Poincaré and Vaihinger both differ from James and Peirce, and how Poincaré and Vaihinger differ from each other.

I. Pragmatic Readings of Poincaré

a. France

René Berthelot devoted part of his inquiry into the pragmatist movement to "Poincaré's fragmentary and qualified pragmatism" (Berthelot 1911, 197–413). By "pragmatism" Berthelot meant a view that tends to identify truth with usefulness or convenience. Poincaré is said to promote a partial and qualified pragmatism, for "sometimes he opposes, sometimes he identifies convenience and truth" (*ibid.*, 264). Other French commentators from the early 20th century see Poincaré as a pragmatist in a similar sense. For example according to André Lalande, Poincaré reduced science to a biologically conditioned discourse towards an efficient action (Lalande 1906). According to Désiré Roustan, Poincaré supersedes the word "truth" with "convenience" (Roustan 1914, 631). It should be noted that these writers wrongly ascribe to Poincaré the thesis that the marker of truth is utility.

b. Great Britain

In Great Britain, two influential writers proposed a pragmatic reading of Poincaré's philosophy: William James and Ferdinand Schiller. James sees Poincaré as an almost-pure pragmatist, just a hair's-breadth away from pure pragmatism (James 1909, 62). Poincaré is not a pure pragmatist because unlike James, he maintains a distinction between what is useful and what is true, while taking account of practical constraints in the choice of scientific conventions. Ferdinand Schiller, on the other hand, considers Poincaré to be a full-blooded pragmatist, even if (1) he distinguishes what is true from what is useful, and (2) he refuses to subordinate science to action. In *Studies on Humanism*, Schiller defends a "humanist" pragmatism, recognizing the value of ideas useful to mankind, but considering the supreme goal of human existence to be the development of human faculties by disinterested high culture. For Schiller, Poincaré illustrated humanist pragmatism, in that the Frenchman subordinated worldly action to disinterested contemplation of truth (Schiller 1907, xi).

c. A Recent View

Among the great French mathematician's more recent readers, Gerhard Heinzmann calls for Poincaréan pragmatism still in a different sense. For him, Poincaré is "pragmatist" because he saw all known objects as never being independent of their constitutive operations or actions (Heinzmann 1995, 39). More recently, he has linked Poincaré's pragmatism to that of Peirce. Following a description of the stages of the process of knowledge in Poincaré's view, Heinzmann writes: "We call such a process pragmatic, with reference to Charles Sanders Peirce" (Heinzmann 2006, 406).

A view of pragmatism differing from that of James and Peirce is developed by Hans Vaihinger[2] in his *Philosophie des Als Ob*. I think that this view comes closer to capturing Poincaré's philosophy. Moreover, in a letter addressed to Hans Vaihinger of May 3, 1919, Einstein claimed that Henri Poincaré adopted an epistemological position "closely similar" to Vaihinger's (Einstein 1919 [1979], 319). While attempting to specify the nature of such a position, I want to argue for the following: Vaihinger and Poincaré develop the same brand of pragmatism when illustrating the operational value of fictions (Vaihinger) and hypotheses (Poincaré) in science, a pragmatism distinct from that of both James and Peirce.

2 Hans Vaihinger (1852–1933), who founded *inter alia* the journal *Kantstudien* in 1897, published his definitive work in 1911: *Die Philosophie des Als Ob. System der theoretischen, praktischen und religiösen Fiktionen der Menschheit auf Grund eines idealistischen Positivismus. Mit einem Anhang über Kant und Nietzsche.* This book corrects and supplements Vaihinger's habilitation thesis from 1877 written in Strasbourg under the direction of the positivist philosopher Ernst Laas: *Logische Untersuchungen, 1. Teil: Die Lehre von der wissenschaftlichen Fiktion.*

II. Intersections of Vaihinger's and Poincaré's Thought

1. "Every Generalization is a Hypothesis"[3]

a. Pragmatism, Hypothesis, and Fiction

Did Vaihinger and Poincaré advance a "pragmatist" interpretation of generalization in mathematics and physics? Vaihinger's basic view, which also governs Poincaré's analyses, is the following: scientists need not be *constrained* by either logical necessity or empirical truth. On the contrary, they may have to admit views that are inadequate due to a logical flaw, or the absence of an empirical verification. Scientists introduce views that, while "false" (i. e., logically or empirically invalid), nevertheless hold because of their "opportunity." Vaihinger employs the term "opportunity," *Zweckmässigkeit*, in order to justify holding just such views. This is the very term used by Louis Rougier for an anthology of Poincaré's writing, *L'opportunisme scientifique* (Poincaré 2002). The term provides an apt description of Poincaré's views on science.

Vaihinger defines the main point of his pragmatism as follows: "There are some presentations that are assumed to be not true, from a theoretical standpoint, but which justify themselves by their services to thought. They have thereby a practical value" (Vaihinger 1911 [1920], xv). These "presentations" are called "fictions" by Vaihinger. For Poincaré, however, they are hypothetical "constructions," and some of them are not only useful, but theoretically valid.

In *La science et l'hypothèse*, Poincaré defines hypothesis quite broadly. There are "verifiable hypotheses" (Poincaré 1902 [1905], xxii) or physical hypotheses that, once expressed in mathematical language, express the law-like structure of nature. There are also hypotheses "useful for fixing ideas" or even for providing an illustration of physical law. Poincaré calls the latter "indifferent hypotheses" (Poincaré 1902 [1905], 153). Such hypotheses are laws to be formulated in the current language of scientists. They serve only to present laws in concrete terms, as when matter is assumed "continuous," for example, or when some phenomena are supposed to be the "cause" of others. These ways of picturing physical processes are indifferent to the law itself, as expressed mathematically; they do not purport to say anything about the nature of

3 Poincaré 1902 [1905], 150.

things. Finally, Poincaré identifies hypotheses that are "conventions." These are decrees "imposed on *our* science," i.e., they make science possible, but "they are not imposed on Nature" (Poincaré 1902 [1905], xxiii). For example, in physics, to act *as if* nature were subject to invariable laws is to adopt a conventional hypothesis, albeit one that forces itself on scientific practice: in order to predict events, one must admit that natural phenomena unfold in a lawlike manner. Conventions are neither true nor false, but preconditions for scientific practice. Poincaré also makes use of the term hypothesis in a way he does not explicitly endorse, namely, as fiction. When referring to purely imaginary entities used in thought experiments, Poincaré refers to both "fiction" and "hypothesis" (Poincaré 1902 [1968], 116).

While Poincaré employs the term "hypothesis" to cover "hypotheses in physics" as well as "indifferent hypotheses," "conventions," and "fictions," Vaihinger limits the extension of this term to physical hypotheses alone. For instance, the atomic hypothesis is a fiction for Vaihinger, not an indifferent *hypothesis.* Likewise, Euclidian space is a convenient "fiction," not a conventional *hypothesis.* According to Vaihinger, a clear-cut distinction ought to be made between "hypotheses" that carry truth and may be verified, and "fictions" that do not carry truth, but provide a "means" by which an operation succeeds (problem solving, discovery, proof, prediction, etc.). I will return later to these distinctions.

Vaihinger divides fictions into two classes: 1) contradictory constructions, or "pure fictions," and 2) imaginary constructions, or "semi-fictions" (Vaihinger 1911 [1920], 24). While lacking objectivity, both sorts of fiction find use in science. The idea is to act "as if" their object were real, just as long as the concerned scientific operation (generalization, calculus, demonstration, etc.) obtains. Does valorization of "as if" processes signify Poincaré's influence on Vaihinger? In every chapter of *La science et l'hypothèse*, Poincaré underlines the act of simulation, when we act "as if" in order to realize a given scientific operation. For instance, one must proceed "as if" bodies were in geometrical space in order to reason about them (Poincaré 1902 [1905], 57). One must proceed "as if" there were an absolute space, in order to simplify laws and calculi (Poincaré 1902 [1905], 90). On must proceed "as if" nature were simple and uniform in order to generalize and predict, and so forth. To justify the employment of views with no truth-value, but which render scientific activity possible is the characteristic feature of Vaihinger and Poincaré's "pragmatism." Such a feature clearly appears in the way both writers handle generalization, inasmuch as this involves illogical inference.

b. Generalization in Mathematics

The way in which Vaihinger and Poincaré construe reasoning by recurrence exemplifies the significance of the "as if" process (or simulation) in discovering new properties of numbers. Irreducible to logic, mathematics exploits the fiction of an actual infinite, i. e., a logically contradictory notion (Vaihinger 1911 [1920], 530). Every mathematical operation involving the actual infinite infringes the principle of non-contradiction, in that it considers completed something (a quantity, an operation) that cannot actually be so. To conceive of the actual infinite is, according to Vaihinger, to conceive of a "clear contradiction," namely something that is both endless and complete (Vaihinger 1911 [1920], 87).

Nonetheless, without such a contradictory notion or "pure fiction" of an actual infinite, reasoning by recurrence is impossible. Indeed, in order to generalize a property p to the infinite series of numbers one must proceed *as if* the infinite operation verifying that property were completed. Let the verifying operation be as follows: since p holds for 1, and, given that if it holds for 1 it holds for $n+1$, then it does hold for 2; since p holds for 2 and for $n+1$, then it holds for 3, and so on. Such an operation cannot ever be completed within N, given that it concerns an infinite group of elements. Consequently, the justification of reasoning by recurrence is illogical. The principle of contradiction rules out the expansion of p to some infinite set of elements, since p can be verified only for a finite number of elements. It is only the recognition of an indefinitely repeated operation $n+1$ that grounds reasoning by recurrence, or even "imagining" this infinite operation as complete. Such an imagination, Vaihinger points out, embodies a "logical impossibility," but one must proceed "as if such an impossibility were possible" (Vaihinger 1911 [1920], 531). When I imagine that any operation proceeding indefinitely is completed, I commit a logical error, of course, but gain an advantage from it. The advantage at hand is heuristic fecundity,[4] namely the discovery of some new property of numbers in

4 Mathematical fecundity is to be obtained, according to Vaihinger, through notions or operations embracing a logical contradiction (infinitesimal, imaginary numbers, reasoning by recurrence, and so on). Thus Vaihinger writes: "The

the case at hand. This example shows that a logically irrelevant view, especially that of a completed infinite series, assumes some *practical* value insofar as it gives rise to a fruitful process: in particular, generalization of a property. A pragmaticist epistemology consists in endorsing such views. The great merit of Poincaré, according to Vaihinger, is "to have established the utility and adequacy of fictional (or even contradictory) concepts in mathematics" (Vaihinger 1911 [1920], xvii).

In the same vein as Vaihinger, Poincaré calls for reasoning by recurrence in order to establish the irreducibility of mathematics to logic. In reasoning by recurrence, he writes, "the view of a mathematical infinite already plays a prominent role." Expanding p beyond what is verifiable shows that mathematical reasoning is of a sort differing from logical deduction: one passes with it from a finite number of cases to a generalization that touches any number. In sum, one passes "from the finite to the infinite" (Poincaré 1902 [1905], 11). Such an induction or generalization, Poincaré points out, is grounded upon an "intuition," that of an indefinitely repeatable operation within a human mind serving as a basis for reasoning. Poincaré explains why the validity of reasoning by recurrence is "imposed upon us with such an irresistible weight of evidence," although it is unsound:

> It is because it is only the affirmation of the power of the mind which knows it can conceive of the indefinite repetition of the same act, when the act is once possible. The mind has a direct intuition of his power, and experiment can only be for it an opportunity of using it, and thereby of becoming conscious of it (Poincaré 1902 [1905], 13).

This understanding of reasoning by recurrence shows the significance of *as if* in the reasoning process. On the one hand, it is by acting *as if* the operation $n+1$ had been repeated indefinitely that one can generalize the property p. On the other hand, this example shows that a mathematical object (the updated general property) cannot be dissociated from its constitutive operation. This is a "pragmatic" feature. In mathematical induction, the action or operation produced by the knower is involved in the final result. Indeed, the indefinite repeatability of the operation

bias according to which only the non-contradictory is logically fecund has not to be prompted" (Vaihinger 1911 [1920], 92). Poincaré, on his part, is opposed to logicism as subjecting mathematics to the principle of non-contradiction. He holds that "the rule of reasoning by recurrence is irreducible to the principle of contradiction" (Poincaré 1902 [1905], 12).

$n+1$ is part and parcel of the generalization and justification of the property *p*.

For Poincaré, the infinite is a mere "methodical" fiction, employed in the construction of mathematical objects. Such a pragmatic view is opposed to realism, which posits the independence of mathematical objects from details of discovery. In *Dernières pensées*, Poincaré insists that unlike pragmatists, the champions of mathematical realism (like Cournot) do not see infinity as an operation of "becoming" because actual infinity "pre-exists the mind that discovers it." In assigning some independent existence to mathematical entities, realists "do believe in an actual infinite." Unlike pragmatists, they separate mathematical objects from human cognitive process or activity (Poincaré 1913, 94).

When it comes to generalization in physics, the aim is no longer to *discover* a new truth, but to be able to *predict* by means of a plausible hypothesis.

c. Generalization in Physics

A physical hypothesis, once verified, is "generalized." To generalize a hypothesis is to proceed "as if" it could be extended to the infinitely many unverified cases. Now both writers take this to be a *tour de force*. According to Poincaré, even a well-verified hypothesis remains uncertain, given that nothing ensures that circumstances under which it was verified will be reproduced in the future. On the contrary, "some of the circumstances will always be missing. Are we absolutely certain that they are unimportant? Evidentlynot!" (Poincaré 1902 [1905], xxvi). Hence "the importance of the role that is played in the physical sciences by the law of probability" (*ibid.*, xxvii). Vaihinger notes in turn that it is impossible for a hypothesis to cover the entire set of unobserved cases, so that modifications of the "general part" of a hypothesis are to be expected (Vaihinger 1911 [1920], 147). He also confers to physical hypotheses the status of plausibility, *Wahrscheinlichkeit* (*ibid.* 152).

Poincaré and Vaihinger acknowledge that a physical hypothesis, although it cannot be held to be strictly true, displays a somehow practical justification at the least. It preserves its entire relevance as long as it helps to make easy and convenient predictions. So Poincaré defines physical hypotheses not in virtue of their theoretical truth but their "fertility" (Poincaré 1902 [1905], xxii). Vaihinger points out in turn that although every hypothesis is unclear as a matter of fact (it neglects some secondary causes as implied in the phenomena), one can proceed *as if* it were clear, so long as it efficiently drives calculi (Vaihinger 1911 [1920], 456). Let us keep in mind here that the criterion of validity for a hypothesis is not truth or knowledge of reality in a strict sense, but the efficiency of operations that are based on them.

Up to this point, one might say that Poincaré and Vaihinger are "pragmatists," in that they place the operational criterion before that of truth, in modes both formal (logical non-contradiction) and material (strict correspondence with given reality). In both cases of generalization at hand, the first requirement to suffer damage is logical non-contradiction (reasoning by recurrence is irreducible to the principle of non-contradiction), while the second is that of sticking to the facts (generalization goes beyond facts).[5] Such an account of some views in virtue of their operational value is also verified in the way Vaihinger and Poincaré handle conventional hypotheses.

2. Conventional Hypotheses

a. Euclidean Space as a Convention

Poincaré and Vaihinger justify the use of "conventions" (Poincaré) and "conventional fictions" (Vaihinger), understood as decrees. In mechanics, Poincaré explains, it is assumed by convention that bodies move in a Euclidean geometrical space. This convention simplifies operations of measurement and calculation. This does not mean that physical space corresponds to Euclidean space. Furthermore, Poincaré and Vaihinger both resort to the phrase "as if" in order to point out that this is a mere simulation. "We reason about bodies," says Poincaré, "as if they were situated in geometrical space" (Poincaré 1902 [1905], 57). Vaihinger writes in turn that we proceed in mechanics "as if" bodies were located in a Euclidean space but without claiming that it is really the case (Vaihinger 1911 [1920], 497).

5 We are far from a writer like Claude Bernard, who writes: "Reasoning will always be correct when applied to accurate notions and precise facts; but it can lead only to error when the notions or facts on which it rests were originally tainted with error or inaccuracy" (Bernard 1865 [1927], 2). On the contrary, both writers show that the use of views that are inexact both logically and factually can be justified at the same time. To recognize the operational relevance of views that are inexact in themselves is a typical feature of pragmatism, in Vaihinger's sense.

The way Poincaré and Vaihinger account for the adoption of a Euclidean space, among other possible spaces, gives rise to a new characteristic feature in their pragmatism: the introduction of human subjects and their concern for adapting to a given environment.

Poincaré holds that Euclidean geometry is the most convenient one because "it agrees sufficiently with the properties of natural solids, those bodies which we can compare and measure by means of our senses" (Poincaré 1902 [1905], 50). Natural solids are not rigorously invariant and cannot be transferred without distortion like solids in geometry, but their properties approximate those of geometrical solids, and this is sufficient for us.

According to Poincaré, the essential properties we assign to space proceed from associations relevant for the defense of organism. Such associations are so ancient in the history of mankind that they seem indestructible. To be sure, one might assume another space in place of the Euclidean space, in order to state some connections between objects within our environment, but that space would be less favorable for our adaptation: "We could conceive of thinking beings, living in our world, whose distribution board would have four dimensions, who would, consequently, think in hyperspace. It is not certain, however, that such beings, admitting that they were born, would be able to live and defend themselves against the several dangers by which they would be assailed" (Poincaré 1908 [1914], 114).

A similar analysis is to be found in Vaihinger. Like Poincaré, Vaihinger holds that the question for us is not which sort of space corresponds to physical space. The real question is: which is the most convenient one for us, that is, the one best suited to the world in which we live and measure? If Euclidean space happened to be selected among other possible ones, Vaihinger points out, this is by virtue of a functional criterion and not a theoretical one. It is so because it was best suited to the natural objects we encounter:

> It is while trying to discover why three dimensional space has been selected and maintained, that we find the causes and reasons which have led human thought to elaborate it: it is the best adapted to natural objects. It is the unique survivor of other possible spaces and, by selection, it has proven to be the most convenient one (Vaihinger 1911 [1924], 55).

For Vaihinger as for Poincaré, it is a practical requirement (adaptation) that guides the selection of our formal tools for knowledge. Poincaré writes that "[by] natural selection, our mind has *adapted* itself to the conditions of the external world, it has adopted the [geometry] *most advantageous* to the species: or in other words *the most convenient*" [Poincaré 1902 [1968], 108].[6] Generally, the genesis of the forms of human knowledge cannot be dissociated from the requirement to act with efficiency within some given environment. Scientific knowledge itself depends on the essential purposes of human organisms. In response to external circumstances, the human mind creates the forms that will make it possible to master such circumstances theoretically and to act more efficiently:

> In the course of its growth, mind creates its organs of its own accord in virtue of its adaptable constitution, but only when stimulated from without, and adapts them to external circumstances (Vaihinger 1911 [1924], 2).

A pragmatic criterion of adaptation thereby explains the adoption of Euclidean space. To account for the adoption of a geometry in accordance with knowers as *human* beings related to some given environment is one characteristic feature of pragmatism for both writers. Such an account does not occur in Peirce, for instance, as the community of scientists is not a strictly *human* community *de jure* (*CP*, 5.408).

Consideration of geometrical space as a convention certainly distinguishes the present writers from Kant, given that the latter took Euclidian space to be the only one possible. On the other hand, it seems that Vaihinger and Poincaré are closely related to Kant when acknowledging the regulative value of views without any theoretical value. Kant even appears to be the common source of their pragmatism.

b. Some Other Conventional Hypotheses: Kant's Heritage

Poincaré points out that, in order to explain and predict, we must proceed as if the laws of nature (as relations between "an antecedent" and "a consequent") were invariant, or as if the world were ruled by an absolute determinism. Poincaré underlines that this is "a *practical* rule" (Poincaré [1905 [1958], 134; my italics). Similarly, Vaihinger justifies the "fiction" of determinism as making the scientific function of explanation and prediction possible. In order for any explanation to hold, it must be assumed that the same causes yield the same effects, and that all processes are subject to known laws of nature. Poincaré thus claims that the hypothesis of determinism is indispensable to science: "Without determinism, science could not even exist" (Poincaré 1913, 45). Vai-

6 My translation. This phrase is missing in (Poincaré 1902 [1905], 88).

hinger, in his commentary on Kantian antinomies, notes in turn that the determinist thesis does not have any value as knowledge, but only as a precondition for the working scientist. It holds as a "regulative principle" (Vaihinger 1911 [1920], 639), and only has a "practical reality" (*ibid.*, 640). By such an expression, Vaihinger means that one must proceed "as if" determinism were real, in order for prediction to be practically possible. Poincaré and Vaihinger thus appear in line with Kant, including in their use of the words "as if."

In the *Critique of Pure Reason,* Kant insists that some assumptions (the view of a teleological organization of the world, an overall unity of nature, a universal determinism, and so on) count as fruitful "regulative principles." Although they have no objective validity, such assumptions serve as guidelines for scientific research. Concerning finality, Kant thus insists how relevant it is "to see any world-ordering *as if* it originated from the purpose of some supreme reason" (A 686/B 714; my translation). He motivates this view in an indirect way. Acting *as if* nature were organized according to some purpose entails that "we could make a host of findings about the shape of the earth, mountains and seas, for example" (A 687/B 715).

Similarly, unity of nature "is nothing but a regulative principle of reason" (A 697/B 725). Kant exhorts the scientist to behave "as if" such a suggestion were true, or for the duration of his investigation.[7] Kant insists that it is not *theoretically* legitimate to advance that nature has some systematic unity. Such an assumption is to be motivated only by the "advantages" it brings: "It should be always in our advantage, and so without being ever harmful, to conduct our examination of nature according to this principle" (A 701/B 729).[8] But what "advantage" is there? Let us take an example quoted by Kant: to assume that every faculty of the soul turns on *one* ultimate faculty can lead one to find some totally new connections between these faculties. To act *as if* they derived from a common source can be accounted for only as a way to stimulate research in psychology (A 649/B 677). It will be shown later that Vaihinger and Poincaré are opposed concerning the

7 Kant writes the following: "The regulative law of a systematic unity within nature directs us to investigate the latter *as if* some systematic and final unity in the greatest possible variety indefinitely occurred therein" (A 700/B 728). Also: "We must consider everything that can only resort to the series of possible experience, acting *as if* the latter constituted some absolute unity" (A 672/B 700).

8 See also the following: "The [allegedly] systematic unity is advantageous for the empirical knowledge of understanding" (A 681/B 709).

status to be conferred to these sorts of suppositions: are they necessary or contingent?

Following Kant, they also acknowledge the value of suppositions in the areas of law and ethics. According to Kant, metaphysical suggestions, although they turn out to be without any theoretical objectivity, must be posited as requirements for moral judgment and action. Thus, the idea of human freedom as an absolute starting point is justified as a precondition for assessing human actions. Kant claims that in order to judge any human act, one must see the latter "*as* completely unconditioned with respect to the former state, *as if* the agent thereby introduced a series of consequences" (A 555/B 583). In this way, Kant goes on, "the effectiveness of freedom is not to be established." This proposition "is just taken here as a transcendental idea," as a precondition for some operation (e.g., judging a defendant or facing up to one's responsibilities).

A similar analysis is to be found in Vaihinger and Poincaré. Both hold that even if human freedom does not exist (an undecidable point), nevertheless one must act "as if" it did for a number of moral and social reasons. In other words, the assumption of free will is justified not theoretically, but practically, as a precondition for social life. This leads Poincaré to justify two contradictory theses as mere operative means: "It is equally impossible not to act *as a free* man when one acts and not to reason *as a determinist* when one does science" (Poincaré 1913, 46; my italics).

Freedom and determinism are on a par, being both undecidable, and count only as preconditions for operations. Thus, doing science one must posit universal determinism in order to be able to predict, for instance (*ibid.*, 45). When acting, on the other hand, one proceeds *as if* free; otherwise, the notions of merit and demerit that ground moral and social life would be meaningless: "The idea of merit and demerit should vanish or change" (*ibid.*, 46).

Vaihinger takes a similar stance when he holds that in a trial the prosecution must assume free will (just as defense must posit determinism). Prosecution would otherwise be impossible since the defendant could not be held responsible for his acts. The idea of free will too is justified not theoretically, but *practically* (as a precondition for prosecution). The prosecuting attorney "does not say that man is free, if he is careful, but the following: man must be treated *as if* he were free, at least in legal affairs and moral judgment" (Vaihinger 1911 [1920], 167).

Given the features just listed, what distinguishes Vaihinger and Poincaré's pragmatism from that of their contemporaries Peirce and James? Is this a *peculiar* sort of pragmatism?

III. Vaihinger, Poincaré, and Anglo-Saxon Pragmatism

a. Utility Versus Truth

Vaihinger qualifies his affiliation to pragmatism while he sees himself as indebted to Anglo-Saxon pragmatism. When he notes that "Pragmatism has done something to prepare the ground for [his] fictionalism," he writes also of a "crucial difference" between pragmatism and fictionalism:

> [My] fictionalism does not admit the principle of Pragmatism which runs: an idea which is found to be useful in practice proves thereby that it is also true in theory, and the fruitful is thus always true (Vaihinger 1911 [1924], viii).

Contrary to James's pragmatism, Vaihinger's pragmatism holds that an idea is not "true" because it is "useful." Conversely, a false or "fictional" idea, which is without correlate in being, does not become useless for knowledge:

> An idea whose theoretical untruth or incorrectness – and thereby its falsity, is admitted, is not for that reason practically valueless and useless; for such an idea, in spite of its theoretical nullity may have great practical importance (*Ibid.*).

According to Vaihinger, mental constructions that enable one to obtain an arbitrary positive result (predicting correctly, discovering a property of numbers, establishing a thesis, and so on) must be assumed fictional, that is, false. In other words, the way to obtain truth cannot be said to be true in turn. Otherwise, it must be said that some fictions (the atom, the infinitesimal, any artificial classification) are "true," and express something actual, which is absurd. Arthur Fine aptly sums up Vaihinger's position on this point:

> Vaihinger regards the inference from utility to reality as fundamentally incorrect. Thus, despite his pragmatic emphasis on thought as a tool for action, he wants to distinguish his position from the Jamesian form of pragmatism that regards truth to be whatever turns out to be 'good' by way of belief (Fine 1993, 8).

Before returning to the difference with the "Jamesian form of pragmatism," let us note that Poincaré adopts the same position as Vaihinger. It must be conceded that, unlike the latter, Poincaré talks about "conventions" (neither true nor false) rather than "fictions" (falsities). Before returning to this significant discrepancy, let us observe meanwhile that both writers are against a certain trend in pragmatism, namely, to hold as true any idea the practical consequences of which are acknowledged as useful and satisfactory. As Jules Vuillemin writes: "[According to Poincaré], science is not true because of its usefulness" (Poincaré 1905 [1970], 8). In the same line, Vaihinger: "From the usefulness of a psychical and logical construction one cannot conclude its truth" (Vaihinger 1911 [1920], 100).

Both writers thus distance themselves from W. James's pragmatism. For James holds that the word "true" designates what is seen to be good for some reasons to be assigned (James 1907 [1975], 42). Truth attaches to an idea provided that the latter turns out to be good and efficient with respect to some given practical purpose. In other words, truth attaches to an idea through consequences that are not separated from it but appear as its development.

In a sense, James supports a more original and innovative view of truth than Vaihinger and Poincaré. In physics, they still think about truth in accordance with some order (the lawlike structure of nature) that ideas (differential equations) are supposed to express. Even if this order is still the order *as seen by human knowledge*, by way of its finite capabilities, it remains that some natural order (to which one has partial access) pre-exists the discovery one is making. When evoking perception of natural laws's harmony (see infra), Vaihinger and Poincaré still maintain a static view of truth as the discovery of some pre-existing order.

From James's viewpoint, the idea that there is some natural order that thought finds, step by step, is to be abandoned. From his perspective, to consider that the statement of a law or system of laws should give a direct access to truth, as expressing some objective structure of nature, amounts to considering this statement only through *pre*formatted ideas, previously experienced beliefs, or habits that are no longer questioned. Noting that "truth gets to an idea," James supports a *dynamic* view of truth. He means that the truth of a statement does not obtain in isolation from some active process of verification: "Truth [of a sentence] is the process of its validation" (James 1907 [1975], 97). As we will see, Poincaré and Vaihinger are much more contemplative when calling for a "pure perception" of the structure or harmony of nature.

b. Antirealism: A Difference with Peirce

With Poincaré and Vaihinger, as seen above, a scientific idea can be picked out when it favors the adaptation of mankind to its milieu. So it goes for Euclidean space, selected for its correspondence to the motion group of natural solids found in our environment. The ends of mankind (adaptation, conservation, avoiding pain) are involved in justifying the choice of an idea. Criteria for selecting an idea are thus not independent of the subject's expectations *qua* human subject.[9] With Peirce, however, the word "pragmatism" does not rely on any purpose appropriate to human action. Objective knowledge involves a community of minds that may not be human. Such knowledge is connected with the human mind only *contingently* (CP 4.550).

A second significant difference concerns the meaning of the term "pragmatic," and the view of truth. For Peirce, every proposition is pragmatic in the sense that it acts or launches action. The first of these is the act of affirmation, by which a speaker makes a truth claim liable to dispute. An affirmation acts in delivering a truth-claim, for which it claims to obtain the agreement of its audience (CP 2.344). Now it also includes a pragmatic dimension, in the sense that it prompts action or results in interaction within a community of interpreters. For Peirce, objective scientific propositions are about states of the world and cannot exist without some intersubjective interaction. Every proposition we make about natural phenomena entails claims to validity while making them subject to confirmation or contestation by each member of the investigating community in the course of some indefinite process in which truth, taken to be an agreement between each member of the community, holds as an ideal goal of inquiry.

Poincaré and Vaihinger are not "pragmatists" in the same sense. Of course, both men see the collective advance towards truth as involving an interaction among members of an intellectual community. Neverthe-

9 One pervasive distinction is found in Vaihinger and Poincaré between properly *human* knowledge, as closely related to given phenomena and meant to apply to them with efficiency, and some sort of *divine*, disembodied knowledge that would have a direct access to things themselves. Both men consider objective knowledge to imply human knowledge of phenomena and the interconnections of phenomena.

less, as we will see in more detail later, there is also the view of a direct and "certain" access to truth (or the structure of reality) by an isolated subject, at the end of the process of elaborating natural laws. The point is then disinterested contemplation, separated from any action, without any pragmatic concern.

There are also important differences concerning the relationship between language and reality. Neither Poincaré nor Vaihinger attach to the *referent* the central role Peirce gives to it. According to Peirce, an assertion is, as a claim to truth, always connected with "something" in the world. Signs point at things, allowing one to "say something about" them (CP 2.231). Poincaré and Vaihinger attach another value to signs: symbols or signs as used by the scientist have a *meaning* but do not require a *referent*. They do not express things, but their interrelations. This brings out a specific feature in the pragmatism of both writers: both hold the possibility of "true," "certain" knowledge about mere connections between things.[10]

For the very same reason, the status of mathematical objects differs for Poincaré and Vaihinger. Unlike Peirce, Poincaré and Vaihinger are antirealists in mathematics: mathematics does not consist in *real* entities and connections independent of their knower, but in formal systems relying on conventional axioms. Mathematical language cannot be seen as a language of objects (entities, substances), i. e., as ontological. Talking about substances or mathematical entities involves the notions of "permanence" and "numerical identity," which are nothing but metaphysical *fictions*.[11]

According to Poincaré and Vaihinger, geometrical axioms are conventions, the adoption of which is directed by the environment, and geometrical objects such as "lines" and "surfaces" are defined by axioms. According to Peirce, on the other hand, "it is not correct [to speak] as if lines and surfaces were something we make. The lines and surfaces are *places* which are *there*, whether we think them or not. ... They are there, in this sense that we can think of them as being there without being

10 Thus Poincaré writes: "The aim of science is not things themselves, as the dogmatists in their simplicity imagine, but the relations between things" (Poincaré 1902 [1905], xxiv). Vaihinger holds the very same position: "[Science] has to do with reality only as establishing connections of invariable succession and coexistence" (Vaihinger 1911 [1920], 97). "Real is constant connection [between phenomena], that is, whatever is nomological" (Vaihinger 1911 [1920], 98).

11 On the non-ontological character of mathematical language in Poincaré, see Ly (2004).

drawn into any absurdity" (CP 2. 387). While emphasizing the independence of mathematical objects with respect to the human subject who thinks of and constructs them, Peirce seems to take a realistic stance in mathematics –. For unlike Poincaré and Vaihinger, Peirce holds that mathematical objects and propositions about time and space are not formed by the subject under the effect of experience, but they belong to it in some "innate" way (CP 4.92).

In the same line, Peirce attaches a realist meaning to concepts Vaihinger and Poincaré take as mere convenient and subjective notions, e.g., "chance." Peirce writes: "When I speak of chance, I only employ a mathematical term to express with accuracy the characteristics of freedom or spontaneity" (CP 6.201). As Philip H. Hwang has shown, in following Aristotle Peirce held a realistic view of chance (Hwang 1993, 262–276). However, Vaihinger and Poincaré interpret "chance" subjectively, linking it to our ignorance of certain physical causes. For an omniscient being with unlimited predictive capacity, there is no such thing as a chance occurrence (Poincaré 1908 [1914], ch. IV and Vaihinger 1911 [1920], 74).

Nonetheless, while both Vaihinger and Poincaré involve the human knowingsubject in the constitution of the object of study, they do not adopt a skeptical stance with respect to scientific truth.

c. A Non-skeptic Pragmatism

Poincaré and Vaihinger hold that our mathematical equations give us partial but genuine knowledge of the lawlike organization of nature, or even of its "harmony." In other words, the way to truth is not constantly postponed as is the case with James and Peirce, who take truth as the ideal limit of scientific inquiry. Vaihinger notes for instance that "science increasingly tends to collapse every event into some purely mathematical connections" (Vaihinger 1911 [1920], 319).

An occasion is given here to make more precise the distinction which both writers make between physical "hypothesis" and physical "law." According to Poincaré, the mathematical language of differential equations has the advantage of avoiding the metaphysical notions of "force" and "causality" that are involved in explanatory hypotheses: "When we say force is the cause of motion, we are talking metaphysics" (Poincaré 1902 [1905], 98). Vaihinger in turn presents causality as an "analogical fiction" of a metaphysical sort. In other words: to speak in terms of causality is to project on things some spontaneous force similar to the one we seem to experience through our voluntary acts (Vaihinger 1911 [1920], 45). Laws, according to both writers, are sufficiently verified hypotheses that are translated into the conventional language of differential equations (Poincaré 1905 [1958], 93 and Vaihinger 1911 [1920], 319). Unlike hypotheses, laws do not purport to "explain" physical processes, but only to state mathematically "constant ratios" that stand between physical parameters.

However, even if the knowledge we gain about nature can be said to be true, it is always finite knowledge, meaning that it is true *for us* and relativized to some provisory state of our science. According to Vaihinger and Poincaré, we have a partial, incomplete view of the lawlike structure of nature. We would not have the very same presentation of such a structure if we were acquainted with other, not yet discovered laws. Vaihinger distinguishes between *our* knowledge of lawlikeness and some other plausible ways to construe this very lawlikeness of nature (Vaihinger 1911 [1920], 289). A broader human knowledge that includes more laws than known today would result in a reading of natural order different from our own.

Poincaré develops a similar thesis in much of his philosophical work. In *La science et l'hypothèse*, he enjoins us to imagine being in a specific universe with creatures who know the same laws as ours, but for whom certain facts would be hidden. These beings would not have the same view of physics as we do, although their partial theory would be consistent with observation. The same holds for us: our physics cannot embrace everything, and will be modified upon observation of some new facts (Poincaré 1902 [1905], 114–115). Nevertheless, we are not thereby deprived of access to a true or objective knowledge of nature: "The harmony expressed by mathematical laws ... is the sole objective reality, the only truth we can attain" (Poincaré 1902 [1905], 14).

Hence Poincaré also argues that law-like knowledge is the only truth *we could* ever reach. In *La valeur de la science*, Chapter X, he clearly distances himself from any skeptical pragmatism that renounces truth in the scientific arena. This kind of pragmatism (that Poincaré associates with Édouard Le Roy) accords to physical laws as a mere practical value which boils down to a "rule for action" (Poincaré 1902 [1905], 15). Against this kind of pragmatism, Poincaré repeatedly states (like Vaihinger) that science does not content itself with directing our actions in the world but reveals the real *structure* of nature, or partially so at the very least. To be sure, symbols as used in our differential equations do not designate things themselves, but they allow us to see *constant* connections among various physical parameters. They prepare us to grasp "the harmonious order of its parts" by means of "pure intelligence" (Poincaré 1908 [1914], 22). According to both writers, we reach true knowledge only of the relative, or relations. Vaihinger writes: "We have only knowledge about the relative relations and invariable laws of phenomena" (Vaihinger 1911 [1924], 78). According to Poincaré: "Beyond those relations [between quantities] there is no knowable reality" (Poincaré 1902 [1905], xxiv).

IV. Hypotheses and Conventions Versus Fictions: The Differences Between Poincaré's and Vaihinger's Pragmatisms

a. Hypothesis and Fiction

Vaihinger clearly departs from Poincaré in drawing a distinction between hypothesis and fiction. Given that Vaihinger read Poincaré, one may ask if the two chapters[12] he devotes to the distinction between fiction and hypothesis concern the French mathematician directly, in that Poincaré uses both terms loosely.

In *La Science et l'hypothèse*, as previously noted, Poincaré distinguishes three sorts of hypotheses: verifiable physical hypotheses, indifferent hypotheses that serve to fix thought (indifferent to mathematical results), and apparent hypotheses, which are "disguised conventions." Poincaré also calls conventions hypotheses "in appearance" (Poincaré 1902 [1905], xxii) and does not count them as genuine hypotheses. Whether they occur in mathematics or physics, indeed, conventions, unlike the other sorts of hypotheses, do not concern reality.

It remains to say that Poincaré retains the noun "hypothesis" and the adjective "hypothetical" as referring not only to "conventions" but to "fictions" that serve to support thought experiments. With respect to Riemannian geometry, Poincaré invites his readers to "imagine" a two-dimensional world devoid of thickness, inhabited with infinitely flat animals, in order to establish that a non-Euclidean geometry can be intuited. He thereby makes use of the word "hypotheses" to refer to "fictions" that compose such an imaginary world: "While we are

12 Cf. Vaihinger (1911 [1920]): Chapter 21 of Part One and chapter 28 of Part Two deal with differences between fiction and hypothesis.

making these hypotheses it will not cost us much to endow these beings with reasoning power …" (Poincaré 1902 [1905], 38).

In the rest of the book, the word "hypothesis" is also sometimes employed for "fiction." In the chapter on classical mechanics, for instance, Poincaré assumes that given a body subjected to no force its position would not change, but its speed would. Poincaré calls such an assumption a "hypothetical law," before he decides to call it a "fiction" a couple of lines later (Poincaré 1902 [1905], 93). One could ask whether this expresses a vacillation on his part, or whether he takes fiction and hypothesis as synonymous here.

The same floating case can be noticed some lines later when Poincaré launches a thought experiment, inviting us to imagine the solar system "traversed by a body of great mass and immense velocity, arriving from distant constellations" (Poincaré 1902 [1905], 94). In two successive sentences concerning this same assumption, he first uses the word "fiction": "The fiction I previously exposed …," and then the word hypothesis: "Such a hypothesis is …" (Poincaré 1902 [1968], 116; my translation).

Poincaré also talks about "hypothesis" with respect to physical determinism (Poincaré 1902 [1905], 134). We recall that Vaihinger takes determinism to be a mere fiction that cannot be verified and only serves as a precondition for the working scientist. Fiction, unlike hypothesis, is a representation without any reality claim, such as determinism. Vaihinger opposes the floating use scientists make of such words as fiction and hypothesis, pointing out the main differences:

1) Hypothesis is meant to be fixed once and for all (i. e., to express truth durably), while fiction, without any claim to truth, is meant to be deleted as soon as it performs its duty within an arbitrary calculus or operation: "Hypothesis remains, but fiction disappears" (Vaihinger 1911 [1920], 149). Thus, from Vaihinger's standpoint, a fiction that serves to support a thought experiment is not a "hypothesis" at all, contrary to Poincaré's view. Unlike hypotheses, fictions do not claim to express reality and tend to be deleted at the end of a demonstration.

2) The connection between fiction and physical hypothesis is that between means and end: "Hypothesis is the result and fiction the means that leads to it" (*ibid.*). Vaihinger quotes the Goethean fiction of a primitive animal, with respect to which any other animal would be a modification. This is mere fiction according to his line of reasoning, without any claim to reality. According to Vaihinger, Goethe urges us to proceed as if all animals would derive from one and the same prim-

itive animal. This fiction leads us to imagine that all species may be derived from other species. Once this point is corroborated by geological findings, one is led to entertain a "new classification of living species." According to Vaihinger, the Goethean fiction is a heuristic one that served as a presage to Darwin's assumption, which claims to express reality (Vaihinger 1911 [1920], 145).

b. Fiction and Convention

By his use of the word fiction, Vaihinger encounters significant difficulties. To talk about "fiction" presupposes in effect that truth is known. For instance, to say with Vaihinger that Euclidean space is fictional is to assume that the genuine nature of physical space is known, and that Euclidean space does not square with physical space. As pointed out by Poincaré, it is absurd to say that one geometrical space is more or less true than another, given that the only question is which of the various geometries turns out to be the most convenient one for mankind (Poincaré 1902 [1968], 108). Just as Euclidean space (the one best adapted to our measurement tools and our environment) cannot be said to be true, it cannot be seen as fictional or false, since that requires knowledge of the "true" space. In a nutshell: once the view of a "true" geometry as meaningless is admitted, it can no longer be said that Euclidean space is "true" or, with Vaihinger, that it is "fictitious," i. e., false (inadequate for representing physical space). One just agrees to uphold the most convenient space.

It is well known that Poincaré eludes the difficulty noted above, by using the term "convention" instead of "fiction." A conventional idea, he repeatedly points out, is neither true nor false. It is an idea that the scientists, guided by experience, have chosen in virtue of its convenience. It must be admitted that, in a few passages, Poincaré makes indifferent use of the words "fiction" and "convention." He says, for instance, that the definite axes that serve as co-ordinates for planets have a "purely conventional existence," and then adds that such co-ordinates are "a convenient fiction" (Poincaré 1913, 103). But it remains that, within this context, fiction (of co-ordinates) is not, as with Vaihinger, synonymous with falsity. Poincaré entertains here a convenient tool, just as when he calls on the "fiction" of infinitesimals. Generally, by a privileged use of the term "concept," Poincaré avoids the very difficulty Vaihinger encounters.

Furthermore, by using the term fiction (= falsity) rather than convention, Vaihinger appears to contradict himself. In order to say that an idea is a "fictitious" one, i. e., that the idea is false, one must already know what the case is. This assumes knowledge of reality prior to framing reality. Thus we are led to a contradiction, given that, from Vaihinger's pragmatist and constructivist perspective, reality does not exist independent of a human mind's activity, which organizes sensations by means of its fictional categories. Finally, Walter Scholz notes that Vaihinger – but not Poincaré – often confuses fiction and convention, for example, when he presents the 0-degree mark on a thermometer, or the idea of some absolute point that is at rest with respect to all bodies in the Universe (body Alpha) as so many fictions (ideas inappropriate to the real). In reality, Scholz argues, they are mere conventions, that is, convenient landmarks without any claim to truth, serving to measure relative intensities or distances (Scholz 1924, 256–257). Body Alpha refers to Carl Neumann's convention by which every motion in the universe is related to an imaginary and absolutely fixed body: the "body Alpha." By relating motion to the sole body Alpha, it becomes possible to consider all motion as absolute motion. Body Alpha thus serves to facilitate the presentation of an "absolute motion," which Neumann took as a necessary assumption for Galileo's and Newton's theories (Neumann 1870). According to Poincaré, however, absolute space and motion are useful "conventions." Scientists proceed "as if" these existed in order to formulate the laws of mechanics more simply (Poincaré 1902 [1905], 90). According to Vaihinger, absolute space and motion are "fictional concepts" that serve to simplify the formulation of laws and the calculus of motion (Vaihinger 1911 [1920], 105–106).

Admittedly, "fictions" and "conventions" have, as a common feature, to be introduced by the human mind, and, unlike physical "hypotheses," cannot be seen as discoveries about nature. But whereas convention is assumed as a pure expedient, distinct from truth and falsehood, fiction embraces the view of a falsity, a contradiction with reality. In view of such motives, the concept of fiction is much more problematic from the pragmatist perspective to be developed here than the concept of convention.

c. Contingency or Necessity of Some Assumptions?

Vaihinger takes notions or fictional propositions that enter into the scientific process to be "arbitrary" and "contingent" constructions. Such propositions are liable to be altered or superseded in the course of history by other, more efficient or convenient fictions. They hold so long as nothing better is found. For instance, the Goethean fiction of a primitive animal from which any other would descend was merely tentative: it guided and favored the Darwinian thesis of the evolution of species by natural selection. Such a fiction may be renounced once its heuristic job is done. Likewise, fictions introduced into mathematical calculus (infinitesimals, imaginary numbers, and so on) serve to unblock a situation and move toward the final result (Vaihinger 1911 [1920], 202 sq.). The same, according to Vaihinger, goes for assumptions introduced in the economical sciences. Thus Adam Smith, in his *Inquiry on the Wealth of Nations*, states as an axiom that human actions are driven by a desire for profit, hence selfishness alone. From such a fictional assumption, he derives the laws supposed to rule trading relations between human beings. According to Vaihinger, this is a "provisory assumption" to be deleted once the laws in consideration are derived. By no means does Adam Smith confer it a truth-value. The assumption only served as a starting-point for inquiry. Fictions serve to make discoveries, to state laws, but they should be renounced once their job is done. There are very few cases in which Vaihinger admits the *necessary* character of assumptions or fictions. I have recorded just one, with respect to law: the defense attorney *must* presuppose determinism in order to prepare his client's defense, contrary to the prosecuting attorney who *must* presuppose the defendant's free will (Vaihinger 1911 [1920], 167).

Unlike Vaihinger, Poincaré stresses the entirely *necessary* and *indestructible* character of unverifiable assumptions in the scientific process. One of the original features in Poincaré's pragmatism is its capacity to identify propositions that make scientific activity itself *possible*. In Kantian terms, these propositions would be said to have a transcendental character. In many passages, Poincaré accounts for the attitude of acting *as if* some unverifiable propositions were true, showing that without this attitude the scientist would be unable to secure at least three of his main purposes: to discover new laws, to generalize, and to make inferences about future and past states of the world.

As for scientific *discovery*, Poincaré notes that one must *necessarily* act as if nature is unified. We must bend Nature "to our ideal of unity"

(Poincaré 1902 [1905], 173). Such an assumed unity of parts in nature is a necessary guide for discovering new laws. For instance, "the sentiment of the unity of Nature" eventually led scientists to relate venerable fluids of caloric, electricity and so on with each other. Assuming the unity or interdependence of parts in nature, they came to see "intimate relations" between these invisible fluids (Poincaré 1902 [1905], 169). In other words, scientists ceased to regard the time-honored fluids as isolated substances and began to see them instead as manifestations of one and the same sort of process.

Similarly, in *Savants et Écrivains*, Chapter X, Poincaré emphasized that it was the assumption of unity in nature that led the chemist Marcelin Berthelot to affirm *continuity* between organic and mineral matter, between laws ruling the former and those ruling the latter. The view of unity, or interdependence among natural substances led him to discover that such elements as oxygen, nitrogen, hydrogen and carbon, when combined with others, enter into the formation of the variety of animal and vegetal substances. Therefore, the thesis of organic matter as essentially distinct from the inorganic should be renounced (Poincaré 1910, 163–164).

d. Science for Science's Sake; Science for Action

Poincaré stresses much more than Vaihinger the disinterested character of scientific research. In this he distances himself from what could be called a radical pragmatism, which would make science dependent on action. Poincaré holds in effect that science sees action neither as an end nor as a foundation. Science aims at nothing but itself, at disinterested knowledge, at science for science's sake: "If we wish ever more to free man from material cares, it is so that he may be able to employ the liberty obtained in the study and contemplation of truth" (Poincaré 1905 [1958], 11). Poincaré recalls very often that the ultimate purpose of human activity lies in contemplation. And conversely, it is only as an exceptional case (I record only one) that Vaihinger considers knowledge for knowledge's sake (and not for efficient action). The point is that, according to Vaihinger, it is a "luxury" when science takes its own achievements to be an end in itself: "If science goes further and considers its construction as an end in itself, it is no longer concerned with improvement of its apparatus, it then becomes a luxury, properly speaking" (Vaihinger 1911 [1920], 95).

Luxury certainly brings pleasure and is worthwhile as such, but it is not indispensable, or essential to human existence. Efficient adaptation and insertion of our action into the world are more substantial goals for Vaihinger: "The ultimate and genuine goal of thought is action and making action possible" (*ibid.*). Science for science's sake, depicted by Vaihinger as a luxury one could forgo, is conversely depicted by Poincaré as the ultimate goal of scientific knowledge (Poincaré 1910, 138, 171).

Therefore, the relationship between science and action is not the same for both writers. Unlike Vaihinger, Poincaré holds that the essential purpose of science is not action, or possible efficient prediction, but knowledge itself:

> It should not even be said that action is the goal of science; should we condemn studies of the star Sirius, under pretext that we shall probably never exercise any influence on that star? To my eyes, on the contrary, it is knowledge which is the end, and action which is the means (Poincaré 1905 [1958], 115).

Poincaré reverses here the relationship introduced by Vaihinger between knowing and acting. He holds that action upon the world may be the means to scientific knowledge. The supreme end of labor and industrial development is, in effect, to provide the scientist with the means to experiment and, subsequently, to the expansion of scientific knowledge: "If I congratulate myself on industrial development, … it is above all because it gives the scientist faith in himself and also because it offers an immense field of experience …" (*ibid.*).

Finally, Poincaré, who sets disinterested contemplation of truth above utility, connects such contemplation as offered by science with an ethics of disinterestedness. Science for science's sake prepares us to rise beyond the crude pursuit of awards and privileges, while morally enlightening us:

> Whoever tastes, or sees the splendid harmony in natural laws, even if only slightly, will be prepared better than anyone to pay scant attention to his egoistical concerns; he will have an ideal he will cherish more than himself, and that is the only ground on which an ethics could be built up. Towards such an ideal, he will work without sparing any effort or accepting any crude awards, which are the only issue for some people; and once he gets into the habit of being thus disinterested, such a habit will follow him everywhere; his entire life will be somehow flavored by it (Poincaré 1913, 36; my translation).

Poincaré's connection between science and ethics is completely lacking in Vaihinger's analysis. Only Poincaré situates the supreme values of humanity beyond practical utility and practice.

V. Conclusion

In the light of Vaihinger's version of pragmatism, Poincaré can no longer be associated with "partial pragmatism" (Berthelot 1911, 197). Berthelot saw pragmatism as the sacrifice of truth in favor of convenience and usefulness, and he judged Poincaré to be a pragmatist in geometry and mechanics only where axioms are selected for the sale of convenience, but not in other areas of science (Berthelot 1911, 222). The Vaihingerian "pragmatism" contains a number of original features that help cast Poincaré's epistemology as a *globally* pragmatist one – in spite of certain differences between Poincaré and Vaihinger. Let us sum up the essential traits of this pragmatism:

1) In order to perform *operations*, scientists should not keep to logic and experience alone, but should accept the introduction of some illogical views (e.g., the completed infinite for mathematical generalization), or views not validated by experience (e.g., the conventional hypothesis of an absolute determinism for predictions).

2) What conveys value to a view is primarily the action or scientific operation it makes possible. A view obtains depending upon what it helps to accomplish. This feature is proper to pragmatism, which, to recall, is derived from the Latin "*pragmaticus*," meaning skillful, efficient.

3) The criterion for adopting a "conventional" (Poincaré) or "fictional" view (Vaihinger) may be to optimize action (in the broad sense) by living creatures within their environment. This is, according to both writers, with regard to Euclidean space: such a space is retained not only because it corresponds to physical space, but because it is better suited to help humans in their activities.

4) The introduction of imaginary items (conventions, fictions, hypotheses) in the empirical sciences is not an obstacle to reaching the truth (the perception of the lawlike structure of Nature). In this sense, one can talk about a non-skeptical pragmatism.[13]

In closing, let us ask whether Poincaré's epistemology motivated the Vaihingerian version of pragmatism. It does seem to be so. In fact, Vaihinger pays a strong tribute to Poincaré for having given a "practical"

---

13 Poincaré gives a good example in the following passage: "If you put the question to me: Is such a fact true? I shall begin by asking you, if there is occasion, to state precisely the conventions, by asking you, in other words, what language you have spoken ; then once settled on this point, I shall interrogate my senses and shall answer yes or no" (Poincaré 1958, 118).

justification of some views without any theoretical validity in science. In the same passage, he presents *La Science et l'hypothèse* as a fundamental, *grundlegend* work (Vaihinger 1911 [1920], xvii). Such a practical justification of the hypothetical (in a broad sense) in science, which permeates Poincaré's entire oeuvre, became the guiding line of Vaihinger's philosophy, at the cost of some adjustments. The main difference is that Vaihinger regards all assumptions that make science possible as useful fictions, while Poincaré regards some of them (like conventional principles of physics for exemple) as theoretically credible.

One last point requires clarification: why should Poincaré and Vaihinger be considered "pragmatists," if they did not declare themselves as such? They are pragmatists insofar as they believe that from the positive practical implications of certain ideas, the value of these ideas may be determined, such implications being conceived in terms of operational convenience and of fruitfulness. Seizing upon the practical implications of ideas in order to establish their value is an essential trait of pragmatism, to be found in its various historical manifestations. Furthermore, when read in the light of pragmatism, the epistemologies of Poincaré and Vaihinger show what I see as their unity: following the practical, rather than the strictly theoretical value of ideas is the guiding thread of their justification of the hypothetical in the scientific attitude.

Bibliography

Bernard, Claude (1865 [1927]), *An Introduction to the Study of Experimental Medicine.* New York: The Macmillan Company. (First French ed. 1865).

Berthelot, René (1911), *Un romantisme utilitaire.* Vol. 1: *Le pragmatisme chez Nietzsche et chez Poincaré.* Paris: Félix Alcan.

Einstein, Albert (1919 [1979]), "Ein Brief Albert Einsteins an Hans Vaihinger vom Jahr 1919", *Zeitschrift für allgemeine Wissenschaftstheorie* 10: 318–319.

Fine, Arthur (1993), "Fictionalism", *Midwest Studies in Philosophy* 18: 1–18.

Heinzmann, Gerhard (1995), *Zwischen Objektkonstruktion und Strukturanalyse. Zur Philosophie der Mathematik bei Jules Henri Poincaré.* Göttingen: Vandenhoeck & Ruprecht.

— (2006), "Henri Poincaré et sa pensée en philosophie des sciences", *L'héritage scientifique de Poincaré.* Ed. Éric Charpentier, Étienne Ghys, and Annick Lesne. Paris: Belin, 404–423.

Hwang, Philip H. (1993), "Aristotle and Peirce on chance", in *C. S. Peirce and the Philosophy of Science.* Ed. Edward C. Moore. Tuscaloosa and London: University of Alabama Press, 262–278.

James, William (1907 [1975]), *The Meaning of Truth: A Sequel to Pragmatism.* New York/London: Longmans, Green.
— (1909), *Pragmatism: A New Name For Some Old Ways Of Thinking.* Cambridge and London: Harvard University Press.
Lalande, André (1906), "Pragmatisme et pragmaticisme", *Revue philosophique de la France et de l'étranger* 61(2): 121–146.
Ly, Igor (2004), "Identité et égalité. Le criticisme de Poincaré", *Philosophiques* 31/1: 179–212.
Neumann, Carl (1870), *Über die Prinzipien der Galilei-Newtonschen Theorie.* Leipzig: Teubner.
Peirce, Charles Sanders (1931–1958), *The Collected Papers of C. S. Peirce* (abbrev. CP), 8 vols. Ed. Charles Hartshorne and Paul Weiss. Cambridge, Mass: Harvard University Press. (Quoted by volume and paragraph).
Poincaré, Henri (1902 [1968]), *La Science et l'hypothèse.* Paris: Flammarion.
— (1902 [1905]), *Science and Hypothesis.* London and Newcastle-on-Tyne: Walter Scott. (Transl. of the foregoing).
— (1905 [1970]), *La Valeur de la science.* Paris: Flammarion.
— (1905 [1958]), *The Value of Science.* New York: Dover Publications. (Transl. of the foregoing).
— (1908 [1914]), *Science and Method.* London/Edinburgh: Thomas Nelson.
— (1910), *Savants et Écrivains.* Paris: Flammarion.
— (1913), *Dernières pensées.* Paris: Flammarion.
— (2002), *L'opportunisme scientifique.* Compiled by Louis Rougier, ed. Laurent Rollet. Basel/ Boston/ Berlin: Birkhäuser.
Roustan, Désiré (1914), "La science comme instrument vital". *Revue de Métaphysique et de Morale*, Année 22: 612–643.
Schiller, Ferdinand C. S. (1907), *Studies in Humanism.* London: Macmillan.
Scholz, Walter (1924), "Kritischer Konventionalismus und Philosophie des Als Ob". *Annalen der Philosophie und philosophischen Kritik* 4 (4–5): 253–263.
Tiercelin, Claudine (1993), "Peirce's realistic approach to mathematics: Or, can one be a realist without being a Platonist". In *C. S. Peirce and the Philosophy of Science.* Ed. Edward C. Moore. Tuscaloosa and London: University of Alabama Press, 30–48.
Vaihinger, Hans (1911 [1920]), *Die Philosophie des Als Ob. System der theoretischen, praktischen und religiösen Fiktionen der Menschheit auf Grund eines idealistischen Positivismus.* Mit einem Anhang über Kant und Nietzsche. Leipzig: Felix Meiner. (First ed. 1911).
— (1911 [1923]), *Volksausgabe der Philosophie des Als Ob.* Leipzig: Felix Meiner.
— (1911 [1924]), *The Philosophy of As If. A System of the Theoretical, Practical and Religious Fictions of Mankind.* Transl. C. K. Ogden. London: Routledge & Kegan Paul.

Werner Heisenberg's Position on a Hypothetical Conception of Science

Gregor Schiemann[1]

Abstract:

Werner Heisenberg made an important – and as yet insufficiently researched – contribution to the transformation of the modern conception of science. This transformation involved a reassessment of the status of scientific knowledge from certain to merely hypothetical – an assessment that is widely recognized today.

I examine Heisenberg's contribution in particular by taking his conception of "closed theories" as an example according to which the established physical theories have no universal and exclusive, but only a restricted validity. Firstly, I characterize the historical process of hypothetization of claims to validity. Then, secondly, I reconstruct Heisenberg's conception, as far as it can be derived from his popular writings, relating it to the process of hypothetization. Finally, I touch on the history of its reception and compare it with conceptions of science that emphasize the significance of the hypothetical for the modern theories of natural sciences. Compared to these conceptions, Heisenberg's contribution turns out to be rather independent.

Werner Heisenberg made an important – and as yet insufficiently researched – contribution to the transformation of the modern conception of science. This transformation involved a reassessment of the status of scientific knowledge from certain to merely hypothetical – an assessment that is widely recognized today. The beginnings of this process can be traced back to the nineteenth century (e.g. John Herschel, William Whewell and Hermann von Helmholtz).[2] Taking Heisenberg as an example, I would like to investigate the influence of the foundation of quantum mechanics, which shaped his conception of science, on the relativization of claims to truth.

Heisenberg's conception of science itself, however, underwent a transformation and is not free of contradictions.[3] By restricting his ma-

1 Slightly revised and translated version of Schiemann 2007.

2 On John Herschel and William Whewell see Snyder 2009, on von Helmholtz see Schiemann 1997.

3 I have presented Heisenberg's conception of science in the context of a reconstruction of the central elements of his thought in Schiemann 2008.

trix mechanics to the calculation of measurable quantities, in 1925 Heisenberg tried to give quantum mechanics a foundation as free of hypotheses as possible. Later, questions concerning the reality of theoretical entities and about the truth of atomic theories, as well as their relation to other physical theories and to the concept of knowledge in the natural sciences – without impugning the formal structure of the foundation – led in the late 20s to a partial hypothetization of claims to truth.

Heisenberg's efforts towards a solution to these theoretical problems are closely linked to *practical contexts*. His enquiry into the question of claims to truth was influenced by exchanging ideas with other quantum physicists. It appeared to be part of Heisenberg's individual career strategy; within physics, it helped to implement a certain understanding of physics and to organize research in physics. Last but not least, it reacted to the political attacks that Heisenberg and other scientists had been facing since the Nazis came to power.

My contribution focuses on Heisenberg's conception of science, as far as it can be derived from his *Popular Speeches*. In her much-noticed book on the justification of quantum mechanics, Mara Beller referred to the general context-dependency of these texts (Beller 1996, 196). Heisenberg's biographer, David C. Cassidy, also supported this opinion (Cassidy 1992, 255). Inconsistencies that may be found between the contents of the various speeches are due, they claim, to the fact that Heisenberg intended to achieve different effects with different audiences. In my view, however, the interpretative possibilities of Heisenberg's conception of science depending on practical contexts are clearly of minor importance. The fundamental elements of his conception persist from his early lectures in the 1930s to his late lectures of the 60s and 70s. They are characterized on the one hand by an ambivalence which emphasizes the certainty of scientific findings, while at the same time, admitting that their claim to truth is limited on systematic and historical grounds.[4] On the other hand, Heisenberg constantly struggles to overcome this conflict. His striving for a unified solution is most clearly expressed in his "conception of closed theories." Following its first programmatic formulation in his speech "Recent Changes in the Founda-

4 The ambivalence of Heisenberg's position on the question of the truth of scientific knowledge has not been discussed in secondary literature. Discussion has focused one-sidedly either on his orientation towards absolute claims to validity (Beller 1999, Carson 1995) or on his rejection of the classical ideal of objectivity (Chevalley 1988).

tions of Exact Science" ("Wandlungen in den Grundlagen der Naturwissenschaft"), which he held at the general meeting of the Association of German Natural Scientists and Physicians in 1934 (Heisenberg 1934), Heisenberg stood by his conception for all his life and without substantial changes, despite its varied theoretical and practical contexts.[5]

My paper is structured as follows: first, I am going to characterize the historical process of hypothetization of claims to truth. Secondly, I will then reconstruct Heisenberg's conception, as far as it can be derived from his popular Writings, relating it to this process of hypothetization. In the third part, I want to touch on the history of its reception and compare it with conceptions of science that emphasize the significance of the hypothetical for the modern theories of natural sciences. Compared to these conceptions Heisenberg's contribution turns out to be rather independent.

1. The Process of Hypothetization of Claims to Truth

Hypotheses have been known in science ever since its origins in antiquity. And yet it is only in the last 200 years, I contend, that they have begun to constitute a distinguishing feature of knowledge and self-conception in the natural sciences. The historical process of hypothetization of claims to truth is closely connected with emergence and decline of the mechanical worldview of classical physics. There are *two ways* – not always sharply distinguishable from each other – in which a statement can be said to be hypothetical: its truth is either assumed to have not yet been established, or it is considered not to be verifiable at all. In the first case, a provisional assumption can be confirmed, disproved or corrected in the course of further research. The second type of hypothesis, by contrast, is not expected to lose its nature of being open to truth. Both types can exist side by side in a given theory. A theoretical supposition about empirical objects can be assumed to be converted into truth, whereas a metaphysical assumption of the same theory can be believed to remain permanently hypothetical. But a competitive situation may just as well develop between the two types. This occurs, for instance, if the justification of one of the types is fundamentally dis-

5 Heisenberg's most important presentations of this conception are Heisenberg 1934, Heisenberg 1948, Heisenberg 1959, ch. VI, and Heisenberg 1973. On the secondary literature cf. section 4.

puted by some scientists. In this case – as some scientists claim – the possibility of being able to transform certain hypotheses into truth is challenged or, vice versa, the uncatchable openness to truth of all hypotheses – as assumed by others – is rejected.

I presume that both definitions of the hypothetical were useful for a long historical period, although the concept of truth during that period was subject to considerable transformations. The historical process of *hypothetization of claims to truth* that has occurred can be very briefly described as follows:

Since the beginning of the early modern era, i.e. the 16th and 17th centuries, *provisional assumptions* have become more and more important in the pursuit and formulation of knowledge in the natural sciences. The question of the claim to knowledge was discussed from the very beginning. Johannes Kepler's, René Descartes's, Isaac Newton's and Gottfried Wilhelm Leibniz's concepts of hypothesis are vivid examples of this.[6] The hypothetization that soon set in was spurred on by questions arising in practical research about observability – e.g. in the use of optical instruments –, and the role of models – e.g. mechanical models of submicroscopic processes. Provisional assumptions became the basis for the hypothetico-deductive model of theory construction as it is standard today. They are also found in explanations that reduce phenomena to processes as yet uninvestigated but assumed to exist, in predictions, in calculations using models and simulations – to give just a few examples.

The second form, however, the *abandonment of claims to truth*, was for a long time categorically rejected by modern natural science. The mechanistic worldview, prevalent in the classical physics of the 18th and 19th century, still postulated final, non-hypothetical knowledge and restricted the use of hypothetical statements to the first kind of hypotheticity. It was David Hume's empiricism that first brought a great deal of attention in the English-speaking world to the second kind of hypotheticity. According to Hume, only inductively secured knowledge could be gained from experience. Induction, however, does not hold by logical necessity. Thus, he argued, empirical knowledge can at best have a hypothetical character.

In the second half of the 19th century, the hypothetical character of the natural sciences increasingly caught the attention of natural scientists

6 On Kepler, Descartes, and Newton see McMullin 2009.

and philosophers in Germany, France and Britain.[7] During the 20th century, a hypothetical conception of the natural sciences was established, which differed from previous conceptions of science in its positive evaluation of the concept of hypothesis, as well as in the historical consciousness related to it. It rejects the absolute claim to validity of classical physics and is convinced that all knowledge in the natural sciences, including the fundamental theories, may be subject to a revision. There are mainly two approaches in the philosophy of science that express this view: Karl R. Popper's falsificationism and Pierre Duhem's and Willard van Orman Quine's thesis of the empirical underdetermination of theories.[8]

2. Heisenberg's Conception of Closed Theories[9]

Heisenberg's writings on the history and theory of physics are relevant to the two meanings of the hypothetical on account of his ambivalent position regarding scientific claims to truth. On the one hand he proceeds from the assumption that theories are hypothetically conceived structures which lose their openness to truth in the course of the continuing experimental investigation of the theories relating to empirical data. Thus, in a speech delivered in 1946 and entitled "Science as a Means of International Understanding" ("Wissenschaft als Mittel zur Verständigung unter den Völkern"), he says:

> I learnt "that in science the question of what is right and what is wrong is ultimately always answered: that it is not a matter of faith or worldview or hypothesis, but that one particular assertion simply turns out to be correct and another false; the decision about what is right is made ... by nature or if you will, by God, but certainly not by man" (Heisenberg 1946, 386).[10]

7 One example for Germany is presented by Pulte 2009, one for France by Heidelberger 2009.

8 On Duhem and Popper see Bartels 2009.

9 Cf. Schiemann 2008, Chap. III.2.

10 Ich lernte, "daß man nämlich in der Wissenschaft schließlich immer entscheiden kann, was richtig und was falsch ist, daß es sich hier nicht um Glauben oder Weltanschauung oder Hypothese handelt, sondern daß schließlich eine bestimmte Behauptung eben einfach richtig ist und die andere unrichtig; und welche richtig ist, darüber entscheidet ... die Natur oder, wenn Sie so wollen, der liebe Gott, jedenfalls nicht die Menschen."

What Heisenberg means by "right" is truth that is not a subject in non-scientific discourses, which are concerned solely with "belief," "world-view" or "hypothesis." In science, hypotheses arise in the early stage of the development of theories and in their application to new fields. In the structures of truth derived from theories one is confronted, he says, with "pure and unconcealed truth" (Heisenberg 1946, 393). This absolute conception of truth can be regarded as an expression of his Platonism.[11] According to this, science develops from hypotheses to knowledge which permits recognition of the fundamental laws of the world existing independently of this knowledge.

However, the historical change of physical theories at the beginning of the last century has unsettled the faith in the validity of their truth and thereby lent them a hypothetical character. Scientific claims to the truth have proved to be partly wrong and partly incompatible with one another in terms of content.[12] The ideal of objective knowledge, i. e. knowledge independent of the circumstances of observation, has proved to be realizable only to a limited degree.[13] It has become questionable whether a unified understanding of nature can be attained.[14] The partial unification of physical theories requires a high degree of abstraction that increasingly diverges from the concrete appearance of phenomena and it is capable of relating to the phenomena only hypothetically.[15] What is more, the pragmatic orientation of experimental research counteracts a truth-oriented process of knowledge acquisition:

> As practical activity became [in our century] the focal point in our view of the world, so the fundamental [scientific] thought patterns lost their absolute significance … In science one became increasingly aware that our understanding of the world cannot proceed from any certain knowledge, cannot be founded on the rock of such knowledge, but that rather all knowledge is suspended as it were above a bottomless abyss. (Heisenberg 1946, 391).[16]

11 Cf. Liesenfeld 1992, Schiemann 2008 Chap. III.3.

12 Cf. below the example of quantum mechanics.

13 E.g. Heisenberg 1954, 399 ff.; Heisenberg 1959, 166 ff.

14 Heisenberg 1934, Heisenberg 1959, 96.

15 Cf. Heisenberg 1948, 338 f.; Heisenberg 1973, 418.

16 "In dem Maß, in dem das praktische Handeln [in unserem Jahrhundert] in den Mittelpunkt des Weltbildes rückte, verloren die grundlegenden [wissenschaftlichen] Denkschemata ihre absolute Bedeutung … In der Wissenschaft wurde man sich immer mehr dessen bewußt, daß unser Verständnis der Welt nicht mit irgend einer sicheren Erkenntnis beginnen kann, daß es nicht auf dem

We have thus a divergence between Heisenberg's orientation towards an absolute concept of truth and his awareness of the inevitable hypotheticity of scientific knowledge. The way he tries to resolve this conflict is made clear by his critique of the claim to validity of classical physics.[17] It is this critique that leads him to the conception of closed theories. The theories of classical physics, Heisenberg claims, have *no universal, but only a limited validity* and the limited fields of application are determined by the fundamental concepts of the theories concerned. Heisenberg attributes this function of fundamental concepts to the fact that experiments make them closely rooted in the kind of experience that is specific for the particular fields of application. The differences between the fields of experience reflect the differences between the concepts of the respective theories. In regard to the relations between the concepts of different theories, Heisenberg tolerates contradictory definitions. However, the differences between the experiences or fields of application do not prevent points of contact and overlap. The realm of one theory can also include the realms of other theories. Theories, therefore, have *no exclusive validity*.

Heisenberg specifies this structure of theories with his conception of closed theories. In this context, the term "closed" has a systematic and a historical meaning. The *systematic* meaning refers to the relation between coexisting closed theories. A closed theory forms a system that is closed in itself (Heisenberg 1934, 100), in which concepts are distinguished from laws and interconnected through a consistent system of axioms. The connection between these concepts is "so tight that it is generally impossible to change any of these concepts without destroying the whole system at the same time" (Heisenberg 1959, 81). This determination leads over to the *historical* meaning of the term "closed." Closed theories cannot be improved by minor modifications, i. e. modifications of their laws (Heisenberg 1973, 417). Major modifications, i. e. changes of concepts, lead to new concepts and theories (Heisenberg 1934, 100; Heisenberg 1959, 84; cf. Scheibe 1993, 252). Therefore,

Fels einer solchen Erkenntnis gegründet werden kann, sondern daß alle Erkenntnis gewissermaßen über einer grundlosen Tiefe schwebt"

17 For Heisenberg, the validity of a theory is founded upon its truth content. I follow here this understanding of the concept..

theories are also closed, because their historical development has come to an end.[18]

Heisenberg considers *four theories of physics* to be closed: Newton's mechanics, electrodynamics, including the special theory of relativity, thermodynamics, including its statistical version, and quantum mechanics. From the fact that they are closed, he draws the conclusion that they are "valid forever": "wherever experiences can be described with the concepts of these theories, be it in the most distant future, the laws of these theories will always prove to be right" (Heisenberg 1948, 339). This formulation is circular, as far as only those experiences are concerned to which the concepts are applicable. As there is always, however, a difference between concepts and experiences, the formulation is not tautological. Correspondingly, it extends to "all times," thus referring to validity that is not timeless, but invariant for all circumstances that are conceivable at a certain time – and therefore of a final nature.

The view of a *far-reaching stability* of reliable physical theories that is hereby expressed shows considerable plausibility; this is what makes Heisenberg's concept really interesting. Newton's mechanics, electrodynamics and thermodynamics continue to be of central importance to the engineering sciences where they find their most important application today; quantum mechanics has proved its validity in an unprecedented way during the 80 years since it was founded. Heisenberg's idea that theoretical stability is grounded in the close relation between concepts and experience is reflected also in present philosophy of science, such as, for example, in Ian Hacking (1992, 30).

By claiming a final validity for particular theories Heisenberg formulates a *non-hypothetical feature* of physical knowledge, which is only in part still linked to the traditional claims to truth – a relation which Heisenberg does not discuss explicitly. The mechanistic worldview of classical physics aimed at the unity of content-related knowledge. Heisenberg contests this aim by renouncing the universal validity of the theo-

18 Heisenberg's most important presentations of his conception are Heisenberg 1934, Heisenberg 1948, Heisenberg 1959, ch. VI, Heisenberg 1969, ch. 8, Heisenberg 1970 and Heisenberg 1973. In Heisenberg 1948, 338 f. he mentions four characteristics of closed theories: 1. inner consistency and axiomatization of concepts, 2. representation of experience within a certain area of application, 3. missing knowledge about the exact boundaries of the application areas, 4. basis for further research. On the secondary literature cf. section 2. On the characteristics of Heisenberg's closed theories cp. Scheibe 1993, Frappier 2004, Bokulich 2006.

ries. The closed character of the theories, which cannot be anymore improved upon, could oppose their unity. If the natural sciences consisted entirely of closed theories, they would acquire a plural character:

> [T]he edifice of the exact natural science" "can therefore scarcely become ... a coherent unity such that one could get from one point in it to all other rooms of the building simply by following a prescribed route. Rather, it is composed of various parts, each of which, although linked to the others in manifold ways, ... still represents an integral unit in itself. (Heisenberg 1934, 101).[19]

This view of scientific knowledge, akin to William James's pluralism (James 1925), does not necessarily contain a renunciation of claims to validity. The closed theories after all remain valid "for all times" in their particular fields of application. Heisenberg leaves room for doubt with regard to the claim to unity of knowledge, but not with regard to its final validity in individual areas. To put it differently, he saves the temporal dimension of the classical claim to truth by separating it from the claim to unity. For Heisenberg, it is still an objective of science to search for an all-embracing theory; but the attainment of such a theory may not be possible any more. The distancing from universal claims to validity could be attributed to the fact that Heisenberg doubted their finality and tried to save this attribute of traditional concepts of truth at least for particular theories.

Heisenberg discusses the anti-hypothetical content of his concept of truth in various places, for instance, when he investigates the question of the consequences of the *emergence of quantum mechanics* for the claim to truth of classical theories (Heisenberg 1936, 110 f.). Heisenberg believes that, against the background of modern theories, classical theories are not false, but remain true in their fields of application. It was wrong, however, that they went beyond their realm of application (e.g. by means of mechanical models of the atom). But only in the light of new theories does this inadmissible transgression become visible. Heisenberg did not doubt that, in the future, the applicational limits of quantum mechanics could similarly be determined by new theories,

19 "Das Gebäude der exakten Naturwissenschaft kann also kaum ... eine zusammenhängende Einheit werden, so, daß man von einem Punkt in ihm einfach durch die Verfolgung des vorgeschriebenen Weges in alle anderen Räume des Gebäudes kommen kann. Vielmehr besteht es aus einzelnen Teilen, von denen jeder, obwohl er zu den anderen in den mannigfachsten Beziehungen steht, ... doch eine in sich abgeschlossenen Einheit darstellt." Cf. also Heisenberg 1959, 96.

just as the limits of classical theories in the past. Through this, the conditions of a final validity would also be more specified for quantum mechanics.

The fact that the demarcations between the realms of application of closed theories can only be determined in the historical process refers to a first aspect of *the irreducible hypothetization* of these theories in Heisenberg's conception. The *determination of demarcations* can change all previous assumptions about the extension of the realms of validity. This process could be accomplished only with an ultimate closed theory. Prior to such a final circumstance, whose future realization Heisenberg does not rule out (Heisenberg 1970, 390 ff.), the limits of all closed theories cannot be determined with certainty. Therefore, statements regarding phenomena acquire a hypothetical character, which relates to Heisenberg's *historical conception of knowledge*, i. e. to his conviction that the validity of certain aspects of knowledge are subject to historical change. Knowledge yielded by closed theories does not expand without repercussions for its already existing elements. The degree of their certainty grows with the increasing determination of their limits without being required to reach final exactness. Consequently, the finality of the theories does not rule out that their statements are of a hypothetical nature – whether provisionally or in principle.

Against this background Heisenberg establishes that:

> The closed theory does not contain any statements that are entirely reliable of the world of experiences. It remains – strictly speaking – doubtful and simply a question of success up to what extent the phenomena can be dealt with by the concepts of this theory … Despite this uncertainty, the closed theory remains a part of our scientific language (Heisenberg 1948, 339).[20]

In Heisenberg's conception there are two more aspects of a hypothetization of scientific knowledge which, albeit to a different extent, are likewise attributed to the historicization of scientific knowledge. While the aforementioned aspect concerns the objects of experience that were wrongly counted as lying within a field of application, the second aspect is about the *relation between experience and theory* within the

20 "Die abgeschlossene Theorie enthält keine völlig sichere Aussage über die Welt der Erfahrungen. Denn wie weit man mit den Begriffen dieser Theorie die Erscheinungen greifen kann, bleibt im strengen Sinne unsicher und einfach eine Frage des Erfolgs … Trotz dieser Unsicherheit bleibt die geschlossene Theorie ein Teil unserer naturwissenschaftlichen Sprache."

boundaries of a field of application. According to Heisenberg, the axiomatization of theories guarantees their consistency and determines certain meanings of the concepts. He assumes, however, the experience in the particular field of application to be changeable. This transformability is covered by the meanings of the concepts only to a limited degree (Heisenberg 1948, 338 f.; Heisenberg 1973, 418).

This aspect of the hypotheticity of closed theories is associated with Heisenberg's *romantic reference to Plato's theory of forms*. In his speech "On the History of the Physical Interpretation of Nature," given in 1933, he differentiates with Plato "four levels of knowledge": knowledge of essence (episteme), insight (dianoia), belief (pistis) and conjecture (eikasia).[21] Whereas Plato relates episteme only to the world of ideas, Heisenberg understands it as a form of knowledge of nature separate from dianoia (Heisenberg 1933, 54). According to him, episteme means an understanding of nature that is immediate, clear and qualitative in character, and is therefore directly connected to experience; in contrast to this, dianoia signifies a quantitative description of nature, which is accomplished through increasing axiomatization. He states that natural science has augmented the amount of dianoia and increasingly drifted away from episteme (Heisenberg 1933, 56). The immediate understanding of the nature of episteme, of which Goethe's conception of nature is paradigmatic in Heisenberg (Heisenberg 1967, 405), allows an awareness of the change of the concrete diversity of phenomena and thus forms a critical instance against the claim to truth of the mathematical knowledge of nature. Heisenberg's Platonism therefore allows a justification not only of absolute claims to validity, but also of the diversity of phenomena.

Like the first one, also the third aspect of the hypothetical in Heisenberg's conception concerns the relations between closed theories. New theories emerge, because old ones fail when expanding their fields of application. Putative areas of application of an old theory turn out to be those of a new theory. Consistent relations between the concepts of the old and the new theory are not necessary. Yet the laws of the old theory can emerge as borderline cases from the laws of a new theory. As borderline cases, the phenomena of an old theory can be understood by means of the concepts of the new theory. The emergence of new theories leads therefore to the possibility of empirically equivalent descriptions of objects by theories that might be conceptually incompati-

21 Plato, Pol. XI 509d–511e.

ble. The *empirical underdetermination* of closed theories not discussed by Heisenberg is expressed in this feature. As long as the unity of knowledge remains a part of the definition of truth, underdetermination results in the hypotheticity of scientific knowledge.

To summarize, I wish to claim that, with the concept of closed theories, Heisenberg took a step towards a conception of science that recognized the hypothetical character of scientific knowledge. Paradoxically, with this step he tried to save the classical claim to the final nature of knowledge. He abandoned the necessary classical conditions of the universality and exclusivity of truth and conceded that there were uncertainties in the relations between closed theories as well as between one theory and its corresponding realm of experience. In terms of the historical development sketched in the first section, Heisenberg stands at the transition from an early modern claim to truth to a hypothetical conception of science.[22]

3. The Reception of Heisenberg's Conception

Despite its plausibility, Heisenberg's conception has not received significant attention in the philosophy of science up to now. It has mainly been reflected in the works by Carl Friedrich von Weizsäcker (1974) and in the circle of his former colleagues (Scheibe 1993, Böhme et al. 1973, Böhme 1970 f.). Today, his conception attracts interest above all because of its historical significance for the understanding of quantum mechanics (Beller 1999, Chevalley 1988, Bokulich 2006) and sometimes also in the philosophy of science (cp. the aforementioned Hacking 1992). Up to now, little attention has been paid to the fact that closed theories are a suitable object to study the process of transformation of claims to truth within the sciences.

I have already mentioned Popper's falsificationism and the thesis – attributed to Duhem and Quine – of empirical underdetermination as paradigms of the hypothetical in the *theory of science.* In the *history of science,* it was Thomas S. Kuhn who laid some of the groundwork on a hypothetical conception of science. To conclude, I want briefly to outline the relation of Heisenberg's conception to these paradigms of the hypothetical.

22 In this classification there is a similarity with Hermann von Helmholtz's conception of science, which is analyzed in Schiemann 1997.

In order to characterize the relation to Popper's philosophy of science in one sentence one could say that Heisenberg's conception of closed theories keeps a clear distance from *falsificationism*, because the claim to finality excludes the possibility of refutation. In his later writings, however, Heisenberg retracts his proposition that closed theories are irrefutable (Heisenberg 1973, 418). He is able to refer to his understanding of finality, which he conceives as not being absolute and timeless, but rather as extending over a relatively long period of time. The non-holistic character of closed theories, in which no concept "in general" is changed "without the whole system being destroyed" (Heisenberg 1959, 81, see above), also weakens any immunity to refutation.

There are different versions of *Duhem's and Quine's underdetermination thesis*. I choose one that appears to me to be the most appropriate to be compared with Heisenberg's conception. Accordingly, a theory is underdetermined if its empirical evidence is not sufficient to confirm or invalidate it. If empirical evidence were the only criterion for assuming or refuting a theory, no decision could be made to decide between logically incompatible theories that relate to the same object. Similarly, Heisenberg's conception leads to the possibility of logically incompatible closed theories in a certain field. He does not, however, reduce the criterion of theory selection to empirical evidence. The concepts of closed theories remain more appropriate to their fields of application than the concepts of those theories whose borderline case can be calculated.

The domain of a closed theory can also lose any relevance. In his first discussion of the concept of a closed theory in 1934, Heisenberg compares the relation of closed classical theories to quantum mechanics with the relation of disc and sphere theories of the earth. The disc theory was replaced by another conceptual system, where parts of its domain as well as related questions no longer occurred (Heisenberg 1934, 98 and 100).

The emergence of new theories, separated by a conceptual gap from old ones, shows *similarities to Thomas S. Kuhn's theory* of the development of science. Heisenberg's conception has frequently been discussed as a precursor of that theory.[23] However, the succession of Kuhn's paradigms does not necessarily lead to progress in knowledge. In contrast, according to Heisenberg, new theories expand knowledge since they

23 On the relationship of Heisenberg and Kuhn cf. also: Beller 1999, Bokulich 2006, Van Dyck 2003.

deal with phenomena hitherto not yet investigated, or not correctly investigated, that differ from the phenomena of previous theories. While, according to Kuhn, the selection of theories depends considerably on non-scientific influences, Heisenberg believes that new closed theories emerge from interdisciplinary contexts. With respect to the conditions of the origin of quantum mechanics, he partly emphasizes the significance of the discussions held among physicists and of their intuition (Heisenberg 1969, 86), and partly claims that the new theory has been "imposed by nature" (Heisenberg 1934, 96).

A comparison of the three hypothetical conceptions of science with the conception of closed theories confirms that Heisenberg continues to orientate himself towards unconditional claims of truth. He (mostly) contests the possibility of any fundamental revision of scientific knowledge, the possibility of its equivalent representations and its historical relativity. The development of some established theories reflects advances in an improvement in knowledge which leads to knowledge within limited fields of application. The concepts of these theories are so well-adapted to experience that they are valid forever.

4. Conclusion

In conclusion I summarize my claims: according to Heisenberg, some established theories of physics function as so-called *closed theories*. Closed theories have a limited realm of application. Their concepts are particularly well adapted to the pattern of experience of their realm of application, within which they are valid for all conceivable circumstances. On the one hand, Heisenberg endorses the claim to final knowledge with this conception as it was typical of physics at the beginning of the early modern age. On the other hand, he relativizes claims to truth within the scope of his conception. The final certainty of physical knowledge does not go beyond closed theories and, as a consequence, ceases to be necessarily directed at the target of a uniform description of nature. There is then uncertainty about validity, *firstly* regarding the limits of realms of application, *secondly* regarding the concepts which, being fixed by axiomatized theories, only conditionally relate to the changing experience they refer to, and *thirdly,* regarding those realms of application that are covered by several closed theories at the same time.

Heisenberg's conception possesses *considerable plausibility* and does justice to the stability of established physical theories better than Popper's falsificationism. In systematic terms it displays similarities with the thesis of the empirical underdetermination of theories, and in historical terms with Kuhn's conception of the development of theories.

Bibliography

Bartels, Andreas (2009), "Hypotheticity and Realism – Duhem, Popper and Scientific Realism", in this volume.

Beller, Mara (1996): "The Rhetoric of Antirealism and the Copenhagen Spirit", *Philosophy of Science* 63(2): 183–204.

— (1999), *Quantum Dialogue: the Making of a Revolution*. Chicago: Chicago Press.

Böhme, Gernot (1979 f. [1980]), "Wie kann es abgeschlossene Theorien geben?", *Zeitschrift für allgemeine Wissenschaftstheorie* 10: 343–351. (Transl. as "On the possibility of closed theories", *Studies in History and Philosophy of Science* 11 (2): 163–172).

— (1973), "Die Finalisierung der Wissenschaften", *Zeitschrift für Soziologie* 2: 128–144

Bokulich, Alisa (2006), "Heisenberg meets Kuhn: Closed theories and paradigms", *Philosophy of Science* 72: 90–107

Carson, Cathryn L. (1995), *Particle physics and cultural politics: Werner Heisenberg and the shaping of a role for the physicist in postwar West Germany*. Ph. D. thesis, Harvard University, Cambridge, MA.

Cassidy, David C. (1992), *Uncertainty. The Life and Science of Werner Heisenberg*. New York: Freeman.

Chevalley, Catherine (1988), "Physical reality and closed theories in Werner Heisenberg's early papers", in: Dederik Batens and J. P. van Bendegem (eds.), *Theory and Experiment: Recent Insights and New Perspectives on Their Relation*. Dordrecht: Kluwer, 159–167.

Frappier, Mélanie (2004), *Heisenberg's notion of interpretation*, Ph. D. thesis, The University of Western Ontario, London/Ontario.

Hacking, Ian (1992), "The self-vindication of the laboratory sciences", in A. Pickering (Ed.), *Science as Practice and Culture*. Chicago: Chicago Press.

Heidelberger, Michael (2009), "Contingent Laws of Nature in Émile Boutroux", in this volume.

Heisenberg, Werner (1933), "Zur Geschichte der physikalischen Naturerklärung", in: Heisenberg 1984–1993, Vol. I, 50–61.

— (1934), "Wandlungen in den Grundlagen der Naturwissenschaft", in: Heisenberg 1984–1993, Vol. I, 96–101.

— (1936), "Prinzipielle Fragen der modernen Physik", in: Heisenberg 1984–1993, Vol. I, 108–119.

— (1941), "Die Goethe'sche und die Newton'sche Farbenlehre im Lichte der modernen Physik", in: Heisenberg 1984–1993, Vol. I, 146–160.

—— (1946), "Wissenschaft als Mittel zur Verständigung unter den Völkern", in: Heisenberg 1984–1993, Vol. V, 384–394.

—— (1948), "Der Begriff der 'Abgeschlossenen Theorie' in der modernen Naturwissenschaft", in: Heisenberg 1984–1993, Vol. I, 335–340.

—— (1954), "Das Naturbild der heutigen Physik", in: Heisenberg 1984–1993, Vol. I, 398–420.

—— (1959), "Physik und Philosophie", in: Heisenberg 1984–1993, Vol. II, 3–181.

—— (1966), "Die Rolle der phänomenologischen Theorien im System der theoretischen Physik", in: Heisenberg 1984–1993, Vol. II, 384–387.

—— (1967), "Das Naturbild Goethes und die technisch-naturwissenschaftliche Welt", in: Heisenberg 1984–1993, Vol. II, 394–409.

—— (1969), "Der Teil und das Ganze. Gespräche im Umkreis der Atomphysik." Ch. 5 (Talk with Einstein), in: Heisenberg 1984–1993, Vol. III, 3–334.

—— (1970), "Abschluss der Physik?", in: Heisenberg 1984–1993, Vol. III, 385–392.

—— (1973), "Die Richtigkeitskriterien der abgeschlossenen Theorien in der Physik", in: Heisenberg 1984–1993, Vol. III, 417–421.

—— (1984–1993), *Gesammelte Werke.* A: *Wissenschaftliche Originalarbeiten* (I: 1922–46, II: 1929–45, III: 1946–72), B: *Scientific Review Papers, Talks, and Books*, C: *Allgemeinverständliche Schriften* (I: 1927–1955, II: 1956–1968, III: 1969–1976). Berlin, New York: Springer. (Cited with series letter and volume; e.g. B III for vol. 3 of series B).

James, William (1914 [1925]), *A Pluralistic Universe.* New York: Longmans, Green and Co.

Liesenfeld, Cornelia (1992), *Philosophische Weltbilder des 20. Jahrhunderts: Eine interdisziplinäre Studie zu Max Planck und Werner Heisenberg.* Würzburg: Königshausen und Neumann.

McMullin, Ernan (2009), "Hypothesis in early modern science", in this volume.

Pulte, Helmut (2009), "From Axioms to Conventions and Hypotheses: The Foundations of Mechanics and the Roots of Carl Neumann's 'Principles of the Galilean-Newtonian Theory'", in this volume.

Scheibe, Erhard (1993), "Heisenbergs Begriff der abgeschlossenen Theorie", in: B. Geyer et al. (eds.), *Werner Heisenberg. Physiker und Philosoph.* Heidelberg: Spektrum, Akad. Verlag.

Schiemann, Gregor (1996), "Descartes' Hypothesenbegriff im *Discours de la méthode* und in den *Principia philosophiae*", in: Allgemeine Gesellschaft für Philosophie (ed.), *Cognitio humana – Dynamik des Wissens und der Werte.* XVIII. Deutscher Kongreß für Philosophie. Leipzig, 1006–1021.

—— (1997), *Wahrheitsgewissheitsverlust. Hermann von Helmholtz' Mechanismus im Anbruch der Moderne. Eine Studie zum Übergang von klassischer zu moderner Naturphilosophie.* Darmstadt: Wissenschaftliche Buchgesellschaft. (English transl.: *Herman von Helmholtz's Mechanism: The Loss of Certainty. A Study on the Transition from Classical to Modern Philosophy of Nature.* Dordrecht: Springer, 2009)

— (2007), "Werner Heisenbergs Position zu einer hypothetischen Wissenschaftsauffassung in seinen populären Reden und Aufsätzen", in: M. Gerhard (ed.), *Oldenburger Jahrbuch für Philosophie*: 9–27
— (2008), *Werner Heisenberg*. München: Beck.
Snyder, Laura (2009), "Hypotheses in 19th Century British Philosophy of Science", in this volume.
Van Dyck, Maarten (2003), "The roles of one thought experiment in interpreting quantum mechanics. Werner Heisenberg meets Thomas Kuhn". philsci-archive.pitt.edu/archive/00001158/. Also published in: *Philosophica* 72: 79–103.
Weizsäcker, Carl Friedrich von (1974), *Die Einheit der Natur*. München: Deutscher Taschenbuch Verlag.

"Instrumentalism" and "Realism" as Categories in the History of Astronomy: Duhem vs. Popper, Maimonides vs. Gersonides

Gad Freudenthal[1,2]

Abstract:

This paper traces the history of the modern philosophical notion of "instrumentalism," popularized through Pierre Duhem and Karl R. Popper, back to the ancient and medieval conflict between two theoretical bodies of knowledge: mathematical astronomy and physical theory. Since Antiquity, the former was admired for its reliability, but could not be accepted as a description of reality that had to conform to the Aristotelian ideal of science. Maimonides, who is at the center of this paper, formulated with particular acuity the idea that mathematical astronomy should be construed as an "instrument" allowing prediction, even though its models violate the laws of physics.

Maimonides used the conflict between these two irreconcilable bodies of knowledge to buttress his skepticism and cast doubt on Aristotle's cosmology: he sought to make plausible the competing, traditional (religious) doctrine, according to which the world was created out of nothing in a determinate past. This conflict between two bodies of knowledge also opened the way for him to make statements concerning the possibility that science might progress in the future, implying that the celestial science of his own day was "hypothetical."

The paper concludes with a short account of the corresponding ideas of Levi ben Gershom (Gersonides, 14th c.), who believed that it is possible to elaborate a theory of the celestial realm that satisfies the exigencies of mathematical astronomical science while also conforming to the ideals of Aristotelian science. Gersonides was a realist who stated his epistemological position with great clarity.

---

1 This article first appeared in: Peter Barker, Alan C. Bowen, José Chabás, Gad Freudenthal, and Y. Tzvi Langermann (eds.), *Astronomy and Astrology from the Babylonians to Kepler: Essays Presented to Bernard R. Goldstein on the Occasion of his 65th Birthday* (= *Centaurus* 45 [2003]), 227–48, and is reprinted here with slight modifications. I am grateful to the editors and publishers for the permission to reprint it.

2 For helpful observations on a draft of this paper I am indebted to Ruth Glasner, Orna Harari, Josef Stern, and particularly Alan Bowen. This paper is a tribute of warm friendship and esteem to Bernie Goldstein, a rare combination of a multi-dimensional erudition and an unflagging readiness to share it with others: I am deeply grateful to him for his unfailing availability for clever, judicious and competent advice and help over many years.

1. How "Instrumentalism" and "Realism" Entered the History of Science

The notion "instrumentalism" has entered the historiography of science fraught with ambiguity: it was simultaneously introduced (I) by philosophers, to denote a general philosophical stance in the theory of knowledge, and (II) by philosophers turned historians of science, to denote a particular view, held by some past philosophers and astronomers, of the aim and means of mathematical astronomy. Subsequently, however, historians of science have failed to distinguish between the two meanings of the term, and the result has been lasting confusion and misunderstanding. In what follows I shall try to clarify this issue and suggest how it may be avoided.

(I) The term "instrumentalism" seems to have been coined by Josiah Royce (1855–1916) in 1908 as a denomination for John Dewey's (1859–1952) pragmatism (Heede 1976). It subsequently made a first "trans-disciplinary migration" and entered the philosophy of science, perhaps through Hans Reichenbach (1891–1953; Morgenbesser 1969, 200), where it acquired a narrower meaning. The late Sir Karl R. Popper's (1902–1994) influential essay "Three Views Concerning Human Knowledge" (1956 [1963]) contributed to giving it wide currency. Popper introduced the notion via some historic examples (see below), but his concern was philosophical: the "instrumentalist view," he noted with regret, "has become an accepted dogma. It may well now be called the 'official view' of physical theory since it is accepted by most of our leading theorists of physics" (Popper 1956 [1963], 100). Popper, who surely reacted to the so-called "Copenhagen Interpretation" of quantum mechanics, defined instrumentalism in general terms "as the thesis that scientific theories – the theories of the so-called 'pure' sciences – are nothing but computation rules (or inference rules); of the same character, fundamentally, as the computation rules of the so-called 'applied' sciences" (Popper 1956 [1963], 111).[3] On this view, which has become standard in subsequent philosophy of science (for representative references see e.g. Lloyd 1978 [1991], 254, n. 3; Barker and Goldstein 1998, 234, n. 3), "instrumentalism" is the thesis that "scientific theories are useful and that scientists are justified in

3 For some interesting references concerning the history of the notion in the early twentieth century see Popper 1935 [1959], 36, n. *4. On the context in which Popper developed his ideas see Hacohen 2000, 202 f.

using them even if the entities they countenance are fictional" (Morgenbesser 1969, 201). Thus, it is opposed to the position called "realism," which holds that scientific theories consist of statements reflecting (more or less precisely) reality, and positing entities that exist.

The positions called "instrumentalism" and "realism" are thus positions in *the theory of knowledge:* they make general philosophical statements about the relationship of scientific theories to reality, specifically about how scientific statements bear on reality. They are *meta-theoretical* or *epistemological* statements. They concern one's *general* position vis-à-vis the very nature of scientific knowledge and its relation to reality: one is not a realist with respect to, say, thermodynamics, but an instrumentalist with respect to the theory of relativity. Although to be sure some practitioners of science have their own epistemological views, still, scientists *qua* scientists may go (and as a rule go) about their work without worrying about how scientific theories relate to the world.

(II) The notion (although not the term) of instrumentalism was introduced into the history of science by the physicist turned historian and philosopher of science Pierre Duhem (1861–1916). Duhem's *Sozein ta phainomena* is a historical narrative of views on the epistemological status of astronomy, putatively expressed by philosophers and astronomers from Plato to the seventeenth century. Duhem identified in the history of epistemological thought about astronomy two traditions, instrumentalism and realism, and made the (meta-historic, judgmental) claim that astronomy was at its best when it was instrumentalist and not realist.[4] Even in this historical work Duhem did not conceal his own instrumentalist convictions (see e.g. Duhem 1908 [1982] 37, 136), a stance directly related to his very conservative political positions: the instrumentalist philosophy of science was to undermine the menace which a realistically-interpreted science exerted on the doctrinal positions of the Catholic Church (Martin 1991).

Half a century later, this second "trans-disciplinary migration" of the notion of instrumentalism, namely from the philosophy to the history of science, was completed by Popper, who was acquainted with Duhem's

4 Duhem (1908 [1982]) occasionally uses "realism": e.g. "le réalisme des astronomes arabes" (32); "réalisme illogique" of the Copernicans (136). Duhem does not use the term "instrumentalism," but describes the Averroist position as one according to which astronomy must not use hypotheses that are "purement conventionnelles et fictives" (56).

Sozein, but whose epistemology was opposite to Duhem's.[5] In "Three Views Concerning Human Knowledge" Popper set out from the conflict between the Church and Galileo Galilei and wrote that the former had "no objection to Galileo's teaching the mathematical theory, so long as he made it clear that its value was instrumental only; that it was nothing but a 'supposition', as Cardinal Bellarmine put it; or a 'mathematical hypothesis' – a kind of mathematical trick, 'invented and assumed in order to abbreviate and ease calculations'."[6] Popper (1956 [1963] 98, n. 2) considered Osiander and Cardinal Bellarmine among the "founding fathers of the epistemology which … I am going to call 'instrumentalism'."

Duhem and Popper thus joined hands in recognizing as instrumentalists some historical figures. But whereas for the former Osiander and Bellarmine were the "good guys," the latter hastened to contrast them with his own positive hero, Galileo Galilei: "Galileo himself, of course, was very ready to stress the superiority of the Copernican system as *an instrument of calculation*. But at the same time he conjectured, and even believed, that it was *a true description of the world*; and for him (as for the Church) this was by far the most important matter" (Popper 1956 [1963] 98, italics in the original). Note that whereas for Duhem the "founding fathers" of instrumentalism were to be sought among the ancient Greek philosophers and scientists, Popper viewed the Greeks as realists (Popper 1956 [1963] 99, n. 6), and held that instrumentalism emerged only as a (reactionary) defensive attitude on the part of the Church toward Copernicanism. Duhem and Popper naturally tried to each enlist the best past minds into his own camp.

Both Duhem and Popper were thus philosophers of science who sought, and unsurprisingly found, precursors for their respective positions in history. But by so doing, both lost from view an essential point. Both failed to appreciate that Osiander and Bellarmine, say, did not at all make general statements about the cognitive value of *science* as such. Rather, they made statements about the views held by historical

5 Popper was acquainted with Duhem's "famous series of papers" (i. e. *Sozein*): Popper 1956 [1963] 99, n. 6.

6 Popper 1956 [1963], 97–98. The first inserted quotation comes from Cardinal Bellarmine, the second from Francis Bacon. Popper similarly describes Bishop Berkeley as holding the view that the Newtonian theory "could not possibly be anything but a 'mathematical hypothesis', that is a convenient instrument for the calculation and prediction of phenomena or appearances" (Popper 1956 [1963], 98–99).

actors on the cognitive value of one *specific* theoretical body of knowledge (or discipline), viz. (Copernican) astronomy: the role of science, they held, was to describe reality; only that this astronomical theory (and perhaps astronomy in general) *was not real science.*[7]

Introducing epistemological notions into the historiography of science is not, *per se*, illegitimate or unfruitful. In the present instance, however, this move was fraught with an original sin: *a term denoting a general epistemological attitude toward knowledge was applied to a historically contingent view on the aims, tools and limits of one particular body of knowledge,* viz. *specific theories of mathematical astronomy*. The philosopher of science's instrumentalism is a general epistemological attitude that has nothing to do with the contents of this or that theory; theories come and go, but the philosopher's view of their possible relationship to reality remains unaltered.[8] By contrast, what Duhem and Popper identified as instrumentalism is something quite different: it is a view of the cognitive value of mathematical astronomy – or rather of different theories of mathematical astronomy – as judged by contemporaries against certain ideals of science. Although Osiander and Bellarmine, say, possibly had an historical influence on the emergence of instrumentalism as a theory of knowledge, they cannot legitimately be regarded as already holding that general epistemological position.

2. How the Epistemology of Astronomical Theories Emerged as a Problem for Greek Philosophers

The very problem concerning the epistemological status of astronomical theories has grown out of what can be regarded as (almost) a historical accident, namely the interaction of two fairly independent theoretical traditions: Greek logic and natural philosophy, especially in their Aristotelian version, on the one hand; and mathematical astronomy of Babylonian-Greek ancestry on the other. They overlapped in two domains:

---

7 [Note added to the present reprinting:] Put differently: Osiander and Bellarmin commented on the cognitive value of one specific astronomical theory (Copernicanism); their judgment was grounded in traditional general theory of science, according to which a theory whose sole merit was predictive capacity was not really science.

8 In some historical contexts, though, inner theoretical problems of an important scientific theory may favor instrumentalism; this was arguably the case with quantum mechanics in the 20th century.

that of their respective *ideals of science*[9] and that of the *substantive* descriptions of the heavenly ("supralunar") realm; the first is meta-scientific, the second inner-scientific.

The salient meta-theoretical ideals of science are of two kinds: (I) Those specific to mathematical astronomy; and (II) Aristotelian general ideals of science.

(I) Early Greek astronomy is associated with Eudoxus's innovative two-sphere cosmology, whose underlying philosophically-inspired ideals are the "perfect" spherical forms of the earth and of the rotating sky, and the regular rotatory motion of the latter around the former (Goldstein and Bowen 1983). These ideals were upheld by Aristotle, whose cosmology integrated Eudoxus's theory, and they inform the synthesis of Greek and Babylonian astronomy that began to take form in the late second century and early first century BCE, in which the Babylonian quantitative data, and the notion of quantitative prediction, were integrated into the qualitative ideals deriving from Greek philosophy (Goldstein and Bowen 1983, 339; Bowen 2001). (The historian imbued with Aristotelianism would be tempted to describe this synthesis as consisting of Babylonian matter and Greek – mathematical – form.) This development reached completion in Ptolemy's astronomy which dominated posterity until the time of Copernicus and Kepler.

(II) Aristotle set out his ideals of science in his *Posterior Analytics*, where he distinguishes between an "understanding of the *fact* and [an understanding of] the reason *why*" (*An. post.* 13, 78a21; see Barnes 1975, 149). In the latter case, not only the fact *that*, but also the *cause* of the *explanandum* is required. (For Aristotle, as Josef Stern has usefully reminded us, a *cause* "is not, as we post-Humeans think nowadays, a prior event that brings about its later effect. Rather, an Aristotelian cause (of any of the four kinds) is an explanatory factor, an answer to the why-question, the clause that follows the connective 'because' in statements of the form '*P* because *Q*'" [Stern 2001]). In order to have an explanation of the "why," the *explanandum* must be syllogistically deduced from primary principles that are "true and primitive and more familiar than and prior to and explanatory of the conclusion" (*An. post.* I, 2, 71b20 f., Barnes 1975), with the resulting argument providing a *dioti* demonstration (demonstration of the reason why). Conversely, in a *oti* demonstration (demonstration of the fact) one proceeds from the perceptible phenomena – which are better known to man – to establish

9 I borrow this notion from Funkenstein 1986, 18–22; Funkenstein 2003.

principles which are better known "in themselves." The later Aristotelian tradition called the first a *demonstratio propter quid*, with the resulting argument called a "synthesis," and the second a *demonstratio quia*, which was referred to as an "analysis." When the principles from which a phenomenon is derived, instead of being "better known in themselves" and thus (in principle) shared by all humankind, are accepted on the basis of a large consensus (*endoxa*), then the resulting argument is a dialectical demonstration. The question of what Aristotle himself thought about whether and to what extent the different kinds of his own scientific research and theories match his ideal of *dioti* demonstrations is a much discussed issue (Barnes 1969 [1975]; Harari-Eshel 2001) into which we need not enter.

Now the cosmology outlined in *De caelo* was in conformity with both the substantive and the meta-theoretical ideals of science: it was consistent with contemporary astronomy, for whose "facts" it could thus provide *dioti* demonstrations (Aristotle may have had some doubts concerning parts of the theory, however; cf. *De caelo* 2, 12, 291b24–28, and Maimonides, *Guide*, II, 19, text and trans. Qafah 1972, 334; trans. Pines 1963, 307; and Kraemer 1989, 56–61). This implies that the substantial affirmations bearing on, say, the center of the earth as the natural place of the two heavy elements, or the one concerning the existence and properties of the "fifth substance," were henceforth taken to be established demonstratively. From the perspective of our *problématique*, Aristotle's is the golden age – that of a unity of physical explanation and astronomical-cosmological theory. As long as that unity prevailed, the idea of an "instrumentalist" interpretation of astronomical knowledge could arise only with respect to an external, namely the Babylonian, body of knowledge. It is indeed essential to keep in mind that the very nature of the demonstrations required and offered in Aristotelian science excluded any idea of construing its theories along instrumentalist lines: the posited eternal "fifth substance" endowed with a natural circular motion, say, was unproblematically taken to exist no less than the four sublunar elements, whose instantiations could be perceived by the senses. In other words: *a realist epistemology was an intrinsic ingredient of the Aristotelian theory of science.*

The primeval harmony of (Aristotelian) physics and (Eudoxan) mathematical astronomy ended with the introduction of epicycles and eccentrics (Goldstein 1980, 135). True, as noted, Ptolemy's mathematical astronomy integrated, and conforms to, ideals deriving from natural philosophy, e.g. the idea that the earth is immobile at the center of the world, the ideal of accounting for the observed planetary motions by positing circular regular motions (Goldstein 1997a). Still, the coherence was no longer complete, inasmuch as Ptolemaic mathematical astronomy does not satisfy the Aristotelian standards of science on two main counts: (i) On the *meta-theoretical* level, it falls short of the Aristotelian *desiderata* because it does not proceed from principles that are "better known in themselves" and thus does not provide causal explanations of astronomical facts.[10] (ii) On the *substantial* level, it conflicts with some established principles of Aristotelian science, inasmuch as it notably posits circular motions that are not uniform about the center of the world (e. g. Lloyd 1978 [1991], 250; Goldstein 1980, 136).

The question how the two ideals of science and the corresponding traditions related to each other has been much discussed, notably by and as a reaction to Duhem. He identified the Greeks as "instrumentalists," who ingeniously did not care about the reality of the entities they posited, but has been taken to task by historians who have shown that he misinterpreted the relevant texts (Mittelstrass 1962, 140 ff.; Lloyd 1978 [1991]; Goldstein 1980; Lerner 1996, 287, n. 2). For our purposes here it is sufficient to note that it is only in the Arabic tradition, beginning in the ninth century, that we find a new type of scientist, one who, in conformity with a general trend to mathematize natural phenomena, endeavored to satisfy the *desiderata* of both (Aristotelian) natural philosophy and mathematical astronomy[11] (later, though, a movement in the opposite direction took place, in an effort to secure the intellectual independence of mathematical astronomy; see Ragep 2001). In the East, Thâbit Ibn Qurra (826–901), Ibn al-Haytham (965–c. 1041) and their successors tried to create a unified theory of the heaven, integrating physics and advanced mathematical astronomy. In the West, we have an at least partly different tradition: the philosophically-inspired criticism of Ptolemy seems to have been launched in the twelfth century by Aristotelian philosophers such as Ibn Bâjja (d. 1138), Ibn Tufail (d.

10 The importance of considering the instrumentalism/realism issue in Greek, medieval and Renaissance astronomy from the viewpoint of Aristotle's scientific methodology has been pointed out in Jardine 1988, followed by Barker and Goldstein 1998. See, in addition to the above, e. g. Hyman 1989; Kraemer 1989; Stern 2001.

11 The causes for this historically highly significant paradigm shift have not yet been investigated; see Ragep 2001. Perhaps also social changes were involved that led to what has been described as role-hybridization; see the model of scientific innovation suggested in Ben-David 1991.

1185), and Averroes (1126–1198), who aspired a "reunification" of astronomy and natural philosophy (Gauthier 1909; Sabra 1984). This tradition gave rise to al-Bitrûjî's astronomy (written around the turn of the twelfth century), which, although disappointing as a work of mathematical astronomy, had great influence both on the Hebrew and Latin traditions (Goldstein 1971, 1, 3, 40–45). Maimonides (1137/8–1204), too, is an heir to this tradition on whom I will concentrate in what follows.

3. Maimonides's Instrumentalism Revisited

Ever since Duhem, Maimonides is justly considered as having formulated with particular clarity (what has been called) the instrumentalist stance on astronomy. Duhem considered the instrumentalist stance to be Maimonides's own view, and barely hid his surprise in face of the "prudent skepticism" expressed by this representative of "Semitic Peripatetism" who, alone among the writers in Arabic, succeeded in "elevating himself up to the doctrine formulated by the Greek thinkers," whose doctrine he expressed in words "almost identical" with theirs (Duhem 1908 [1982] 36–38; on Duhem's attitude to the Arabs [including, for once, the Jews], see Ragep 1990). Practically all subsequent authors on instrumentalism have also commented on Maimonides and in what follows my aim is to clarify further Maimonides's attitude – or rather attitudes – toward the science of the heaven in general and toward astronomy in particular, as expressed in the *Guide of the Perplexed.*[12] My further aim is to shed some light on the usefulness of the notion of instrumentalism.

In the well-known chapter 24 of the second part of the *Guide*, Maimonides gives a long list of incompatibilities between Aristotelian natural science and the received mathematical astronomy. Maimonides's discussion, it should be realized, is written from the *standpoint of an Aristotelian natural philosopher.* He does not argue from the vantage point of a meta-theorist, an epistemologist, say, who as a neutral observer points out that two theoretical positions are incompatible. Rather, he takes certain Aristotelian principles of natural philosophy (e. g. the impossibil-

12 The question to what extent the *Mishneh Torah* reflects Maimonides's personal views is much debated (for one view see Kellner 1991), and hence I will leave this work out of consideration; I believe, though, that it tends to corroborate, rather than weaken my interpretation.

ity of a circular motion about a center other than the center of the earth, the impossibility of void; *Guide*, II, 24; text and trans. Qafah 1974, 352–354; trans. Pines 1963, 322–324) to have been established through "reason" (*qiyâs*), i.e. demonstratively, and to "have been made clear" (*tabayyana*) by natural science (*Guide*, II, 24; text and trans. Qafah 1974, 351; trans. Pines 1963, 322; see Hyman 1989, 40–42 on Maimonides's salient logical terminology and Stern 2001 on Maimonides's different notions of "demonstration"), so that "nothing can be added" to them (*Guide*, II, 24; text and trans Qafah 1974, 352; trans. Pines 1963, 323). Even Thâbit Ibn Qurra's affirmation of the existence of an inter-spherical (fluid) body is considered as having been demonstrated and proven (*bayyana*, *barhana*; *Guide*, II, 24; text and trans. Qafah 1974, 354; trans. Pines, 325). Other principles of natural philosophy pertaining to the supralunar realm, have been established (by dialectical syllogisms) in conformity with the less stringent criteria which, according to Alexander of Aphrodisias, are applicable in this domain (*Guide* II, 3, 22; text and trans. Qafah 1974, 296, 348; trans. Pines, 254, 320; Hyman 1989, 43; Kraemer 1989, 64–68). It is from the viewpoint of the natural philosopher, who considers some essential parts of Aristotelian natural science as having been adequately demonstrated, that Maimonides takes cognizance of the impressive realizations of the astronomers: the fact that "what is calculated" by astronomers "is not at fault even by a minute," that their calculations concerning eclipses are always precise, etc. (*Guide*, II, 24; text and trans. Qafah 1974, 355; trans. Pines, 326). It is the fact that although astronomy is a *non-science* (namely because it is not a proper/Aristotelian science, not being demonstrative) it yet possesses such striking exactness (we would say: predictive power) that constitutes for Maimonides the "true perplexity" (see Langermann 1991). Indeed, Aristotle himself, had he been acquainted with the claims of mathematical astronomy, would have regarded them as "established as true" (*sahha lahû*) and "would have become most perplexed" (text and trans. Qafah 1972, 356; trans. Pines 1963, 326), a speculation with which one feels inclined to concur.

This is the context in which Maimonides puts forward his famous statement concerning the astronomer's role:

> [A]ll this does not obligate (*yulzimu* or *yalzamu*) the master of astronomy. For his purpose is not to tell us in which way the spheres truly are, but to posit an astronomical system [or: configuration; *hay'a*] in which it would be possible for the motions to be circular and uniform and to correspond to what is apprehended through sight, regardless of whether or not

> things are thus in fact. (Text and trans. Qafah 1974, 355; trans. Pines 1963, 326, slightly modified).

Maimonides's statement, it should be stressed, must not be interpreted as *normative*, i. e. as reflecting Maimonides's views on what the knowledge of the celestial realm *should* be, nor on what it can *possibly* be (cf. Langermann 1999). Rather, it is a purely *descriptive*, matter-of-fact statement, pragmatically taking note of an existing social division of intellectual labor between the natural philosopher and the astronomer, and of the brute fact that natural philosophy and astronomy *are* "compartmentalized" (Goldstein 1980, 139). The astronomer, inasmuch as he is not committed to the Aristotelian ideals of science, simply need not care about what the philosophers (including Maimonides himself), *qua* philosophers, perceive as "the true perplexity"; he, *qua* astronomer, has no reason whatsoever to be perplexed. In this statement (but others are to follow) Maimonides, perhaps echoing statements by Greek philosophers (namely those that, rightly or wrongly, could be understood as "instrumentalist"), gives a classic statement of the position which was to be dubbed "instrumentalist"; he is thereby implicitly rejecting the scientific ideal of, say, Thâbit Ibn Qurra, Ibn al-Haytham, or even al-Bitrûjî, who abolished the separation of astronomy and natural science, and sought to create a unified theory of the heavens. In fact, Maimonides's purpose was to induce in his readers a dose of skepticism, not with respect to mathematical astronomy, but rather with respect to what Aristotle said *on the heavens* and, in a further move, on the issue of the eternity of the world. His target was metaphysics, not mathematical astronomy (see e.g. Langermann 1991, 167–168). This reading is confirmed, and refined, by another well-known passage:

> Know with regard to astronomical matters mentioned that if an exclusively mathematically-minded man reads and understands them, he will [or: may] think that they form a cogent [i.e. decisive] demonstration (*burhân qât'i*) that the form and number of the spheres is as stated. Now things are not like this, and this is not what is sought in the science of astronomy. Some of these matters are indeed founded on the demonstration [*umûr burhâniya*] that they are that way. Thus it has been demonstrated that the path of the sun is inclined against the equator. About this there is no doubt. But there has been no demonstration whether the sun has an eccentric sphere or an epicycle. Now the master of astronomy does not mind this, for the object of that science is to suppose an arrangement [or: configuration; *hay'a*] that renders it possible for the motion of the star to be uniform and circular ... and have the inferences necessarily following from the assumption of that motion agree with what is observed. (*Guide* II, 11; text

and trans. Qafah 1974, 296–297; trans. Pines 1963, 273–274; see Stern 2001, 57 f. on *burhân qâti')*.

Maimonides here repeats that, in general, mathematical astronomy does not produce demonstrations: it does not conform to the Aristotelian ideal of science and it is misguided to identify (as the astronomer is allegedly prone to do) the proofs offered in astronomy with demonstrations as they are required in Aristotle's *Posterior Analytics*. It is important to note, however, that Maimonides concedes that *some* of the statements made by astronomers are sustained by true (presumably dialectical) *demonstrations* and are, therefore, embraced as true by the natural philosopher too. This is notably the case with the statement that the sun moves along an inclined ecliptic, a fact, indeed, that Aristotle himself had already integrated in his cosmology. Similarly, Maimonides also considered certain astronomical data on the planetary distances, which he had adopted from the *Treatise on Distances* by al-Qabîsî (d. 967) (*Guide* II, 24; text and trans. Qafah 1974, 355; trans. Pines 1963, 325; Goldstein 1980, 138), as "proven," or "made clear" (*tabayyana*; *Guide* III, 14; text and trans. Qafah 1974, 496; trans. Pines 1963, 456), thus holding that "the latter-day scientists have proven (*barhanu*) through a correct demonstration (*al-burhan al-sahîh*), regarding which there is no doubt," of the eccentricities of the purported eccentric spheres (allowing one to conclude that they are impossible) (*Guide* II, 24; text and trans. Qafah 1974, 353; trans. Pines 1963, 324 [modified]). Again, Maimonides grounded his own (idiosyncratic) four-sphere model of *physical* cosmology on astronomical information bearing on the relative places of the spheres of the planets Mercury and Venus (*Guide* II, 9; see also Kraemer 1989, 82).[13]

Maimonides's position is thus more complex than it may have seemed: he holds that the astronomer need not worry about the real constitution of the world, but at the same time he considers some of the information he supplies as providing physical science with reliable premises. When Maimonides writes that "regarding all that is in the

---

13 Averroes, we may add in passing, was even more severe than Maimonides regarding astronomy, for in his Epitome of the *Almagest* he wrote that astronomy provides neither "[*quia*] proofs" (*re'ayot*), nor "[*propter quid*] demonstrations" (the posited "prior" principles, notably the epicycles, being impossible), so that his exposition of celestial science makes no claim to constitute scientific knowledge and will simply follow what is accepted by general the consensus (Lay 1996, 55).

heavens" man "grasps [or: fully understands; *'ahâta bi*]" only a little, which is mathematical (*Guide* II, 24; text and trans. Qafah 1974, 356; trans. Pines 1963, 326, modified), he refers precisely to these few things that mathematical astronomy has firmly established and which can be used as premises in natural science. This is why Maimonides could state that "matters pertaining to the astronomy [or: configuration; *hay'a*] of the spheres," just as those pertaining to natural science, are "necessary for the apprehension of the relation of the world to God's governance as this relation in truth is and not according to imaginings (*'alâ al-haqîqah la' bi-hasb al-khayâlât*)" (*Guide* I, 34; text and trans. Qafah 1974, 77; trans. Pines 1963, 74): that they are relevant to that apprehension means that they depict reality as it really is.

Once introduced into the "true perplexity," Maimonides's reader will want to know whether there are prospects of seeing it dispelled one day: can one hope for a guide for that perplexity (too)? In other words: will the astronomer's and the natural philosopher's roles (again) coincide one day, with all science consisting of *dioti* demonstrations (with the instrumentalist construal of astronomy becoming vacuous)? Maimonides's answer, I believe, consists of several complementary parts: (I) a complete knowledge of the celestial realm will forever remain beyond human ken, for both (i) ontological reasons, namely because the heavens bear the marks of having been created by a Particularizer and not through natural necessity, and (ii) epistemological reasons, namely owing to their distance and rank. (II) Still, some progress of astronomy and/or celestial physics, leading up to a physically and philosophically sound astronomy, may come about.

I. (i) In chapter II, 19 of the *Guide* Maimonides elaborates his objections to the necessitarian cosmology he ascribes to Aristotle. The multitude of the particular details of the heavenly motions and of the locations of the heavenly bodies, Maimonides urges, cannot have come into being through necessity. These arrangements must therefore be due to "a being that has individualized"; the world has been created in a finite past by an intelligent agent, who acted out of free choice.[14] Given this stance, which is ontological in the sense that it bears on the world as it actually *is*, the following point, which seems not to have been appreciated sufficiently, must be strongly emphasized. Given the Aristotelian theory of science, the Maimonidean view that

14 My view of Maimonides's much debated stance on creation follows Davidson 1979.

the world in part issued out of particularization *excludes* the possibility of *ever* finding a scientific theory of the heavens: the results of the deity's particularization by definition escape natural science, inasmuch as the latter apprehends necessities only. In other words: since demonstrations as required by the Aristotelian ideal of science can be given only with respect to segments of nature governed by necessity, and since, moreover, Maimonides holds that necessity is operative only in the sublunar realm, it follows that the kind of explanations available for the sublunar world is, and will remain, beyond reach with respect to the supralunar one.

Maimonides contrasts the knowledge attainable of the sublunar and of the supralunar realms in the following way:

> if Aristotle had been able – as he thought – to give us the *cause* [*'illa*] of the differences between the motions of the spheres [i.e. of those aspects of reality that Maimonides considers as having resulted from particularization], so that these should be in accordance with the order of the positions of the spheres with regard to one another, this would have been extraordinary. In this case, what holds of the cause [*'illa*] of the particularization characterizing this diversity of motions would have been the same as that which holds of the cause [*'illa*] of the differences between the elements with respect to their different positions between the encompassing sphere and the center of the earth" (*Guide* II, 19; text and trans. Qafah 1974, 336; trans. Pines 1963, 308–309, modified after Munk 1856–1866, II, 159).

The generation of the elements, Maimonides reasons, has been accounted for by *dioti* demonstrations as resulting from the motions of the sphere (see Glasner 1996), and a genuine science of the supralunar realm would require analogous demonstrations allowing to deduce the details of the spheres's motions from some established principle(s). Alas, the particularization makes this goal unattainable.

Again, Aristotle's aim in *De caelo*, Maimonides states, has been to "bring order" into the celestial realm, just as he "brought order" into the sublunar realm, his goal being "that the whole [universe] should exist in virtue of natural necessity, and not," he urges, "in virtue of the purpose of one who purposes according to his will … and the particularization of one who particularizes in whatever way he likes" (*Guide* II, 19; text and trans. Qafah 1974, 333; trans. Pines 1963, 306). This is precisely what Maimonides takes to be impossible: inasmuch as (according to his assumption) the world came to be through "the particularization of one who particularizes in whatever way he likes," he can conclude with confidence that "this task has not been accomplished by [Aristotle], *nor will it ever be accomplished*" (*Guide* II, 19; text and trans. Qafah 1974, 333; trans. Pines 1963, 306).[15] It is in this sense that Maimonides writes that "the deity alone *fully* knows the true reality, the nature, the substance, the form, the motion, and the causes of the heavens" (*Guide* II, 24; text and trans. Qafah 1974, 356; trans. Pines 1963 327). The principled unknowability of the number of the heavenly bodies (*Guide*, I, 31, text and trans. Qafah 1974, 68; trans. Pines 1963, 66) presumably also belongs to the same category.

(ii) Maimonides buttresses this reasoning by an epistemological argument: man cannot acquire scientific knowledge about the celestial realm, namely because "it is impossible for us to get hold of evidence allowing to draw conclusions about the heavens; for the latter are too far away from us and too high in place and rank" (*Guide* II, 24; text and trans. Qafah 1974, 356; trans. Pines 1963, 327 modified): the word translated as "evidence," *asbâb*, which often means "causes," here refers to empirical evidence which, had it been available, would have allowed one to "draw conclusions about the heavens," namely to infer principles from which, in turn, one could deduce the heavenly phenomena, thereby explaining them through *dioti* demonstrations. Maimonides here echoes a passage from *Parts of Animals* in which Aristotle states that although "the grasp of the eternal things is but slight, nevertheless the joy which it brings is, by reason of their excellence and worth, greater than that of knowing all things that are here

15 It is true, as H.A. Davidson (1979, 29 f.) remarks, that Maimonides considers that he has here offered an argument, not a demonstration. Still, I do not share Professor Davidson's view that Maimonides "recognized ... that a future philosopher or scientist might yet detect an underlying regularity in the structure of the heavens; and the argument would then collapse" (29 f.). This interpretation seems to be excluded by the sentence just quoted, according to which Aristotle's project has not been accomplished "*nor will it ever be accomplished.*" Maimonides admittedly qualifies what he has said as providing merely a non-demonstrative argument (*dalîl*, Qafah 1972, 337) for the existence of an intention, but he immediately afterwards refers to it as "a correct proof which is not exposed to doubt" and, moreover, seems to enlist the prophets in support of the same view (Qafah 1972, 337–338, Pines, 310–311). It is also true that Maimonides does not exclude the possibility that a new astronomy be found, not beset by "the true perplexity" (see below), but there is no indication that Maimonides believed that that astronomy would account for the arrangement of the heavenly realm as the result of necessity. It is difficult to imagine that Maimonides countenanced the eventuality of relinquishing a belief as fundamental as that of God as a Particularizer.

below" (1:5, 644b31 f.; Peck 1961).[16] But Maimonides, who was acquainted with an Arabic translation of the work, reverses Aristotle's point and uses the excellence of the heavens as an argument against their knowability (Freudenthal 2003), a line of argument which is in continuity with the emphasis he put (contrary to Averroes, for one) on the difference between sublunar and celestial matter (Glasner 2000). As has repeatedly been pointed out, this epistemological argument is part and parcel of Maimonides's general view of the limitations of the human intellect, however this view is construed (Pines 1979 [1997], Kraemer 1989, 61–62, 82–84; Stern 2004).

In sum, scientific inquiry will never furnish full demonstrative knowledge of the heavens, because necessity does not prevail in them, and because of their high rank: human science has its limit where natural necessity ends. Accordingly, Maimonides thinks that Aristotle's theory of the heavens is in part, although not entirely, mere "guessing and conjecturing" and thus subject to doubt (*Guide*, II, 22; text and trans. Qafah 1974, 348; trans. Pines 1963, 320; see also *Guide* II, 19, 24). In Maimonides's view, the most important thing celestial science (to be distinguished from mathematical astronomy) can teach us about the heavens is that it is the outcome of particularization, its main purpose being to indicate the existence of God (Kraemer 1989, 79–80; Langermann 1999, 2–5; Davidson 2000). Since knowing the heavens would mean knowing the deity's particularization, the only way man could obtain such knowledge is through prophecy, indeed prophecy of the highest rank, that of Moses: "let us give over the things that cannot be grasped by reasoning [*qiyâs*, i. e. demonstration] to him who was reached by the mighty divine overflow so that it could be fittingly said of him: "With him do I speak mouth to mouth" (Num. 12:8)" (*Guide* II, 24; text and trans. Qafah 1974, 356–357; trans. Pines 1963, 327; Pines 1979 [1997], 92–93; Kraemer 1989, 61 f., n. 32).[17] Maimonides indeed more than once singled out Moses's prophecy, and affirmed that Moses saw things "as they really are" and not, as other prophets, through parabolic apprehension (*Mishneh Torah, Hilkhot yesodey ha-Torah* VII, 6; see also *Guide* II, 45; *Comm. Mis-*

16 This argument is still repeated in the sixteenth century, e. g. by Nicodemus Frischlin (1586), who also seems to echo this Aristotelian passage; see Jardine 1988, 700 f.; Barker and Goldstein 1998, 248 (cf. n. 25).

17 In the sixteenth century much the same belief was held (Barker and Goldstein 1998, 248–252).

hnah, *Sanhedrin*, chap. 10 [the seventh of Maimonides's thirteen principles]).

(II) But Maimonides's denial of the possibility of fully knowing the heavens scientifically does not imply that any improvement of the existing demonstrative knowledge is excluded, and that the "true perplexity" will forever be an inherent part of human condition. "It is possible," Maimonides writes, "that someone else may find a *demonstration* (*burhân*) by means of which the truth (*haqîqah*) of what is obscure to me will become clear to him"(*Guide* II, 24; text and trans. Qafah 1974, 357; trans. Pines 1963, 327 slightly modified; cf. Langermann 1991, 166). Indeed, we have seen that Maimonides held that man already possesses some true knowledge of the celestial realm: not only "the little which is mathematical" (above, p. 281), but also some knowledge about the physical constitution of the heavens, as e. g. the cosmological tenet posited by Thâbit Ibn Qurra (above, p. 278). Maimonides more than once comments that the science of astronomy has made considerable progress between Aristotle's and his own days, and he is likely to have thought that that progress would continue and that an improved mathematical astronomy could develop (Kraemer 1989, 80 ff.). Here Maimonides reflects the influence of his Arabic milieu, in which, as mentioned, attempts to devise a physically sound mathematical astronomy were made.

Maimonides's view of astronomy thus integrates beliefs derived from two traditions: the stance (ascribed to some Greeks) that, as things are *hic et nunc*, astronomers need not care about demonstrations and the reality of their mathematical circles; and the hope, perhaps even confidence (revived by the Arabs), that a different situation may prevail one day and that the "true perplexity" will give way to a true, physically valid astronomy (cf. Langermann 1991, 170). Maimonides's epistemological pessimism, particularly with respect to celestial science, let me add, is presumably one of the reasons why, contrary to his Andalusian counterparts, Maimonides did not issue a call for a reform of existing astronomy.[18]

Does the latter consideration imply that, in contradiction to what has been said before, Maimonides thinks that one day astronomy and physical science will both have made such progress as to produce a demonstrative, physically sound, mathematical astronomy? That he be-

18 Another reason is Maimonides's general view that the sciences are but propaedeutic disciplines, preparing man to accede to metaphysical knowledge, which alone can lead up to the immortality of the soul.

lieved that man will after all join God and Moses in "fully know[ing] the true reality, the nature, the substance, the form, the motion, and the causes of the heavens" (above, p. 283)? Probably not: what Maimonides apparently intended was that a future cosmology, based on (dialectical) demonstrations, will be found that would be in contradiction neither with the principles of (Aristotelian) science, nor with those of mathematical astronomy. That future cosmology would have good predictive power and yet avoid physically impossible constructs. Still, it would not provide *dioti* explanations of *all* celestial phenomena: the knowledge of the particularization of the heavens would still be left to Moses. Since Maimonides thought that neither Aristotelian celestial science nor actual mathematical astronomy were demonstrative from top to bottom, it seems reasonable to assume that he thought that this new cosmology would involve changes in either natural philosophy, or in mathematical astronomy, or in both. This hope is entirely consistent with Maimonides's "ontological" beliefs discussed above.

4. Gersonides: The Self-Appointed Moses

To devise a new astronomy that describes "the true reality, the nature, the substance, the form, the motion, and the causes of the heavens" – this was the immodest task Levi ben Gershom (Gersonides, 1288–1344) set himself:

> [Astronomy] cannot be split in such a way that part of it would belong to a master of one science and the remainder to the master of the other … It follows that this investigation can only be undertaken in its perfection by one who is at once a mathematician, a physicist and a philosopher, for he can be aided by each of these sciences and take from them whatever is needed to perfect his work. (Goldstein 1985, 304, 323)

Gersonides was a realist in both the philosophers's and the astronomers's sense (Goldstein 1980, 140–142; Hugonnard-Roche 1992), and this outlook determined his entire research program, philosophical and scientific (Freudenthal 1992, 1996; Goldstein 1997b). Here it is his view of Ptolemy that is of particular interest. Gersonides devotes a long chapter (43) of his *Astronomy* to a meticulous and partly innovative criticism of Ptolemy. He concedes that despite his shortcomings, Ptolemy "has helped us enormously" and deserves gratitude. He suggests that Ptolemy failed, but not according to his own standards, for, he says:

> You should know that Ptolemy did not strive to lay down an astronomy that is in conformity with these celestial bodies, but rather has striven to lay down an astronomy that would allow one to determine the positions of the planets at any given moment, as far as this is possible ... He [himself] sensed that what he laid down does not agree with what is observed concerning these motions (Paris, Bibliothèque nationale de France, ms héb. 724, fol. 82a; héb. 725, fol. 60a).

Consider the irony of history: the arch-realist Gersonides anticipates the view of Ptolemy as an instrumentalist to be propagated six centuries later by the arch-instrumentalist (and anti-Semite) Pierre Duhem; only that Gersonides thinks that Ptolemy deserves gratitude *despite* his instrumentalism, whereas Duhem views his instrumentalism as the quintessence of the "génie grec."

5. Conclusion: To Be or Not to Be (Instrumentalist or Realist)?

Pierre Duhem, his many vices (moral and scientific) notwithstanding, was right on one essential point: the meta-theoretical views of scientists may influence the course of their research. Put differently: the historian of science must take into consideration the "second order" views held by the historical actors. This methodological principle has been repeatedly stated in the historiography of science of the second half of the twentieth century, and Duhem was certainly one of the precursors of this insight. Duhem's substantial theses on the history of astronomy were for the most part refuted by later historians, but this process of "conjectures and refutations" itself has confirmed Duhem's historiographic stance. Owing to the debate triggered by Duhem, we now understand better, for instance, Ptolemy's research program as structured in part by the wish to give a true account of the heavenly configuration. The general lesson for which we are indebted to Duhem (among others) thus is that without taking into consideration the astronomers's meta-theoretical beliefs we cannot give a full and adequate account of the historical development of the discipline. Now for these meta-theoretical beliefs, or "ideals of science" (above, p. 274), we need convenient names. In the history of astronomy the relevant ideals are those that have come to be called "realist" and "instrumentalist" (or: fictionalist). Their continued use seems to me warranted, provided we state precisely what we mean by them: they must reflect meta-theoretical beliefs or ideals held by historical actors, not by the historian.

Bibliography

Barker, Peter and Goldstein, Bernard R. (1998), "Realism and Instrumentalism in Sixteenth-Century Astronomy: A Reappraisal", *Perspectives on Science* 6: 232–258.

Barnes, Jonathan (1969 [1975]), "Aristotle's Theory of Demonstration". *Articles on Aristotle*. Ed. Jonathan Barnes, Malcolm Schofield, and Richard Sorabji. Vol. 1, *Science*. London: Duckworth, 65–87.

— (1975), *Aristotle, Posterior Analytics* (transl.). Oxford: The Clarendon Aristotle Series.

Ben-David, Joseph (1991). *Scientific Growth: Collected Essays on the Social Organization and Ethos of Science*. Ed. Gad Freudenthal. Berkeley: University of California Press.

Bowen, Alan C. (2001), "La scienza del cielo nel periodo pretolemaico". *Storia della scienza*. Ed. Vincenzo Capppelletti. Vol. 1: *La Scienza Antica*. Roma: Istituto della Enciclopedia Italiana, 806–839.

Davidson, Herbert A. (1979), "Maimonides' Secret Position on Creation". *Studies in Medieval Jewish History and Literature*. Ed. Isadore Twersky. Cambridge, Mass.: Harvard University Press, 16–40.

— (2000), "Further on a Problematic Passage in Guide for the Perplexed 2.24". *Maimonidean Studies*. Ed. Arthur Hyman, vol. 4. New York: The Michael Scharf Publication Trust of Yeshiva University Press, 1–13.

Duhem, Pierre (1908 [1982]), *Sozein tà phainomena. Essai sur la notion de Théorie physique de Platon à Galilée*. Paris: Vrin.

Freudenthal, Gad (1992), "Sauver son âme ou sauver les phénomènes: Sotériologie, épistémologie et astronomie chez Gersonide". *Studies on Gersonides: A Fourteenth-Century Jewish Philosopher-Scientist*. Ed. Gad Freudenthal. Leiden: Brill, 317–352.

— (1996), "Levi ben Gershom (Gersonides), 1288–1344". *The Routledge History of Islamic Philosophy*. Ed. Seyyed Hossein Nasr and Oliver Leaman. London and New York: Routledge, 739–754.

— (2003), "Some Unnoticed Implicit Quotations of Philosophical Sources in Maimonides' Guide of The Perplexed", *Zutot: Perspectives on Jewish Culture* 2: 114–25.

Funkenstein, Amos (1986), *Theology and the Scientific Imagination. From the Middle Ages to the Seventeenth Century*. Princeton: Princeton University Press.

— (2003), "The Disenchantment of Knowledge: The Emergence of the Ideal of Open Knowledge in Ancient Israel and in Classic Greece". *Aleph: Historical Studies in Science and Judaism* 3: 15–81.

Gauthier, Léon (1909), "Une réforme du système astronomique de Ptolémée, tentée par les philosophes arabes du XIIe siècle". *Journal asiatique*: 483–510.

Glasner, Ruth (1996), "Gersonides' Theory of Natural Motion", *Early Science and Medicine* 1: 151–203.
—— (2000), "The Question of Celestial Matter in the Hebrew Encyclopedias". *The Medieval Hebrew Encyclopedias of Science and Philosophy: Proceedings of the Bar-Ilan University Conference.* Ed. Steven Harvey. (Amsterdam Studies in Jewish Thought, vol. 7). Dordrecht: Kluwer, 313–334.
Goldstein, Bernard R. (1971), *Al-Bitrûjî: On the Principles of Astronomy: an edition of the Arabic and Hebrew versions with translation, analysis, and an Arabic-Hebrew-English glossary.* (*Yale Studies in the History of Science and Medicine,* vol. 7) *Vol. 1: Analysis and Translation; vol. 2: The Arabic and Hebrew Versions.* New Haven: Yale University Press.
—— (1980), "The Status of Models in Ancient and Medieval Astronomy". *Centaurus* 24: 132–147.
—— (1985), "Towards a Philosophy of Ptolemaic Planetary Astronomy". *Ancient Philosophy* 5: 293–303.
—— (1997a), "Saving the Phenomena: The Background to Ptolemy's Planetary Theory". *Journal for the History of Astronomy* 28: 1–12.
—— (1997b), "The Physical Astronomy of Levi ben Gerson". *Perspectives on Science* 5: 1–30.
Goldstein, Bernard R. and Alan C. Bowen (1983), "A New View of Early Greek Astronomy". *Isis* 74: 330–340.
Hacohen, Malachi H. (2000), *Karl Popper – the Formative Years, 1902–1945.* Cambridge: Cambridge University Press.
Harari-Eshel, Orna (2001), "Knowledge and Demonstration in Aristotle's *Posterior Analytics*". *Explanation: Theoretical Approaches and Applications.* Ed. Giora Hon and Sam S. Rakover. Dordrecht: Kluwer, 137–164.
Heede, R. (1976), "Instrumentalismus". *Historisches Wörterbuch der Philosophie.* Ed. Joachim Ritter and Karlfried Gründer, vol. 4. Basel: Schwabe, 424–428.
Hugonnard-Roche, Henri (1992), "Problèmes méthodologiques dans l'astronomie au début du XIVe siècle". *Studies on Gersonides: A Fourteenth-Century Jewish Philosopher-Scientist.* Ed. Gad Freudenthal. Leiden: Brill, 55–70.
Hyman, Arthur (1989), "Demonstrative, Dialectical and Sophistic Arguments in the Philosophy of Moses Maimonides". *Moses Maimonides and His Time.* Ed. Eric L. Ormsby. Washington, D.C.: Catholic University of America Press, 35–51.
Jardine, Nicholas (1988), "Epistemology of the Sciences". *The Cambridge History of Renaissance Philosophy.* Ed. Charles B. Schmitt and Quentin Skinner. Cambridge: Cambridge University Press, 685–711.
Kellner, Menachem (1991), "On the Status of Astronomy and Physics in Maimonides' *Mishneh Torah* and *Guide of the Perplexed:* A Chapter in the History of Science". *British Journal for the History of Science* 24: 453–463.
Kraemer, Joel L. (1989), "Maimonides on Aristotle and Scientific Method". *Moses Maimonides and His Time.* Ed. Eric L. Ormsby. Washington, D.C.: Catholic University of America Press, 53–88.
Langermann, Y. Tzvi, (1991), "The 'True Perplexity': The *Guide of the Perplexed,* Part II, Chapter 24". *Perspectives on Maimonides: Philosophical and His-*

torical Studies. Ed. Joel L. Kraemer. Oxford: Oxford University Press, 159–174.
— (1999), "Maimonides and Astronomy: Some Further Reflections". *The Jews and the Sciences in the Middle Ages*. Aldershot: Ashgate, Essay IV (no continuous numbering).
Lay, Juliane (1996), "L'*Abrégé* de l'Almageste: Un inédit d'Averroès en version hébraique", *Arabic Sciences and Philosophy* 6: 23–61.
Lerner, Michel-Pierre (1996), *Le Monde des sphères*. Vol. I: *Genèse et triomphe d'une représentation cosmique*. Paris: Les belles lettres.
Lloyd, Geoffrey E. R. (1978 [1991]), "Saving the Appearances". *Methods and Problems in Greek Science: Selected Papers*. Ed. Geoffrey E. R. Lloyd. Cambridge: Cambridge University Press, 248–277.
Martin, R. Nial D. (1991), *Pierre Duhem. Philosophy and History in the Work of a Believing Physicist*. La Salle, Ill.: Open Court.
Mittelstrass, Jürgen (1962), *Die Rettung der Phänomene*. Berlin: de Gruyter.
Morgenbesser, Sidney (1969), "The Realist-Instrumentalist Controversy". *Essays in Honor of Ernest Nagel: Philosophy, Science, and Method*. Ed. Sidney Morgenbesser, Patrick Suppes, and Morton White. New York: St. Martin's Press, 200–218.
Munk, Salomon (1856–1866), *Maïmonide, Guide des égarés*. Publié pour la première fois dans l'original arabe et accompagné d'une traduction française et de notes critiques, littéraires et explicatives par Salomon Munk, 3 vols. Paris: A. Franck.
Peck, Arthur L., ed. and trans. (1961), *Aristotle, Parts of Animals* with an English translation by Arthur L. Peck. *Movement of animals; Progression of animals* with an English translation by E. S. Forster. Rev. ed. (Loeb classical library 323. Aristotle in Twenty-three volumes, vol. 12). London: Heinemann.
Pines, Shlomo (1963), Moses Maimonides, *The Guide of the Perplexed:* Transl. with an introd. and notes by Shlomo Pines. With an introductory essay by Leo Strauss. Chicago: University of Chicago Press.
— (1979 [1997]), "The Limitations of Human Knowledge According to Al-Fârâbî, Ibn Badja, and Maimonides". *Studies in Medieval Jewish History and Literature*. Ed. Isadore Twersky. Cambridge, Mass.: Harvard University Press, 82–109.
Popper, Karl R. (1935 [1959]), *The Logic of Scientific Discovery*. Transl. prep. by the author. London: Hutchinson.
— (1956 [1963]), "Three Views Concerning Human Knowledge". *Conjectures and Refutations: The Growth of Scientific Knowledge*. London: Routledge, 97–119.
Qafah [Kafah], Yoseph (1972), *Rabbenu Moshe ben Maimon, Moreh ha-nevukhim. Dalâlah al-ha'îrîn*, 3 vols. Jerusalem: Mossad Harav Kook.
Ragep, F. Jamil (1990), "Duhem, the Arabs, and the History of Cosmology". *Synthese* 83: 201–214.
— (2001), "Freeing Astronomy from Philosophy. An Aspect of Islamic Influence on Science". *Osiris* 16: 49–71.

Sabra, A. I. (1984), "The Andalusian Revolt against Ptolemaic Astronomy. Averroes and al-Bitrûjî". *Transformation and Tradition in the Sciences: Essays in Honor of I. Bernard Cohen*. Ed. Everett I. Mendelsohn. Cambridge, Mass.: Cambridge University Press, 133–153.
Stern, Josef (2001), "Maimonides' Demonstrations: Principles and Practice". *Medieval Philosophy and Theology* 10: 47–84.
—— (2004), "Maimonides on the Growth of Knowledge and the Limitations of the Intellect", *Maïmonide: philosophe et savant (1138–1204)*. Ed. Tony Lévy and Roshdi Rashed. Leuven: Peeters, 143–191.

Postscript (2009)

The article reprinted above was written in 2002 and published in 2003 in a Festschrift for Professor Bernard R. Goldstein, entitled *Astronomy and Astrology from the Babylonians to Kepler: Essays Presented to Bernard R. Goldstein on the Occasion of his 65th Birthday* (= *Centaurus* 45 [2003], 227–48). As its title indicates, it does not directly address itself to the concept of "hypothesis," which is at the core of this volume. Nevertheless, the article has immediate bearing on the ideas we associate with the term "hypothesis." A hypothesis is, roughly, a statement bearing on reality, about which we suspend judgment: as it may be true or may be false, it needs to be tested. It may be (provisionally) "accepted" for some purposes, but it is not regarded as ascertained (confirmed, corroborated) or trustworthy knowledge. A hypothesis is thus a statement about reality which is distinguished by its specific epistemological status, that of a provisional knowledge-item.

The regnant medieval theory of science was that of Aristotle as set out mainly in his *Posterior Analytics*. (Often scientific practice did not follow this normative methodology, but it nonetheless remained the dominant scientific ideology.) As is explained in some detail in this article, Aristotle and his followers regarded truly scientific knowledge as having been *demonstrated:* this was the case when it was (allegedly) deduced syllogistically from premises which were known by reason to be true. Although Aristotle and his followers rarely provided demonstrations of their knowledge-claims (which would satisfy their own standards), they regarded many such knowledge-claims as demonstrated, and it was inconceivable for them that they would be overthrown or refuted. Aristotle also recognized other kinds of knowledge-claims: his concept of dialectical demonstrations (based on premises not derived from reason, but on less secure foundations, such as commonly accepted notions) gradually gave rise to ideas about knowledge-claims that are accepted temporarily in order to be examined.

The present paper does not probe into these developments. Instead, it considers a case in which two established bodies of knowledge collided and asks: how were their respective epistemological statuses negotiated? The example considered here is the well-known clash between mathematical astronomy and Aristotelian physics. The geometrical models (such as epicycles) posited by astronomy allowed for accurate calculations of the heavenly motions, but ran against the most basic principles of Aristotelian physics; conversely, the kinds of motions authorized by Aristotelian physics did not allow the framing of a mathematically sound astronomical system. Many medieval thinkers were deeply troubled by this conflict and reacted to it in various ways. One such way was to posit that mathematical astronomy is "merely" a convenient device – or "instrument" – for calculating astronomical phenomena, but has no bearing on reality, which is described by physical theory. This position, whose roots go back to Antiquity and which (in a generalized form) philosophers of science call today "instrumentalism," was formulated with particular acuity and clarity by Moses Maimonides (1138–1204), the famous Jewish jurist, theologian, philosopher, and physician. Maimonides's views on the epistemological status of astronomical theories are at the center of this paper.

This paper was not written out of an interest in the history of epistemology per se, however. Maimonides drew on the incompatibility of the two bodies of knowledge to argue that the knowledge we have of the celestial realm cannot possibly be demonstrative and true; this argument he then used to cast doubt on Aristotle's cosmology, including the thesis of the eternity of the world, thereby increasing the plausibility of the alternative, traditional (religious) doctrine, which posits that the world was created in a determinate past after not having been existent. It must be noted here that there is no unanimity among scholars as to Maimonides's views on the kind of knowledge about the heavens that man can at all attain: Some scholars understand Maimonides as a sceptic who thought that (among many other things) the heavens are unknowable; they also hold that (contrary to a long-standing tradition) Maimonides denied that the circular motion of the spheres allows man to infer the existence of a Mover. Other scholars reject these claims. It is in this context, in any event, that Maimonides's states (in two passages of his *Guide of the Perplexed*) the "instrumentalist" epistemological position on the nature of astronomical knowledge. Maimonides's formulation needed clarification, and to provide it was the main purpose of the paper.

At the same time, the paper can also be read as implicitly asking whether Maimonides considered some knowledge, astronomical or other, as merely provisional i. e. (in anachronistic terms) whether he regarded some knowledge-claims as hypothetical. Maimonides recognized that astronomical science had progressed since antiquity, and expressed the idea that it may continue to do so ("It is possible that someone else may find a demonstration by means of which the truth of what is obscure to me will become clear to him"; *Guide* II, 24, quoted in the article). He thereby implicitly suggests that (in some general sense) the astronomy of his days is merely hypothetical, that it may one day be replaced by true, demonstrated knowledge. However, as the paper shows, for Maimonides the dividing line was not provisional vs. permanent knowledge, but rather full knowledge vs. partial knowledge of its object (the knowledge being construed as demonstrative in both cases).

The paper also briefly touches upon Levi ben Gershom (Gersonides, 1288–1344), the rationalist Jewish philosopher-scientist who lived in Provence. He tried to achieve precisely what Maimonides considered unattainable: knowledge of the heavens as they really are. In other words, he sought an astro-physical theory that satisfies both the desiderata of physics (viz. Aristotelian physics) and of Ptolemaic astronomy. Epistemologically, Gersonides was therefore a "realist," whose position in matters of epistemology was contrary to that of Maimonides. A savant well ahead of his time, Gersonides fully drew on the notion of hypothesis, although this is not discussed in the present paper. *En passant*, the examination of Maimonides's and Gersonides's ideas about the knowability of the heaven shows how science, meta-science (epistemology), and religious concerns may be entangled: it illustrates how competing epistemological positions emerge and acquire greater precision through their use in such multi-faceted debates.

The paper implicitly touches upon yet another theme that may be of interest in the context of the present volume: what is the usefulness, or otherwise, of categories drawn from the philosophy of science in writing the history of science? It seems to me that this paper illustrates that distinctions and concepts deriving from the philosophy of science – such as "instrumentalism" – can be useful to the historian of science, provided they are not used anachronistically. In the present case, although Duhem's listing of Maimonides as an early instrumentalist was unreflective, it proved useful inasmuch as it drew attention to Maimonides's epistemology, allowing discussions of the extent to which it adumbrates "instrumentalism" in the general, philosophical, sense. Put differently, the philosophical distinctions and conceptualizations may supply fruitful vantage points from which to look at the medieval authors and the problems with which they grappled.

Further Reading:

For more comprehensive and recent treatments of Maimonides's epistemology the reader may wish to consult:

Freudenthal, Gad (2005), "Maimonides' Philosophy of Science". *The Cambridge Companion to Maimonides*. Ed. Kenneth Seeskin. Cambridge: Cambridge University Press, 134–166.

Stern, Josef (2005), "Maimonides' Epistemology", in: *ibid:* 105–133.

"Maimonides on the Knowability of the Heavens and Their Mover (*Guide* 2:24)", "Discussion Forum" by several authors, in *Aleph: Historical Studies in Science and Judaism* 8 (2008): 151–358.

Hypotheticity and Realism – Duhem, Popper and Scientific Realism

Andreas Bartels

Abstract:

In contemporary philosophy of science the hypotheticity of scientific knowledge is often seen as being irrelevant to the question of a realist (or anti-realist) interpretation of scientific theories. I shall explain why the alleged independence of hypotheticity and realism is judged to be self-evident nowadays although, to the impartial observer, this seems far from being the case. I shall investigate why the concept of "realism" has changed so much that any relevance of hypotheticity to realism is denied today. My historical sketch will deal with Duhem and Popper. For both of these authors the relation between hypotheticity and realism was still significant, but today we find it hard to follow them in this respect.

Both Duhem and Popper took hypotheticity as an obstacle for realism, although for quite different reasons. For Duhem, realist interpretations of the most general statements of science fail because the purpose of such statements is not to deliver an explanation proper of the empirical laws of science, but to classify them. Since there are no explanations proper in the realm of fundamental (hypothetical) theories, they cannot be interpreted realistically. For Popper, realism with regard to scientific theories has to be understood as a metaphysical, i. e. non-empirical assumption about science, since it involves belief in the truth of hypothetical scientific statements. Neither Duhem nor Popper have anticipated that realism can itself be considered as an empirical theory. In the end only this further move – the claim of truth itself to be conjectural – made it possible to reconcile hypotheticity with realism.

1. Introduction

Today's debate on scientific realism is not concerned with the hypotheticity of scientific knowledge. That (empirical) scientific statements and theories are "hypothetical" is now taken for granted by almost everyone, but it is, as a rule, not seen as a fact that is relevant for realism. It seems as if, in contemporary philosophy of science, the two questions "Is all of our knowledge hypothetical?" and "Should we interpret scientific theories realistically?" were treated as completely independent of each other. It is as if one could be convinced of the hypothetical status of each statement that occurs as part of a scientific theory (including perhaps even the methodological principles governing modern science), without having to abstain from even a strong ("metaphysical") realist approach to scientific theories. But, it is far from obvious *at face value* that hypotheticity and realism do not conflict with each other. (For instance, undergraduates in philosophy of science courses sometimes ask how one can be a realist with respect to scientific theories and claim at the same time that scientific knowledge is hypothetical; I think that this question is completely legitimate).

The influential philosophical movement that, by denying the existence of a tension between hypotheticity and realism, contributed to the situation described above was *scientific realism*, as represented, for example, by Smart (1968), Putnam (1975), Boyd (1984) and Leplin (1984, 1997). These philosophers claimed that the central realist statements – "that the purportedly referential terms of mature science refer and that the laws of mature science are approximately true"[1] – are themselves *scientific hypotheses* that have to be evaluated mainly according to their explanatory performance (in particular with respect to the predictive successes of science). According to scientific realism, to be a realist with respect to some theory means to believe that the realist hypothesis saying that this theory is approximately true comes off best in explaining the epistemic virtues of the theory. Thus, the natural way for the scientific realist to defend his or her theory is *inference to the best explanation.*

Even if today's realism is not in all quarters a simple continuation of *scientific realism* (e. g., think of Fine's critique of inference to the best explanation as a justification for realism), scientific realism has shaped the general attitude of realists to the effect that hypotheticity is not longer seen as a main obstacle for realism. Even today's anti-realists, such as Bas van Fraassen, do not argue that the hypothetical character of some scientific statements *as such* must preclude one from taking a realist stance towards those statements (whereas Duhem had exactly argued that way). Van Fraassen's point is rather that such a realist stance with respect to some hypothetical theoretical statement (referring to unobservable objects), *prima facie* reasonable as it may be, can never be favoured against possible alternative explanations of the epistemic virtues of that statement (or of the theory entailing it).

In the following, I shall explore the philosophical views which motivated the belief that hypotheticity and realism were incompatible, and which had to be overcome by scientific realism with the result that it now seems hard even to see the relevance of hypotheticity to realism.

---

1 Cf. Leplin (1984), p. 203.

Duhem and Popper are my historical examples, since they represent a strong and a weak version of the view that hypotheticity is an obstacle for realism. While Duhem thought that realism can only reside where hypotheticity is absent (strong version), Popper did not cut down the place for realism, but diminished its status to that of a metaphysical, i. e. non-scientific idea (weak version). But for both these authors the relation between hypotheticity and realism had some significance *at their time* which we find hard to experience *now*.

2. Hypotheticity

Duhem and Popper represent two opposing views concerning the aims of scientific theories. Whereas Duhem considers realism as a misplaced attitude towards scientific theories, Popper is an advocate of scientific realism, insofar as he holds truth to be the aim of science. Going a bit deeper into Duhem's and Popper's conceptions of science, the sharp contrast between them seems to become blurred. For both of them, science is essentially hypothetical. Moreover, their views on what hypotheticity *is* do not differ significantly. A statement is a "hypothesis" in Popper's sense if it is an "anticipation"[2] of, or an assumption about some state of affairs the truth of which cannot be established by any (past, present or future) observational experiences. It tells a *possible story* of how reality could be, given the experience we have, but it need not be falsifiable.

This characterisation fits the conception of hypotheticity that figures in the picture of scientific theories as *hypothetical explanations* that has been rejected by Duhem. Hypothetical statements, according to this picture, are *as if descriptions* of reality used in order to make sure "that all our perceptions are produced *as if* the reality were what it asserts."[3] For Duhem, in order to achieve a correct picture of theories, abstract scientific principles have to be stripped of their alleged descriptive character. They have to be seen then as the fundamental principles of a theory which "do not claim in any manner to state real relations among the

2 Cf. Popper (1994), p. 223.

3 P. Duhem (1974), p. 8. The French edition reads: "elle se contente alors de prouver que toutes nos perceptions se produisent *comme si* la réalité était ce qu'elle affirme." (Duhem 1981, p. 5)

real properties of bodies."[4] Thus, whereas Popper and Duhem have contrary views concerning the question of whether scientific statements *are* hypotheses, they use the term "hypothesis" in the same sense. Furthermore, there is also a difference between Popper and Duhem concerning the extension of "hypothesis." According to Popper, even singular statements about actual experiences entail "anticipations," theoretical assumptions which can not be verified by observations.[5] In this sense, singular statements have "hypothetical character." In contrast to that, hypotheticity is for Duhem restricted to the fundamental principles of scientific theories.

3. Duhem: Realism and Explanation Proper

Both Duhem and Popper think that hypotheticity is an obstacle for realism, although for quite different reasons. For Duhem, realist interpretations of the most general statements of science reveal a misunderstanding of the role of those statements. Since their purpose is to *classify* the empirical laws of a special scientific area, and since there are necessarily many alternative systems of classification that could do the job, a realist attitude with respect to them is at the very least misleading. For instance, one should contend oneself with the logical properties of Newton's law of gravitation allowing for the deduction of approximations to Kepler's laws, without any further pretension to metaphysical speculations about the nature of gravitating bodies from which the law itself could be derived. The purpose of Newton's law is to provide a symbolic scheme for the deduction of approximations to Kepler's laws, but it is not its purpose to provide an *explanation proper.*[6] By this term Duhem refers to explanations by means of entities and mechanisms that can be directly and completely grasped by our senses – without any mediation by means of theoretical inferences. Thus, "explanation proper" stands in strict opposition to "hypothetical explanation."

Why are there no explanations proper in fundamental theories of science, according to Duhem? Explanation proper means, as we have seen above, "to strip reality of the appearances covering it like a veil,

4 Cf. Duhem (1974), p. 20.

5 Cf. Popper (1994), p. 45.

6 Explanations proper pretend "... to strip reality of the appearances covering it like a veil, in order to see the bare reality itself." (Duhem 1974, p. 7)

in order to see the barc reality itself." At least for fundamental physics theories, no theoretical deduction of any empirical law fulfils those characteristics. No fundamental theoretical statement uncovers reality as it is, so it can be no candidate for "truth." What would it be for a scientific statement to uncover reality as it is? Here, Duhem's considerations concerning the theory of acoustics are illuminating:

"This reality whose external veil appears in our sensations is made known to us by theories of acoustics. The latter are to teach us that where our perceptions grasp only that appearance we call sound, there is in reality a very small and very rapid periodic motion ... Acoustic theories are therefore explanations. The explanation which acoustic theories give of experimental laws governing sound claims to give us certainty; it can in a great many cases make us see with our own eyes the motions to which it attributes these phenomena, and feel them with our fingers."[7]

Thus, for Duhem, explanations proper can be exemplified by theoretical statements that reduce the experimental laws of their domains to movements and mechanical interactions which are observable in principle. To describe a domain of reality *as it is* turns out to be the same as to describe its internal processes by means of spatiotemporal movements and mechanical interactions of its parts which are as concrete as descriptions of processes in our common world of experience. The reason why Duhem thinks realism and hypotheticity to be incompatible is that he restricts *realism* exclusively to this preferred mode of *mechanical explanation* which offers directly sensible entities and mechanisms and which can only be realized by such low-level theories such as acoustics.

Realism, for Duhem, is restricted to such fields of science where explanation proper works. A scientific statement can only be interpreted realistically if it is a premise within some explanation proper. Since this condition cannot be met by the statements of any abstract scientific

---

7 Duhem (1974, p. 8). The French version reads: "Cette réalité, dont nos sensations ne sont que le dehors et que le voile, les *théories acoustiques* vont nous la faire connaître. Elles vont nous apprendre que là où nos perceptions saisissent seulement cette apparence que nous nommons le *son*, il y a, en réalité, un mouvement périodique, très petit et très rapide ... les théories acoustiques sont donc des explications (*explanations proper*, A. B.). L'explication que les théories acoustiques donnent des lois expérimentales qui régissent les phénomènes sonores atteint la certitude ; les mouvements auxquels elles attribuent ces phénomènes, elles peuvent, dans un grand nombre de cas, nous les faire voir de nos yeux, nous les faire toucher du doigt." (Duhem 1981, p. 5)

theory, there is no chance for a realist interpretation of abstract theories. There are no explanations proper in science, at least not in the realm of fundamental theories, thus science cannot be interpreted realistically.

Nevertheless, in cases where a theory by its hypothetical constructions yields some order and organization within a field that has formerly been in a state of disorder and intricacy, Duhem admits the appearance of some irresistible inclination of scientists to interpret their theories realistically. The scientist, then, feels that the theory comes close to a "*classification naturelle*," a representation that mirrors the true relations of nature:

"But while the physicist is powerless to justify this conviction, he is nonetheless powerless to rid his reason of it. In vain is he filled with the idea that his theories have no power to grasp reality, and that they serve only to give experimental laws a summary and classificatory representation. He cannot compel himself to believe that a system capable of ordering so simply and so easily a vast number of laws, so disparate at first encounter, should be a purely artificial system ..."[8]

The symptom telling the physicist that his theory has achieved the status of mirroring nature is the unity it yields within a field of science. Thus, the physicist, when striving for the systematic unity of his theories, at the same time searches for a realization of his intuition for truth[9]:

"Thus, all those who are capable of reflecting and of taking cognizance of their own thoughts feel within themselves an aspiration, im-

8 Duhem (1974), p. 27. The French version reads: "Mais cette conviction, que le physicien est impuissant à justifier, il est non moins impuissant à y soustraire sa raison. Il a beau se pénétrer de cette idée que se théories n'ont aucun pouvoir pour saisir la réalité, qu'elles servent uniquement à donner de lois expérimentales une représentation résumée et classée; il ne peut se forcer à croire qu'un système capable d'ordonner si simplement et si aisément un nombre immense de lois, de prime abord si disparates, soit un système purement artificiel ..." (Duhem 1981, p. 36)

9 According to Darling 2003, p. 1132, "the intuition that physical theory is or tends to be a natural classification and the intuition that physical theory should be logically coordinated function together to motivate the physicist". The passage cited in the following (Duhem 1981, p. 152/53) shows that the two intuitions are not mutually independent, but the latter depends on the former: The increase in logical unity provided by some physical theory *indicates* that this theory tends to be a natural classification. Thus, the intuition that "physical theory should be logically coordinated," motivates the physicist only in a secondary way, because of its indicator function for the realization of the primary intuition that "the physical theory is or tends to be a natural classification."

possible to stifle, toward the logical unity of physical theory. This aspiration toward a theory whose parts all agree logically with one another is, moreover, the inseparable companion of that other aspiration, whose irresistible power we have previously ascertained, toward a theory which is a natural classification of physical laws. We, indeed, feel that if the real relations of things, not capable of being grasped by the methods used by the physicist, are somehow reflected in our physical theories, this reflection cannot be devoid of order or unity. To prove by convincing arguments that this feeling is in conformity with truth would be a task beyond the means afforded by physics; how or to what could we assign the characters that the reflection should present when the objects which are the source of this reflection escape visibility? And yet, this feeling surges within us with indomitable strength; whoever would see in this nothing more than a snare and a delusion cannot be reduced to silence by the principle of contradiction; but he would be excommunicated by common sense."[10]

Duhem thus asserts that there is an irresistible disposition, originating from a *sense commun*, to assume that a scientific theory which is epistemically ideal (simple, coherent, and producing unity within a field of scientific facts) should also be *true*, at least in the sense that its structures correspond to structures in reality. Even if it can never be *rational* to interpret any particular theory realistically, there are *extra-rational* forces producing a metaphysical realist attitude with respect to our best theories.

---

10 Duhem (1974), p. 103–104. The French version reads: "... Ainsi, tous ceux qui sont capables de réfléchir, de prendre conscience le leurs propres pensées, sentent en eux-mêmes une aspiration, impossible à étouffer, vers l'unité logique de la théorie physique. Cette aspiration vers une théorie dont toutes les parties s'accordent logiquement les unes avec les autres est, d'ailleurs, l'inséparable compagne de cette autre aspiration, dont nous avons déjà constaté l'irrésistible puissance ..., vers une théorie qui soit une *classification naturelle* des lois physique. Nous sentons, en effet, que si les rapports réels des choses, insaisissables aux méthodes dont use le physicien, se reflètent en quelque sorte dans nos théories physiques, ce reflet ne peut être privé d'ordre ni d'unité. Prouver par arguments convaincants que ce sentiment est conforme à la vérité serait une tâche au-dessus des moyens de la Physique ; comment pourrions-nous assigner les caractères que doit présenter le reflet, puisque les objets dont émane ce reflet échappent à notre vue? Et cependant, ce sentiment surgit en nous avec une force invincible; celui qui n'y voudrait voir qu'un leurre et une illusion ne saurait être réduit au silence par le principe de contradiction ; mais il serait excommunié par le *sens commun*." (Duhem 1981, p. 152–53)

4. Popper on Explanation

In contrast to Duhem's metaphysically charged concept of explanation, Popper, in the *Logik der Forschung* (LdF), presents an early version of the Hempel-Oppenheim model. "Causal explanations," Popper claims, are built up by two different sorts of statements, first *general statements* – hypotheses, laws of natures – and *particular statements* – statements which are only valid in a special case or "boundary conditions." From the general statements, one can, with the aid of the boundary conditions, deduce the explanandum, or the "Prognose," as Popper calls it.[11] Thus, explanations are characterized merely by their deductive logical form; even the word "causal" in "causal explanation" carries no metaphysical connotations, but simply means that explanations provide *predictions*.

Thus, Popper's sceptical attitude with respect to any particular concrete realist commitment, in contrast to Duhem, has nothing to do with an idea of an "explanation proper," of explanations that enable us to see reality "face to face"; in order to explain, scientific statements together with some boundary conditions need only enable us to deduce "predictions" from them. There are no particular "reality criteria" for the premises of an explanation which could restrict our realist commitments to the premises. Thus Popper did certainly not agree to Duhem's strong version of the thesis that hypotheticity precludes realism. On the other hand, Popper was still never willing to admit any whole-hearted realist interpretations for the premises of scientific explanations in the style coined some decades later by the philosophers of scientific realism. According to the account that was established by scientific realism, the success of the predictions deduced from the premises supports, by means of an *inference to the best explanation*, the assumption that the premises are true. No better reason for their truth could be demanded – or so scientific realists claimed. The next section shall explain why Popper never reached the scientific realists's camp.

5. Popper's Reservation Concerning Hypothetical Realism

Popper is known to be a passionate advocate of (metaphysical) *realism*. Most explicitly he argues for metaphysical realism in the first volume of the Postscript to the *Logik der Forschung*, entitled *Realism and the*

11 Popper (1994), p. 32.

aim of science (RAS), written in the 1950s. In contrast to scientific realism, as developed in the 1960s, metaphysical realism holds the central claims of realism – the assumptions that theoretical statements refer and that they provide (approximately) true descriptions of their reference objects – to be *non-empirical* (neither verifiable nor falsifiable) assumptions.

20 years earlier, in his LdF, comments on truth and realism are scarce. In Chapter 84, right at the end of the original 1934 version, Popper declares that we should abstain from attributing "true" or "false" to scientific statements or theories, and instead substitute "true" and "false" by terms such as "corroborated" which refer to the logical relations between scientific statements.[12] Keuth mentions two possible reasons for Poppers reservation: First, the dubious status of the correspondence theory of truth, and second, the possible objection that the word "true" should only be used if a truth criterion is available.[13] Since Poppers views regarding hypotheticity of scientific statements did, *despite* his adoption of Tarski's theory of truth, not significantly change between LdF and RAS, I suspect that the latter reason is the more important and permanent reason for his rejection of "true" and "false." Since we never have conclusive evidence in order to reason that a theory is true, we should abstain from calling it "true." To call a theory "true" would misleadingly suggest that we are certain that the theory will never be falsified. This point is also made explicitly by Popper in RAS.[14]

By then, Popper's theory of verisimilitude had allegedly established a link between the corroboration status and the truth of a scientific theory. Would this be a way to justify realist commitments regarding particular scientific statements? As Popper remarks, "it may ... be thought reasonable to believe that there exists a true law of nature, provided there exists a thoroughly discussed and well tested law of nature."[15] Such a particular "thoroughly discussed and well tested law [better: *law statement*, A. B.]" would have a high degree of corroboration, and therefore a high degree of verisimilitude. Thus, it would be a good candidate for instantiating the existential statement "There are true laws of nature." In this case we would have empirical reasons for

12 Popper (1994), p. 219/220; cf. Ayer's comments on this reservation of Popper against the use of "truth" in "Logik der Forschung" (Ayer 1974, p. 684/85).
13 Cf. Keuth (2000), p. 167.
14 Cf. Popper (1983), p. 72.
15 Popper (1983), p. 79.

that existential claim of realism. Realism then would turn out to be itself an empirical hypothesis.

The following passages, however, considerably weaken this proposal. Popper here draws attention to Hume's scepticism concerning the reality of the physical world. No matter how much empirical evidence we have for the truth of a particular law statement, we are not able to infer, from that evidence, the truth of that law statement or the "existence of the corresponding law." The confidence in the existence of laws will always be unfounded.

Thus, hypotheticity continues to motivate Popper's refutation of hypothetical realism with regard to particular statements. This is also explicitly stated by Popper in an earlier passage: "Our assertion – 'there are true natural laws' – being existential, does not even refer to any particular physical law, but merely asserts that at least one such law is true. … I, for one, would not be prepared to point to any particular law of physics and say: 'This law is true, in its present formulation and interpretation'."[16]

Popper had not changed his attitude with respect to the status of realism. The realist statement does "not belong to science"[17]; methodology (including the theory of verisimilitude) "merely supports our preference for one law or another, and [does] not establish, or support, the view that any one of them is true."[18]

Scientific realism (or hypothetical realism) does neither appear in LdF nor in RAS, but first in Chapter Two of *Objective Knowledge*, entitled *Two Faces of Common Sense*, where Popper discusses arguments for realism. One of these arguments is the argument concerning scientific theories: "We can … assert that almost all, if not all, physical, chemical, or biological theories imply realism, in the sense that if they are true, realism must also be true. This is one of the reasons why some people speak of 'scientific realism'. It is quite a good reason. Because of its (apparent) lack of testability, I myself happen to prefer to call realism 'metaphysical' rather than 'scientific'."[19]

I understand Popper here as saying the following: According to "scientific realism," if someone had a good reason to (hypothetically) accept a scientific theory as true (as a true description of reality), then

16 Popper (1983), p. 72.
17 Popper (1983), p. 73.
18 Popper (1983), p. 73.
19 Popper (1972), p. 40.

he would also have good reason to be a hypothetical realist with respect to this theory. Scientific realists indeed think that there currently exist theories we have good reason to accept as true. But Popper had not changed his mind with regard to his doubts about any "good reasons to accept a theory as true." The passage following the one cited above indicates that he continues to think of realism as a metaphysical thesis. Popper only concedes that, in the days of LdF, he had "identified wrongly the limits of science with those of arguability," but later had changed his mind to the effect that "nontestable metaphysical theories may be rationally arguable."[20] That realism is an arguable, and therefore rational thesis is not a new insight then, but one that had already been elaborated by Popper in RAS.

Why was Popper not able to accept a *scientific realism* in that sense, i.e. to accept the *scientific character of realism?* Why is Popper not ready to accept that a particular realist statement – such as "This law statement 'a' expresses a natural law" – can be rationally supported by the fact that it (hypothetically) explains the empirical facts which can be deduced from it? Why did the acceptance of the hypothetical status of explanation not motivate him to accept *hypothetical realism* as a form of second order hypothetical explanation for the success of first order explanations?

In search for an answer we follow the lines of Popper's arguments in RAS:

On the one hand, Popper has remained sceptical with regard to any particular realist commitment because of hypotheticity ("... I would not be prepared to point to any particular law of physics and say: 'This law is true ...'"[21]).

Obviously, what he has in mind when he rejects calling a particular law "true" are such connotations as "I feel certain that it will never be falsified" which are in conflict with hypotheticity.

On the other hand, Popper is prepared to accept that from a simple assertion of a law statement "a" it follows that "a is true." Thus, to reject "a is true" means also to reject the simple assertion of "a." But no scientist will be willing to sacrifice simple, tentative assertions of law statements. If "a" can be accepted hypothetically, then hypothetical acceptance must also apply to "a is true."

20 Popper (1972), p. 40, Footnote 9.
21 Popper (1983), p. 72.

Yet, while hypotheticity passes on from "a" to "a is true" and further to "there is a true law of nature," "there is a true law of nature" does "not acquire *scientific character* simply by following from a scientific law"[22], according to Popper, because it remains *untestable*. Thus, "there is a true law of nature" does not belong to science. In contrast to the hypothetical premises of scientific explanations, this existential statement can only be the object of our "belief" in an extra-scientific sense.

Popper is right, by his own standards, to reject the scientific character of "there is a true law of nature" (because existential statements are non-scientific in general), but the same does not apply to "a is true," where "a" is a law statement. Since "a" and "a is true" imply each other, "a" is testable if and only if "a is true" is testable. Thus, "a" and "a is true" differ from each other neither with respect to hypotheticity nor with respect to their scientific status. It follows that Popper should not, adhering to his own standards, reject hypothetical realism while accepting hypothetical explanations. Either Popper would have to accept as a scientific hypothesis the statement "a is true" for any law statement "a"; then each statement implied by "a is true," and in particular the statement "there is a true law of nature," would have to be considered as scientific hypothesis (although this would have conflicted with the exclusion of existential statements from science). Or he would have to refuse scientific acceptance passing from "a" to "a is true" (which seems plainly unsettled), in order to preserve the exclusion of existential statements. His old reservations against truth attributions may have moved Popper to choose the second horn of this dilemma.

6. Poppers's Metaphysical Realism

Apart from his continued reservation about realist commitments with respect to particular scientific statements, in RAS Popper explicitly argues for his *metaphysical realism*, which he had only outlined in LdF. His metaphysical realism can be characterized by the following theses:

1) Scientific theories are to be taken literally. They are genuine conjectures about the world.[23] Thus, a scientific theory can be true, even though we can never be sure of its truth.

22 Popper (1983), p. 73.
23 Popper (1983), p. 110.

2) Science aims at true theories about the world, where truth is understood in the sense of the correspondence theory. The aim of science is, in the first place, to find "satisfactory (non-ad-hoc) explanations"[24], but, as Popper thinks, the fulfilment of that aim *implies* the search for universal laws of nature.
3) There are true natural laws.[25]

(1) That scientific theories are genuinely descriptive must be true in order to make it possible that scientific theories can be descriptions of laws of nature. Thus, (1) is a *semantic condition* the fulfilment of which has to be presupposed, if science is to be conceived as an instrument for finding "true natural laws," the existence of which is proposed by (3). In a similar way, (2) can be conceived as a *methodological condition* for epistemic access to "true laws of nature"; if science did not have the aim to detect true laws of nature, those laws would never be uncovered by us.

Thus, whereas the role of (1) and (2) is to guarantee that science makes the real structures of the world, that are proposed by (3), accessible to us, the very core of the realist proposal is entailed in (3).

Apart from its more explicit presentation, Popper's conception of metaphysical realism in RAS does not differ significantly from what he had outlined earlier in LdF. There, Popper had claimed that "... our guesses are guided by the unscientific, the metaphysical (though biologically explicable) faith in laws, in regularities which we can uncover – discover."[26] With similar emphasis, Popper in RAS argues that "clearly our assertion [that there are true laws of nature], being existential, cannot be empirically tested; it is not falsifiable; and it is not verifiable either ... As our assertion is irrefutable, we may certainly describe it as "metaphysical" in the technical sense in which this term is used in L. Sc. D. ... It is "metaphysical" also in the traditional sense of the word, since it deals with subject matters which are regarded as characteristic of metaphysics."[27]

24 Popper (1983), p. 132.
25 Popper (1983), p. 71.
26 Popper (1959), p. 278. In the German edition the passage reads: "... unser Raten ist geleitet von dem unwissenschaftlichen, metaphysischen (aber biologisch erklärbaren) Glauben, dass es Gesetzmäßigkeiten gibt, die wir entschleiern, entdecken können." (Popper 1994, p. 223)
27 Popper (1983), p. 74.

In RAS, in contrast to LdF, Popper thinks he is able to present arguments for his metaphysical realism that turn this realist faith into a rational position: "I share this belief, and think it more reasonable than any alternative of which I know"[28], and he even claims: "... there is an all-important difference between [the irrefutable theses of idealism and metaphysical realism]. Metaphysical idealism is false, and metaphysical realism is true."[29] The arguments Popper offers for metaphysical realism are a bit confused, and in my view not very strong.

First, Popper makes a methodological point. He argues that metaphysical realism "forms a kind of background that gives point to our search for truth. Rational discussion, that is, critical argument in the interest of getting nearer to the truth, would be pointless without an objective reality, a world which we make it our task to discover."[30] The question is how strong this phrase "to give point to" [our methodological practices] is actually meant. Popper emphasizes repeatedly that metaphysical realism is not part of scientific methodology: "... it seems to me that within methodology we do not have to presuppose metaphysical realism. Nor can we derive any help from it, except of an intuitive kind. For once we have been told that the aim of science is to explain, and that the most satisfactory explanation will be the one that is most severely testable and most severely tested, we know all that we need to know as methodologists. That the aim is realizable we cannot assert – neither with nor without the help of metaphysical realism, which can give us only some intuitive encouragement, some hope, but no assurance of any kind."[31] Thus, the connection between metaphysical faith and methodological practice seems to be more of a motivational kind; perhaps we may conceive of it as a sort of *regulative principle*, but in no case has it a necessary status.

Second, there is an argument from *common sense:* "The reality of physical bodies is implied in almost all the common sense statements we ever make; and this, in turn, entails the existence of laws of nature."[32] Since common sense realism, including the belief in the reality of physical bodies, is itself part of metaphysical realism, this argument is very weak. It can at most show that certain realist commitments, for in-

28 Popper (1983), p. 75.
29 Popper (1983), p. 82–83.
30 Popper (1983), p. 81.
31 Popper (1983), p. 145–46.
32 Popper (1983), p. 128.

stance with regard to physical bodies, imply some other realist commitments, for instance with regard to laws of nature; but it cannot show that realist commitments are reasonable in the first place. But that is what Popper is looking for.

Third, Popper offers an argument from *explanation*: "Realism ... explains to us why our knowledge situation is *necessarily* precarious [such that our knowledge can only be a trial-and-error affair]."[33] I find it difficult to grasp the idea that the hypotheticity of human knowledge is necessitated by the fact that we are "animals trying to adjust ourselves to our environment."[34] Shall we believe that a metaphysical assumption is nevertheless explanatory, that it explains an empirical fact about how we gain knowledge about the world? Has Popper not argued that realism is neither falsifiable nor verifiable? How then can some empirical prediction be deduced from it? Or is the statement of the hypotheticity of knowledge itself no empirical statement that can be true or false? If the statement of hypotheticity were itself a metaphysical statement, how could it depend on the truth of an empirical statement, namely the statement proposing that we are animals of a particular constitution? And regardless of all that, is Popper really willing to propose that the hypotheticity of knowledge is a necessary truth relative to our biological nature?

Since the aim of my paper is not an extensive discussion of Poppers arguments for metaphysical realism, I will content myself with these critical annotations.

7. Conclusion

To summarize, concerning their reservations against hypothetical realism, Popper's and Duhem's positions are not as much opposed to each other as one would expect in view of their different personal inclinations in matters of realism. Duhem's strong substantial requirements for realism delimit realist commitments to in principle observable entities figuring in mechanical ("proper") explanation. Thus, for him, scientific hypotheses in general are non-referring statements which cannot be interpreted in a realist manner. In that way, Duhem thought hypotheticity in science to be a phenomenon that strongly restricts the

33 Popper (1983), p. 102.
34 Popper (1983), p. 102.

ground for realist attitudes to science. Yet even Popper's realism is a long way away from the kind of scientific realism dominating the 1960s and 1970s. On the one hand, Popper did not accept any sort of "explanation proper" as opposed to "hypothetical explanation," and thus could not fall a victim to Duhem's strong thesis that there could be only a residual place for realism. According to Popper, "hypothetical explanation" is the only sort of explanation for science in general. On the other hand, he was not willing to go the further step to declare that realist commitments to hypothetical explanations share with them their hypothetical and scientific status. Instead of this, he felt that the hypotheticity of scientific knowledge would permit a general realism for science only as a sort of metaphysical idea – a conception that, as Section 6 has shown, is itself threatened by inconsistency. In contrast to Popper's position, scientific realists declared a hypothetical scientific status for realism itself, and thereby were able to deny any restriction of realist commitments following from hypotheticity.

Duhem and Popper exemplify two different ways in which the discovery of hypotheticity had an influence on the general realist commitment to science. For Duhem, the requirements for explanations proper, which he understood also as requirements for realism, restricted chances for a realist interpretation of theories to a small set of low level theories which make no use of hypothetical explanations. For Popper, the hypothetical character accepted for explanation did not pass on realism. That realism can itself be considered as an *empirical* theory, as scientific realists put it, is neither anticipated by Duhem nor by Popper. Only this move made hypotheticity and realism compatible. If all of our knowledge is conjectural, then the claim to truth can only be maintained if this claim itself is taken to be conjectural in nature. The turn towards an empirical and naturalistic understanding of scientific methodology that has taken place in philosophy of science from the 1960s onwards thus appears not so much as a fashionable desire to naturalize another philosophical domain, but as a necessary move which realists felt obliged to make in light of the insight that scientific knowledge is hypothetical.

Bibliography

Ayer, A. J. (1974), "Truth, Verification and Verisimilitude", in: Paul Arthur Schilpp (ed.): *The Philosophy of Karl Popper*, Book II. Open Court: La Salle, 684–692.

Boyd, Richard (1984), "The Current Status of Scientific Realism", in: Leplin (1984), 41–82.

Darling, Karen Merikangas (2003), "Motivational Realism: The Natural Classification for Pierre Duhem", in: *Philosophy of Science.* Proceedings of the 2002 Biennial Meeting of the Philosophy of Science Association, Part I, Vol.70 (5), 1125–1136.

Duhem, Pierre (1974), *The Aim and Structure of Physical Theory.* Atheneum: New York.

— (1981), *La Théorie Physique. Son Objet – Sa Structure.* Librairie Philosophique J. Vrin: Paris.

Keuth, Herbert (2000), *Die Philosophie Karl Poppers.* Mohr Siebeck: Tübingen.

Leplin, Jarrett (ed.) (1984), *Scientific Realism.* University of California Press: Berkeley.

— (1997), *A Novel Defense of Scientific Realism.* Oxford University Press: Oxford.

Popper, Karl (1994), *Logik der Forschung.* Mohr Siebeck: Tübingen (10th edition).

— (1959), *The Logic of Scientific Discovery.* Hutchinson: London.

— (1972), *Objective Knowledge.* Clarendon Press: Oxford.

— (1983), *Realism and the Aim of Science.* Hutchinson: London.

Putnam, Hilary (1975), "What is Realism?", in: Leplin (1984), 140–153.

Smart, J. J. C. (1968), *Between Science and Philosophy.* Random House: New York.

The Hypothesis of Reality and the Reality of Hypotheses

Alfred Nordmann

> We have to believe that everything has a cause,
> as the spider spins its web in order to catch flies.
> But it does this before it knows
> there are such things as flies.
>
> (Georg Christoph Lichtenberg, H 25)

Abstract:

In a long footnote to an 1868-paper Charles Sanders Peirce remarks upon various meanings of the word "hypothesis." One among these treats the hypothetical as an epistemic qualification of scientific knowledge: "too weak to be a theory accepted into the body of a science." Another meaning of the word associates it with Peirce's pragmatism: a hypothesis is the conclusion of an abduction and, as such, it is a productive anticipation of reality. The conclusion of an abduction creatively posits a reality which might serve to explain why something has occurred. The ensuing process of inquiry articulates and elaborates this posit. It determines the initially vague meaning of the hypothesis and thereby determines also the real itself – since reality is that which corresponds to the true belief that is achieved at the end of inquiry. The most general hypothesis is therefore the hypothesis that there is a mind-independent reality and it is this hypothesis, according to Peirce, that underwrites the scientific method for the fixation of belief. The reconstruction of Peirce's conception of hypothesis shows that he does not associate hypotheticity with fallibilism at all – these two notions play very different roles within his realist metaphysics and epistemology. Accordingly, his views are closer to constructivist and technoscientific accounts of world-making rather than Popperian characterizations of the scientific method.

When Charles Sanders Peirce declared in 1903 that "Pragmatism whatever it may be is nothing else than the true Logic of Abduction" hypotheses took center stage in his philosophy (1903a, p. 224, comp. 1903b, p. 235). After all, the conclusion of an abduction is a hypothesis, and the very term "abduction" succeeded the earlier designation "(method of) hypothesis" as in his 1878 paper "Deduction, Induction, and Hypothesis." It would therefore appear that Peirce fits nicely into the story-line of hypotheticity, that is, of an increasing emphasis on the hypothetical in 19th and 20th century philosophy of science. This impression is bolstered when Peirce is said to anticipate or influence the philosophy of Karl Raimund Popper whose *Conjectures and Refutations* may well represent the apotheosis of this development. Accordingly, Peirce's fallibilism corresponds to Popper's falsificationism and both advance an epistemic view of hypotheses and their role in science as a truth-seeking enterprise and unending quest in which everything remains revisable.

A closer look at Peirce and the logic of abduction reveals, however, that he does not associate hypotheticity and fallibility. Hypotheses do not serve as epistemic qualifiers of belief but as productive anticipations of reality. Their articulation coincides with the settlement of opinion and the determination of reality. Peirce's pragmatism thus points beyond Popper to constructivist accounts of world-making. In these accounts, awareness of the merely hypothetical character of theoretical representations recedes in favor of a pragmatic realism that enables the formation of behavioral and technical habits. Accordingly, Peirce's realism corresponds to a constructivist view of hypotheses and their role in technoscience as an enterprise dedicated to the formation of habits of action, including the acquisition and demonstration of basic capabilities of technical intervention in the world.

The following aims mainly to elaborate the difference between these two conceptions of "hypothesis" and thus also of the two ways of reading Peirce's philosophy. After a reconstruction of the role of hypothesis in Peirce's philosophy, it presents the Popperian and constructivist interpretations, and concludes with a consideration of Peirce's fallibilism. From all this emerges a critical qualification of the claim regarding an ever more pronounced awareness of the merely hypothetical character of scientific knowledge. While it may hold for science conceived as a strictly epistemic enterprise and all the scruples that come with that, it does not hold for technoscientific research which is oriented towards experimental intervention and technological transformation. Here, hypotheses do not signify loss of truth, but are instrumental in the production of truth. And instead of advancing further and further, the general awareness of the conjectural character of all scientific knowledge has been eclipsed by a rise to prominence of epistemically unscrupulous technoscience (Nordmann 2004, 2008). Not all of this can be established here, but a beginning can be made by articulating the notion of "hypothesis" in the work of Charles Sanders Peirce which prepares the parting of the ways between science and technoscience.

1. Peirce's Hypotheses

Charles Sanders Peirce articulated the core intuitions of his philosophy early on. The metaphysical view developed in his 1871 review of Fraser's edition of Berkeley are amended but essentially persist in his Harvard and Lowell lectures of 1903.[1] The pragmatic maxim was formulated early on and assumed greater prominence and scope in his later writings.[2] The following reconstruction of the role of hypothesis in Peirce's philosophy therefore does not need to distinguish between various stages of his intellectual development – it always served not as epistemic qualifier of belief but the productive anticipation of reality.

Even as Peirce replaces the term "hypothesis" with "retroduction" and "abduction," he holds fast to the now-familiar scheme that abduction proposes hypotheses, deduction articulates their consequences, and induction evaluates them (1868b, pp. 31–34; 1908, pp. 441 f.). Peirce presents this succession as a continuous process of reasoning where "just as we say that a body is in motion, and not that motion is in a body we ought to say that we are in thought, and not that thoughts are in us" (1868b, p. 42). To be in thought is for these successions to be nested within each other and to run on whether we are aware of them or not. A physical sensation prompts a perceptual hypothesis. Deduction tells us what to expect if this hypothesis were true. Induction from further perceptions evaluates the original perceptual hypothesis, for example, by confirming it. In the meantime, those further perceptions owe to parallel processes, requiring perceptual hypotheses more or less of their own. Since the coincidence of a series of perceptions also wants explanation and gives rise to more general hypotheses, the formulation of perceptual hypotheses may contribute to a process of

1 The later notion of a "spreading of reasonableness," for example, cannot be found in the Fraser review – but this later amendment leaves quite intact Peirce's critique of nominalism and conception of reality. Indeed, this notion can be said to solve a problem that arose from the Fraser review, namely how to explain that "human opinion universally tends in the long run to a definite form, which is the truth" (1871, p. 89). If reality as conceived by the nominalists cannot provide guidance and if a Darwinian random sporting of hypotheses is insufficient, might a tendency towards reasonableness properly constrain the development of human opinion (see Fisch 1986)?

2 Initially Peirce conceived as a semantic criterion the notion of the "practical bearing" that gives meaning to the objects of our conceptions. Later, that practical bearing would consist open-endedly of all the consequence of an abduction (see note 12 below).

evaluating such more general hypotheses. Aside from massive parallelism, even the formation of fairly basic perceptual hypotheses may thus participate simultaneously in bottom-up (from sensations) and top-down (from general conceptions) chains of reasoning.[3]

This process found its most general and consequential expression in Peirce's epistemological writings on the fixation of belief. The irritation of doubt corresponds to a sensation or percept that needs to be put to rest by being accounted for. With this irritation of doubt, a thought process is started that begins with a hypothesis and terminates in the fixation of belief. There are two prominent sources for the irritation of doubt, namely a clash between what we expect to happen and what does occur, and the social impulse or the disagreement among people. Were it for the first of these sources alone, any manner of explaining the irritable fact would do: as long as that fact appears to us as a matter of course, we can accommodate it and more or less tenaciously put our minds to rest. Since we also have to contend with the social impulse, however, what we need is a method of fixing belief that can draw consensus. A generalized scientific method allows for that. It posits an "external permanency" that serves as a common referent and involves a self-correcting methodology that converges upon it (1877, pp. 120 f.).

On this account, hypotheses do not designate a particular stage in a thought process such that a hypothesis might be proposed to explain definite perceptual facts and such that it eventually ceases to be merely hypothetical but assumes the status of a true theory. Instead, thinking begins with abduction and the prompting of a hypothesis by an irritation of doubt, and thinking ceases with the fixation of belief and the coincidence of opinions and facts. Rather than designate a problematic stage, hypothetical reasoning is coextensive with mind and thought as such. At the one end of process, it is only through error and the irritation of doubt that self and self-consciousness appear (1868a, p. 20; 1868b, p. 55). At the other end, mind becomes crystallized when upon the fixation of belief thinking hardens by taking on the form of habit (1878, p. 129; 1891, pp. 293 and 297).

3 This view underwrites Peirce's rejection of philosophical foundationalism or any human faculty of having immediate knowledge of objects or of oneself: the continuous process of reasoning cannot be traced to an absolute beginning (an intuition) or end, see Peirce 1868a.

Even more importantly, perhaps, the scientific method of fixing belief is initiated by a most general kind of hypothesis.[4] It posits an external permanency as a common referent of all inquiry and all inquirers.

> Such is the method of science. Its fundamental hypothesis, restated in more familiar language, is this: There are real things, whose characters are entirely independent of our opinions about them: those realities affect our senses according to regular laws, and, though our sensations are as different as our relations to the objects, yet, by taking advantage of the laws of perception, we can ascertain by reasoning how things really are, and any man, if he have sufficient experience and reason enough about it, will be led to the one true conclusion. (1877, p. 120)

The fundamental hypothesis thus posits a reality that is prior to and explanatory of our sensations. It directs the formulation of all special hypotheses – starting with the perceptual hypotheses – toward causes of our sensations that are independent of and external to our mind, causes that operate with perfect generality such that anyone in the same position would have experienced the same sensation. The fundamental hypothesis thus directs inquiry to a future time and a reality-to-be (Esposito 1980, p. 220), namely to a discovery of "how things really are." It anticipates reality in a general way by positing an indefinite future time when reality will be known as that which corresponds to the one true conclusion of an unbounded process of inquiry:

> [A]s what anything really is, is what it may finally come to be known to be in the ideal state of complete information, so that reality depends on the ultimate decision of the community; so thought is what it is, only by virtue of its addressing a future thought which is in its value as thought identical with it, though more developed. In this way, the existence of thought now, depends on what is to be hereafter; so that it has only a potential existence, dependent on the future thought of the community. (1868b, pp. 54–55)

On the one hand, the hypothesis of reality posits a mind-independent reality as the cause of our sensations, and on the other hand this reality is to depend on the "decision of the community." What is posited is therefore something that can draw the ultimately unanimous decision of the community. And the community decides on what can be said to be really the case, that is, it decides upon what is independently of and prior to any such decision. This seemingly precarious metaphysical

4 Adopting the term from Stephen Pepper, Andrew Reck suggests that one might consider it a "world hypothesis" (1994, p. 130).

construction is supported by the logic of abduction and thus by the conception of hypotheses as anticipations of reality that are productive in that they inaugurate a self-correcting process of convergence on a limit.[5] The following four propositions are at the heart of this logic of abduction and contextualize the hypothesis of reality as an instrument of discovery:

i) What the hypothesis of reality states is a metaphysically flawed conception of reality, namely, a philosophical nominalism that regards reality as the incognizable cause of mental action.
ii) What Peirce endorses instead is a philosophical realism that regards reality as the normal product of mental action, that is, as a product of hypothetical reasoning.
iii) The scientific method of fixing belief adopts the nominalist hypothesis of reality. This hypothesis posits real things or events as the causes of mental action. These presumed causes are only established as such in the course of reasoning and therefore appear as the product of mental action. The scientific method thus proceeds from a productive "as if": If and to the extent that there is a fixed reality (laws, definite values of variables, limits of series of measurements or experiences), then a self-correcting method that wagers on its existence will reliably establish this reality.
iv) Since that self-correcting method involves multiple chains of hypothetical reasoning (abductive-deductive-inductive), particular hypotheses posit real entities and processes that are then gradually established in the course of inquiry. All the while, the fundamental hypothesis also becomes articulated and the assumption of the real realized.

Peirce develops these four points nowhere more forcefully and cogently than in his 1871 review of an edition of the works of George Berkeley. It contrasts the nominalist and realist philosophical conceptions of reality and begins with the more familiar one:

---

5 This self-correcting process cannot be fully reconstructed here. It found its most succinct formulation in Reichenbach 1961 where the hypothesis of reality appears as a "posit": There is no guarantee that a series of inductions will point towards some truth, that is, that it will converge upon a limit of that series. However, if there is a limit, the method of induction will discover it in the long run as relative frequency approximates objective probability. Thus, if we want to gain truth at all, we have to posit a knowable reality as a limit to the series and hope that this posit proves to be self-fulfilling.

> Where is the real, the thing independent of how we think of it, to be found? There must be such a thing, for we find our opinions constrained; there is something, therefore, which influences our thoughts, and is not created by them. We have, it is true, nothing immediately present to us but thoughts. Those thoughts, however, have been caused by sensations, and those sensations are constrained by something out of the mind. The thing out of the mind, which directly influences sensation, and through sensation thought, because it *is* out of the mind, is independent of how we think it, and is, in short, the real. (1871, p. 88)

What is stated here corresponds to the hypothesis of reality and generally agrees also with positions that are currently known as "scientific realism." Peirce rejects this view, however, as the *nominalist* conception of reality. His reasons for this are at least threefold: i) the nominalist view posits the real as something unknown, perhaps unknowable – it is the we-know-not-what that must be causing our sensations; ii) it assumes that the real is real in virtue of being outside the mind, whereas "the immediate object of thought in a true judgment *is* the reality" and thus in the mind, though not therefore exclusively in the mind (1871, p. 91); iii) nominalism denies the reality or objectivity of universals and holds instead that general conceptions serve only to organize sensations but do not enter into judgment and the process of realizing the real.[6] Another, less familiar conception of reality does not suffer from these three defects:

> All human thought and opinion contains an arbitrary, accidental element, dependent on limitations in circumstances, power, and bent of the individual; an element of error, in short. But human opinion universally tends in the long run to a definite form, which is the truth. Let any human being have enough information and exert enough thought upon any question, and the result will be that he will arrive at a certain definite conclusion, which is the same that any other mind will reach under sufficiently favorable circumstances ... The individual may not live to reach the truth; there is a residuum of error in every individual's opinion. No matter; it remains that there is a definite opinion to which the mind of man is, on the whole and in the long run, tending. On many questions the final agreement is already reached, on all it will be reached if time enough is given ... This final

6 It is due to this third feature, of course, that Peirce chooses the label "nominalist" for this conception of reality: "from this point of view it is clear that the nominalistic answer must be given to the question concerning universals ... the one mental term or thought-sign 'man' stands indifferently for either of the sensible objects caused by the two external realities [i. e., two real men]; so that not even the two sensations have in themselves anything in common, and far less is it to be inferred that the external realities have" (1871, p. 88).

> opinion, then, is independent, not indeed of thought in general, but of all that is arbitrary and individual in thought; is quite independent of how you, or I, or any number of men think. Everything, therefore, which will be thought to exist in the final opinion, is real, and nothing else. (1871, p. 89)

Peirce attributes the passage from the nominalistic to the realistic view of reality to Kant's Copernican turn (1871, pp. 90–91; compare Nordmann 2006b).

> It was the essence of his philosophy to regard the real object as determined by the mind. That was nothing else than to consider every conception and intuition which enters necessarily into the experience of an object, and which is not transitory and accidental, as having objective validity. In short, it was to regard the reality as the normal product of mental action, and not as the incognizable cause of it ... The realist will, therefore, believe in the objectivity of all necessary conceptions, space, time, relation, cause, and the like. (1871, p. 91)[7]

Towards the end of his discussion Peirce briefly reflects the tension that results from the fact that his realist metaphysics and epistemology is fueled by a flawed nominalist conception of reality.

> The realistic philosophy of the last century has now lost all its popularity, except with the most conservative minds. And science as well as philosophy is nominalistic ... On the other hand, it is allowable to suppose that science has no essential affinity with the philosophical views with which it seems to be every year more associated. History cannot be held to exclude this supposition; and science as it exists is certainly much less nominalistic than the nominalists think it should be. (1871, p. 104)

By proposing a somewhat casual ad hoc explanation for the nominalism of science, Peirce clearly does not fully appreciate as of yet that according to his own realism, scientists have to start from the nominalist hypothesis of reality. Instead of providing a realist justification for the only apparent but necessary nominalism of science, that nominalism of science still seems in this passage to be something of an embarrass-

---

7 Peirce goes on to state that "[n]o realist or nominalist ever expressed so definitely, perhaps, as is here done, his conception of reality." Peirce's own definite expression claims Kant for his own "common-sense position" and thereby rejects Kant's notion of an incognizable thing in itself (but see Peirce 1931–35, 5:525 and 1905b, 353–354). And in this, too, Peirce might actually be going beyond or even against Kant. Peirce refers to his own view of reality as inevitably realistic, "because general conceptions enter into all judgments, and therefore into true opinions. Consequently a thing in the general is as real as in the concrete" (1871, pp. 91, 90).

ment to him. According to Andrew Reck, the resolution of this tension required gradual adjustments to his conception of the relation of science and philosophy: "The tension between scientific method, as an empirical method involving hypothesis and experimental observation, and the goal of science, in a systematic formulation of truth such as the philosopher seeks, is somewhat resolved when Peirce defines philosophy as a science of discovery" (Reck 1994, p. 130 [sic]).

The fundamental hypothesis of a foregoing reality sets us on a track towards the discovery of truth. And reality *will be* what corresponds to the eventually established truth. Science thus vindicates the "bad metaphysics" of nominalism in that it gradually realizes a conception of a unified reality as the cause rather than the product of our sensations and their attendant mental actions.[8] The course of science cannot establish the uniqueness of reality, however. "[I]f two groups of inquirers could never compare notes, there could be conflicting sets of perfect knowledge" or conflicting ways in which general conceptions enter into experiences and judgement.[9]

To complicate things even further, the "bad metaphysics" comes with a mechanism of validation – once a nominalist conception of reality is introduced as a hypothesis, the logic of inquiry will articulate it and determine the real accordingly. In contrast, there is no hypothesis of realism. Indeed, Peirce's assertion that "human opinion universally tends in the long run to a definite form, which is the truth" has no clear status. It is partly (historical) observation, partly normative insistence against arguments from tenacity, authority and a prioricity, and partly a consequence of the self-corrective method ("if there is truth or a limit to a series of observations, the method of abduction-deduction-induction is sufficient to discover it in the long run"). Peirce was keenly aware

8 In Nordmann 2006a I describe this as the interplay between metaphysics (that posits a substantial underlying reality) and metachemistry (that deals with the realization of the real). More in line with Peirce's debt to Schelling, this could also be viewed as a dialectic of *natura naturans* and *natura naturata*.

9 Quoted here is Sandra Rosenthal's paraphrase of unpublished ms. 409.112 (Rosenthal 1994, p. 138). To be sure, the social impulse guarantees that this case cannot be sustained in the long run and that we will arrive at perfect knowledge and a corresponding determination of reality on some track, and there will then be no question whether or not there might have been a possibly preferable alternative track: "Wherever universal agreement prevails, the realist will not be the one to disturb the general belief by idle and fictitious doubts" (1871, p. 91).

that the truth of the assertion required constraints on abduction or hypothesis-formation. Rather than being entirely random, the hypotheses that account for our sensations and experimental findings have to continue us on some adopted track towards the truth.[10] Aside from being controlled by the hypothesis of reality, they ought to be consistent with the best of our prior knowledge. Peirce referred to these controls on abduction as reasonableness on the one hand (1903b, p. 235; 1903c), instinct on the other (1913, pp. 464–465).[11]

A final aspect of Peirce's conception of hypothesis emerges when one considers why consistency with background knowledge or previously adopted hypotheses is not enough for a sufficiently powerful logic of abduction that explains how human opinion always tends to the truth. This aspect is closely linked with the pragmatic maxim and thus with the establishment of meaning: Hypotheses are anticipations of reality not in the sense that they specify how things would be if they were true (Rosenthal 1994, p. 135). Instead, they anticipate reality vaguely. By discovering instances that can serve to verify the hypotheses, the process of inquiry simultaneously articulates their meaning. When the pragmatic maxim enjoins us to consider what effects the objects of our conceptions might conceivably have, we are asked to engage in an experimental investigation. The full meaning of our conceptions emerges at the end of inquiry and the determination of reality coincides with the clarification of ideas (1878, 1903a).[12] The concept of "reality" and the concept, for example, of "electricity" therefore have in common that both are first introduced by abduction in a hypothesis and then accrue meaning as they grow from vagueness to determinacy, that is, the power to determine how things really are. Accordingly,

10 "If hypotheses are to be tried haphazard, or simply because they will suit certain phenomena, it will take the mathematical physicists of the world say half a century on the average to bring each theory to the test …" (1891, p. 288). To solve this problem, Peirce's earlier and metaphysically sparser account needed to be amended (see note 1 above).

11 "We call that opinion reasonable whose only support is instinct" (1903a, p. 218).

12 In his Harvard Lectures Peirce revisits the pragmatic maxim. Rejecting a purely linguistic reading of the maxim (whereby we introspectively group conceptions of effects under conceptions of objects), he emphasizes that the meaning of a term is the "entire general intended interpretant" or the sum of consequences of a perceptual nature (1903a, pp. 220, 225, see also 1903b).

the following passage can be read as an account of the history of the term "reality" (as the general account or symbol of the real):

> [T]he universe is intelligible; and therefore it is possible to give a general account of it and its origin. This general account is a symbol; and from the nature of the symbol it must begin with the formal assertion that there was an indeterminate nothing of the nature of a symbol. This would be false if it conveyed any information. But it is the correct and logical manner of beginning an account of the universe. As a symbol it produced its infinite series of interpretants, which in the beginning were absolutely vague like itself ... But every endless series must logically have a limit. (1904, p. 323)

Here, Peirce introduces again the scientific method of fixing belief. It begins with an abduction, the merit of which is not in any information that it conveys but in that it posits a hypothesis or a symbol that sets in motion an unbounded process of inquiry. The endless series converges on a limit and that limit is reality.[13] A hypothesis or a symbol is therefore "essentially a purpose, that is to say, a representation that seeks to make itself definite" (1904, p. 323) – in other words, an anticipation of reality.[14]

2. Interlude: The Popperian Interpretation

Hypotheses or conjectures also take center stage in Karl Popper's philosophy of science. It has therefore been suggested that there is a philosophical kinship between Popper's and Peirce's approaches. And indeed, Popper can be said to provide an interpretation or further development

13 "Reality, therefore, can only be regarded as the limit of the endless series of symbols" (1904, p. 323).

14 Inversely, Peirce speaks of the "Universe being precisely an argument" that is "working out its conclusions in living realities" (1903a, 193–194; compare 1891, 293, 297): "Was ist Wirklichkeit? Vielleicht gibt es so etwas gar nicht. Wie ich wiederholt hervorgehoben habe, ist sie nur eine Retroduktion, eine Arbeitshypothese, die wir ausprobieren, unsere einzige, verzweifelte Hoffnung, etwas zu erkennen ... Aber wenn es irgendeine Wirklichkeit gibt, dann besteht sie, insofern es eine Wirklichkeit gibt, in folgendem: daß es etwas im Sein der Dinge gibt, das dem Prozeß des Schlußfolgerns, daß die Welt *lebt* und *sich bewegt* und *ihr Sein hat,* in der Logik der Ereignisse entspricht. Wir alle stellen uns die Natur syllogistisch vorgehend vor, selbst der mechanistische Philosoph tut das, der so nominalistisch ist, wie es ein Naturwissenschaftler nur sein kann" (this passage from manuscript 439 has only been published in a German translation so far, Peirce 1991, p. 396).

of Peirce's philosophy of science.[15] This interpretation, however, makes three mistakes: It solidifies Peirce's fluid continuity of abduction-deduction-induction, it unwittingly transforms Peirce's realism into nominalism, and it construes hypotheses not as vague anticipations of reality that seek to make themselves definite but takes them to be perfectly meaningful.

In *The Logic of Scientific Discovery* and his later works Popper distinguishes contexts of discovery and justification. He leaves the context of discovery unconstrained and, indeed, allows for guesswork, for general or metaphysical conceptions to enter into the process of hypothesis-formation. Indeed, his stance towards discovery, invention, and creativity is so liberal or respectful that he leaves the domain of abduction entirely unregulated, saying hardly anything about it. What matters for the purposes of science is only that the context of justification is sharply set off from the context of discovery. It is the domain of logic and ruled by the principle of non-contradiction. Here, testable predictions are derived from the hypotheses and subsequently evaluated by way of experiment and controlled observation. This corresponds to Peirce's stages of deduction and induction with the notion of "induction" wide enough to accommodate Popper's anti-inductivism: The evaluation of hypotheses through induction is introduced by Peirce as a "self-correcting" process which is always attended by normal observational error and that can at best corroborate and will often falsify or modify the hypothesis. Indeed, Peirce's fallibilism – "there is a residuum of error in every individual's opinions" (1871, p. 89) – appears to constitute the strongest link to Popper:

> [N]o matter how far science goes, those inferences which are uppermost in the mind of the investigator are very uncertain. They are on probation. They must have a fair trial and not be condemned till proved false beyond all reasonable doubt; and the moment that proof is reached, the investigator must be ready to abandon them without the slightest tenderness towards them. Thus, the scientific investigator has to be ready at a moment to aban-

---

15 Popper called Peirce "one of the greatest philosophers of all time" (1972, p. 212). The most significant intellectual kinship between Popper and Peirce concerns the propensity interpretation of probability and the notion of indeterminacy in physics (which lie beyond the scope of this paper). Larger claims of kinship owe to lax interpretations that find it comparatively easy to understand Peirce as a proto-Popperian. Popper himself fostered this by distinguishing the clear and easy-to-understand Peirce – a good Popperian, of course – from the obscure and speculative Peirce.

> don summarily all the theories to the study of which he has been devoting perhaps many years. (Peirce 1895, p. 25)

As long as we are thinking and have not settled on a final opinion, we are working with hypotheses, Peirce is saying. Popper appears to agree by adding that our best available knowledge is therefore only hypothetical and always on probation. Upon closer scrutiny, however, this agreement proves illusory. It conflates Peirce's conception of hypotheses as anticipations of reality with Popper's notion of the hypothetical as an epistemic qualification of our best available knowledge. After identifying the philosophical differences between Popper and Peirce and suggesting Peirce's affinity to constructivist accounts, Peirce's fallibilism will be revisited and shown to be unrelated to his conception of hypothesis.

The philosophical difference between Popper and Peirce emerges from Popper's motto at the opening of *Logic of Discovery*. Popper quotes Novalis: "Theories are nets: Only he who casts will catch" (1968, p. 11). Novalis, of course, was an idealist poet-philosopher whose views are close to Schelling's (and thus closer to Peirce than Popper), but the one isolated sentence as appropriated by Popper suggests not only a philosophical nominalism but also a nominalist reading of Kant: Reality is out there as the cause of all our sensations but it is shrouded in an inaccessible darkness; we can bring it to light only partially by formulating scientific hypotheses, hoping that they will capture something. What we capture, however, remains tentative because we do not see the things as they are in themselves but only as we brought them to the light of reason. The ways in which we frame our hypotheses structures our scientific experience but hypotheses are not otherwise productive and do not inaugurate a process of clarification of ideas alongside the fixation of belief and the determination of the real. Reality is in no way thought of as "the normal product of mental action" and the real is not what corresponds to a true judgement and not something that is in the mind as much as it is outside it.

This difference becomes more pronounced when one considers the meaning of a hypothesis. According to Peirce, the hypothesis seeks to become definite and requires deduction and induction not as a test of its truth or falsity but – in line with the pragmatic maxim – for the exploration and discovery of the sensible effects that belong to its conceptions. Any hypothesis thus retains a residuum of vagueness (and the criterion of non-contradiction or consistency is therefore insufficient to guide the formation of further hypotheses). In contrast, the virtue of Popper's hypotheses is that they have definite truth-conditions: To understand a hypothesis is to know under which conditions it would be false. Moreover, good hypotheses are very general and therefore could be falsified by a wide range of experimental findings.

Popper's hypotheses are semantically determinate and experimental inquiry does not serve "to make our ideas clear." On the contrary, an experiment can test a hypothesis only to the extent that it is clear already. This is also why Popper views as a static logical sequence rather than as a fluid continuity the succession of abduction, deduction, and induction: Abduction ends when it issues in a hypothesis; deduction and experimental evaluation refer to that hypothesis with the aim of falsifying or else corroborating it. As many of his critics have pointed out, Popper's idealization of this process neglects the formation of auxiliary hypotheses, perceptual judgments and other aspects of discovery, creativity, or abduction in the context of justification.

Popper's idealization is meant to remind us of limits of knowledge and to establish epistemic norms for intellectual honesty. Popper does not "believe in belief" because what distinguishes Einstein from an amoeba is that Einstein can learn from his mistakes and does so by maintaining a healthy distrust of all claims to knowledge (1972, pp. 24 f.).

Like Popper, Peirce views hypotheses as inhabitants of a "third world" of ideas (Popper 1956, p. 156–161). The scientific method of fixing belief requires the possibility of a disagreement between expectation and experience but also of disagreement among inquirers. By being detached from their inventors, hypotheses have a life of their own and become part of a communal process of inquiry, and in this process the attitude of personal belief or disbelief drops out as insignificant – the only opinion that matters is the final opinion that is reached by all inquirers as questioning ceases and knowledge becomes sedimented as a habit of action. But as opposed to Popper, Peirce's understanding of this process gives him something other than criteria for the evaluation of a discontinuous series of theories, where each theory is primarily a linguistic artifact that offers a nominalistic conception of an otherwise incognizable foregoing reality. Instead, an understanding of the process of inquiry provides Peirce with an explanation of reality and a view of the evolution of mind and world (Pape 1991). Even if no individual ever knows the final truth, the hypotheses that are advanced by this individual are productive in that they contribute to the determination of reality as the product of an indefinitely long process of collective mental action.[16]

3. The Constructivist Interpretation

Karl Popper associates the hypothetical with the tentative and always fallible character of scientific representations or descriptions of the world. As such he remains firmly within the confines of a nominalist epistemology, that is, an epistemology that must remain nominalist because the world "out there" is unknowable except indirectly through evidences gathered from observation and experiment. In contrast, Peirce associates the hypothetical with productive anticipations of reality. Abduction and hypothesis are central to his scholastic realism that includes Kantian idealism,[17] and thus central to a view according to which general conceptions enter into true judgements, thereby determining the real as that which corresponds to true judgements.

This conception of the realization of the real is not, of course, a social constructivism but rather more closely akin to Bruno Latour's notion of the world as a construction jointly of human and non-human agents (Nordmann 2006a). Indeed, Latour echoes Peirce's critique of nominalism in his critiques of the purely social or mental constructivism that he attributes to Immanuel Kant and William James. Inverting Peirce's reading of Kant, Latour attributes to him an "extravagant form of constructivism" according to which "everything was ruled by the mind itself and reality came in simply to say that it was there, indeed, and not imaginary":

> Kant invented this science-fiction nightmare: the outside world now turns around the mind-in-the-vat, which dictates most of that world's laws, laws

16 This again might be seen as close to Popper: He praises the demolition of hypotheses as clearing the way for the generation of new and presumably better ones. In this sense, too, the failure of the individual is productive for the whole. To the extent that this is Popper's view, it becomes interesting how little he makes of this. Since each hypothesis is logically distinct from previous and subsequent hypotheses (even if it were to be a mere modification of them), Popper cannot and does not attempt to envision the continuity of the productive process.

17 Peirce designated his own "realism" also as "objective idealism" and "critical common-sensism" (Nordmann 2006b). With reference to Peirce, Ian Hacking contrasts nominalism somewhat misleadingly with "dynamic nominalism" (2002, pp. 48 f.).

> it has extracted from itself without help from anything else (Latour 1999, 5 f.).

Latour's reading differs from that of Peirce because he assumes that Kant's inaccessible things-in-themselves are the "real reality." While Latour takes Kant to be an unreformed nominalist, we saw that Peirce is more generous and views him as the first realist. According to Peirce, Kant does not "think of the mind as a receptacle, which if a thing is in, it ceases to be out of" (1871, p. 91). Peirce maintains that the Kantian "things in themselves" become dispensable once one recognizes that there is no question of a reality beyond the world of experience and that this experience or reality is a product of the joint actions of mind and nature (1931–1935, 5.525, 1905b, 353 f.).

Though a nominalist like Peirce, Latour credits not Kant or Peirce but only William James with debunking the notion that reality is that which lurks behind our experiences, that we can only gather highly mediated evidences about this reality which nevertheless serves to ground and validate our knowledge:

> When a rationalist insists that behind the facts is the ground of facts, the possibility of the facts, the tougher empiricists accuse him of taking the mere name and nature of a fact and clapping it behind the fact as a duplicate entity to make it possible. (James 1907, p. 263, quoted in Latour 1990, p. 64)[18]

But James's critique does not go far enough for Latour, nor would a Peircean realism if all it did was assert that general conceptions enter into true judgements, thereby determining the real as that which corre-

---

18 That the "rationalist" here holds the same position as Peirce's "nominalist" becomes apparent when Latour quotes James another time: "On the pragmatist side we have only one edition of the universe, unfinished, growing in all sorts of places where thinking beings are at work. On the rationalist side we have a universe in many editions, one real one, the infinite folio, or *edition de luxe*, eternally complete; and then the various finite editions, full of false readings, distorted and mutilated each in its own ways" (James 1907, p. 259, quoted in Latour 1990, pp. 78 f.). Here, the *edition de luxe* stands for the incomprehensible reality as it is in truth and knowing subjects are condemned to write books of their own that always fall short of the original. The pragmatist or nominalist, in contrast, engages with all subjects to write the single book that in the course of time determines reality as the normal product of mental action. Reality is that which would correspond not to the original but only to the very final edition which represents the consensus of all and has quieted the social impulse and all other irritations of doubt.

sponds to true judgements. Narrowly construed, such a realism would still amount to merely a social or mental constructivism. And on such a narrow reading, the notion of hypotheses as productive anticipations of reality would reduce to the claim that reality is shaped or constructed by our hypotheses, and that reality is fitted to the requirements of the mind. Insisting that "there is a history of things, not only of science" Latour therefore formulates his critique of James's pragmatism:

> The limit of pragmatism is to be concentrated on man (individual at that). But if essence is existence and existence is action, this pragmatism is to be extended to the things in themselves now endowed with a history. James was ready to "add to reality." He transforms the metaphor of the book one reads to a book one *writes* ... But, he was prepared to do it as you add shape to a shapeless and plastic matter, not as you meet other non-human actors who have also their history. This shift away from human overcomes the other limit of pragmatists. They have no way slowly to withdraw existence out of essence. This withdrawal occurs by shifting the task of maintaining the consensus to non-humans and moving from interactions, talks and controversial practices to a world *in which* we live (1990, p. 66 and 78 f.).

This, to be sure, is a rather cryptic passage. As a critique of James it echoes Peirce's critique of James's all too narrowly conceived pragmatism – leading Peirce to adopt "pragmaticism" to designate his more comprehensive constructivism that goes beyond epistemology to semiotics, philosophy of nature, and a metaphysics of non-human agency (Pape 1991, Reynold 2002). In particular, Peirce's doubt-belief dynamic offers what Latour finds lacking in James, namely an account of the gradual withdrawal of "existence out of essence": In the process of fixing belief and of determining reality, how does one arrive at an essential nature of things that is no longer dependent on what some knowing subjects experience as a currently existing fact? In other words: if science begins with sensations and perceptions, perceptual hypotheses and the like, how does it attribute those finally to a more or less immutable and eternal reality which causes the sensations as signs of something other and prior to these sensations? Where Latour speaks of withdrawing existence out of essence such that the essence remains after everything accidental has been removed, Peirce uses the metaphor of settlement or sedimentation. He refers to the fixation of belief also as a settlement of opinion and thereby captures that this settlement cannot be compared to an explicit human consensus on a hypothesis or on a political issue. Instead, the settlement of opinion is a withering away of discourse and thought as a habit is being formed that becomes a subterranean stratum or a robust way of living in the world.

Latour studied the withdrawal of existence from essence and thus the construction of a robust objective reality through the cooperation of human and non-human actors in the course of his laboratory studies (e.g., Latour 1990). A "laboratory man" himself,[19] Peirce tells a similar story of experimental practice. It begins by bringing together various existences and establishes something that no longer depends on existence but fixes a certain relation that can be invoked at any time:

> What are the essential ingredients of an experiment? First, of course, an experimenter of flesh and blood. Secondly, a verifiable hypothesis … The third indispensable ingredient is a sincere doubt in the experimenter's mind as to the truth of that hypothesis. Passing over several ingredients … we come to the act of choice by which the experimenter singles out certain identifiable objects to be operated upon. The next is the external (or quasi-external) ACT by which he modifies those objects. Next, comes the subsequent *reaction* of the world upon the experimenter in a perception …

Out of all these diverse ingredients that are all more or less arbitrary existences (human bodies, sentences, mental states, chosen objects, actions and reactions) the successful experiment generates an experimental phenomenon that is independent of any particular event:

> When an experimentalist speaks of a *phenomenon*, such as "Hall's phenomenon," "Zeeman's phenomenon" and its modification, "Michelson's phenomenon," or "the chessboard phenomenon," he does not mean any particular event that did happen to somebody in the dead past, but what *surely will* happen to everybody in the living future who shall fulfill certain conditions. The phenomenon consists in the fact that when an experimentalist shall come to *act* according to a certain scheme that he has in mind, then will something else happen, and shatter the doubts of sceptics, like the celestial fire upon the altar of Elijah (1905a, pp. 339 f.).

What Peirce only hints at has been the subject of laboratory studies by Latour and many others: The production of objectivity requires that one controls for experimental artifacts, that the contingencies of the laboratory are reduced, and that the phenomenon can exist outside the laboratory in which it was first demonstrated.

19 His own experience as an experimentalist was to have disposed him towards Kant and the view "that a *conception*, that is, the rational purport of a word or other expression, lies exclusively in its conceivable bearing upon the conduct of life" (1905a, p. 331–3).

More generally, of course, the "withdrawal of existence from essence" refers to the interplay of a realist metaphysics with the nominalist hypothesis of reality. According to realism, we begin with sensations, irritations of doubt, and hypothetical conceptions and find ourselves quite in the realm of existence. That there might be natures or essences "behind" these sensations and existences is suggested only by the hypothesis of reality. To the extent that the process of inquiry adduces evidence (further existences) to this hypothesis – that it confirms or vindicates it – the hypothesized reality or realm of natures emerges as the endpoint and final product of inquiry. At the end of inquiry, there is therefore no longer any reference to particular existence. The work of abduction, deduction, and evaluation has hardened or crystallized into fixed material relations – paraphrasing Latour, the discourses have been weighted down by habits and things (1996a).

These material relations or habits obtain equally between "causes" and "effects" and between "human expectation" and "action in the world."[20] Thus, Peirce (like Latour) treats material things in the world symmetrically along with human thought and action. Both are accounted for semiotically as signs that grow through the articulation of their meaning (that is, their practical effects or bearing on conduct), and both derive from a primordial "law of mind" (1891, p. 292):

> [A]ll mind is directly or indirectly connected with all matter, and acts in a more or less regular way; so that all mind more or less partakes of the nature of matter. Hence, it would be a mistake to conceive of the psychical and the physical aspects of matter as two aspects absolutely distinct. Viewing a thing from the outside, considering its relations of action and reaction with other things, it appears as matter. Viewing it from inside, looking at its immediate character as feeling, it appears as consciousness. (1892, p. 349)

It is common-place within Kantianism to apply these two standpoints to human beings who face the special predicament that they must conceive of themselves simultaneously as determined by nature and as free moral agents. Peirce extends this notion by applying it to all material things (including humans) and concludes that "if habit be a primary property

---

20 Where Peirce speaks of the hardening of habits, Latour refers to trajectories of action that become more predictable as motions are weighted down. For example, in his essay about the heavy key-rings in hotels, he remarks that the behavior of guests (their readiness to leave their keys at the desk) could be regulated by standardizing the guests (for example, through extensive indoctrination) or by making the key so heavy that no one wants to carry it around (Latour 1996a, p. 54).

of mind, it must be equally so of matter, as a kind of mind" (1892, p. 350).[21]

Latour's constructivism boasts that its conception of reality is more robust than that of William James because it complements the social or mental construction of existence with an account of how we can institute an unchanging, eternal, mind-independent nature through the gradual withdrawal of existence from essence. Like Peirce he must therefore account for the peculiar constitution of the modern world and its paradoxical "constitutional guarantee" that "even though we construct Nature, Nature is as if we did not construct it" (1993, 32). He confronts this difficulty in the most direct and philosophically sustained manner in a chapter on the historicity of things: "Where Were Microbes before Pasteur?" (1999, 145–173). In typical Latourian fashion, he first heightens the sense of paradox by suggesting a kind of backward causation: Only our present actions bring into being the world that preceded and enabled these actions. The appearance of backward causation can be dissolved, however, when one considers instead a process of "sedimentation." With new experimental capabilities, people and things assume articulated competencies. In the laboratory, chemical substances can do things that they could not do before, but of this capability it is said that it always existed latently but so far without the chance to manifest itself. As the laboratory experiment is reproduced and varied, the still fairly new behavior of the chemical substance appears more and more to be a property of the substance that does not require specific laboratory conditions for its manifestation. And thus, a newly acquired competence gradually settles into history to reconfigure the past: Its existence in the lab becomes an expression of an immutable essence that precedes existence.

Latour thus shows that not only scientists like Pasteur but also Pasteur's microbes need the laboratory to show what they are capable of. Simultaneously, the new experimental capabilities produce a new past: Controversies about fermentation and spontaneous generation in 1865 produce a year 1864 in which vague, haphazard, and invisible

21 In particular, Peirce applies the two standpoints to the atoms and molecules of protoplasm, and goes on to entertain as a consequence of his theory that collectives (of atoms or of people) can act together, acquiring depersonalized habits of thought and action of their own (Peirce 1892). To be sure, Peirce lacked Latour's sociological background or imagination and never considered the world in which we live as a networked collective of human and non-human actors.

processes appear to be in need of interpretation and experimental investigation. Pasteur's persuasive experimental demonstration in 1867 produced a year 1864 in which vague, haphazard, and invisible processes result from the action of microbes. The present in the year 2009 produces a year 1864 in which, as in most of the past, people had limited knowledge of a reality that is theirs as much as ours: We withdraw existence from essence by declaring contingent states of knowledge as accidental existents in the past – opposing these states of knowledge to a nominalistically conceived, essentially given reality. While Peirce could have explained this production of a new past by saying that Pasteur worked with and articulated the hypothesis of reality, Latour's more paradoxical explanation amounts to the same – in Pasteur's laboratory the microbes were constructed as something that has always existed: "After 1864 airborne germs were there all along" (1999, 168–173).

This complementarity of accounts suggests that Peirce and Latour are engaging in a similar project, albeit with different backgrounds and vocabularies. When present articulations of reality sediment and permeate the past, and when they construct now what has always been, this amounts to the realization of a hypothesis of reality that is assumed to hold already. Pasteur's microbes, in other words, were constructed according to the hypothesis that they existed before their construction and that they are certainly not Pasteur's. The construction thus consists in the withdrawal of temporally bound existence so as to leave behind a purified fixed reality which corresponds to our eternally true beliefs. In this process, human and non-human actors work together to co-construct reality. Where Latour speaks of a complete, rather than merely social constructivism (1990, p. 71), Peirce presents hypotheses as self-fulfilling *and* guided by instinct. According to the pragmatic maxim, the meaning of hypotheses becomes articulated through experimental inquiry, and the clarification of ideas coincides with the determination of reality that owes to matter and mind doing their "collective work in the middle" (Latour 1990, p. 68).

But what is the significance of this complementarity, and what is the significance for the history of hypothesis and hypotheticity of the fact that Peirce's conception has greater affinity to Latour's constructivism than to Popper's critical rationalism? The answer to this question is suggested by a succinct statement that was suggested at the outset but cannot be fully elaborated here: Popper is a theorist of science, Latour of technoscience. Popper's science is an epistemic enterprise that aims to produce theoretical representations of the phenomena. As such it is challenged by the problem of underdetermination and bound up with skeptical questions regarding the relation between representations and their objects. As it deals with these questions, this theoretical enterprise experiences a loss of certain truth (Schiemann 1997); it becomes incapable of marking a threshold between mere hypothesis and true theory – a trajectory that culminates in Popper's philosophy of science. In contrast, technoscience is not an epistemic enterprise (Nordmann 2004, 2008), but constructively pursues hypotheses as productive anticipations of reality. For Peirce and Latour, and for technoscientific research as characterized, for example, by Peter Galison (2006), it amounts to the same whether this reality is conceived of as the one which corresponds to our settled opinions at the end of inquiry or whether this reality is shaped to conform to our technologies and habits of action. The settlement of opinion coincides with formation of habit as we assimilate ourselves to reality and reality to us.

With the rise and increasing prominence of technoscience the epistemic scruples of science fade away. The robustness of technoscientific knowledge does not owe to "confirmation" or "corroboration" but to opinion settling into habit. Accordingly, the notion of hypotheses as epistemic qualifiers of belief may serve to pick out only one strand of 19th and early 20th century "basic science" – of theoretical physics in particular. While it is tempting to place Peirce in this tradition, he is rather the "laboratory man" who constructs phenomena such that something else will happen, "and shatter the doubts of sceptics, like the celestial fire upon the altar of Elijah" (Kant, interestingly but perhaps not surprisingly, is claimed by both traditions as the one who specifies limits of theoretical knowledge and the one who establishes the constructedness of reality – on the one hand offering epistemic qualifications, on the other hand providing an account of experience and reality that is just as knowable as anything that is humanly constructed.).[22]

22 In what is perhaps his most sustained reflection on the word "hypothesis," Peirce distinguishes eight meanings of that term, only the eighth coming close to a Popperian usage: "too weak to be a theory accepted into the body of a science." Peirce himself adopts the seventh sense of "hypothesis": "Most commonly in modern times, for the conclusion of an argument from consequence and consequent to antecedent," that is, what he came to refer to as abduction. He then identifies this use of "hypothesis" in seven authors, including Newton, Mill, Kant, and Herbart (1868b, 34 f.).

4. Coda on Fallibilism

Peirce views hypotheses as productive anticipations of reality and it is therefore that pragmatism is the true logic of abduction. This sets him apart from the Popperian concern that even our very best knowledge is only hypothetical because we cannot with certainty traverse the threshold from hypothesis to certainty of truth. Against this sharp juxtaposition might be objected that Peirce was a fallibilist who insisted that "there is a residuum of error in every individual's opinions" and that "the scientific investigator has to be ready at a moment to abandon summarily all the theories to the study of which he has been devoting perhaps many years" (1871, p. 89, 1895, p. 25). Though not in the name of "hypothesis," one might argue, Peirce's fallibilism expresses some limit of knowledge and marks an unbridgeable gap between mere opinion and certainly true belief. However, a brief consideration of Peirce's metaphysics can show that even his fallibilism does not provide an epistemic qualification upon belief.

In Peirce's categorial scheme, "hypothesis" belongs to Thirdness – it involves the doubt-belief dynamic and thus thought and mind as constructive of reality. In contrast, Peirce's fallibilism belongs to Secondness and thus stands entirely outside the doubt-belief dynamic and all things epistemic.[23] A short elaboration must suffice to make this point.

Thought and mental activity begins with the irritation of doubt; it aims for the fixation of belief and the formation of habit. Indeed, thinking ceases with the formation of habit – Thirdness becomes Secondness. Here, Thirdness refers only to a three-place relation: x is a sign of y for z. The continuous interplay of abduction, deduction, and induction involves such thirdness, if only because percepts and concepts are interpreted and mediated in the course of experimental inquiry. As was pointed out above, mental activity is coextensive with hypothetical rea-

23 Helmut Pape has grappled with this issue: How does Peirce's method of hypothesis arrive at individual things and not just representations of them? Inversely, how can the two-place indexical relation between a person and an individual thing become represented? These questions have bearing on Peirce's fallibilism: On the one hand awareness of the fallibility of all knowledge marks the beginning of all inquiry in that the mind steps into action only once a subject experiences the discrepancy of what one expects and what is. On the other hand, this awareness does not represent anything but a general sense that all our expectations might be frustrated in their brute confrontation with reality (compare Pape 1997).

soning, it subjects hypotheses to the pragmatic maxim, simultaneously clarifying ideas and determining reality. There is no state of belief beyond hypotheses because a habit is not a belief; it does not interpret the world. Instead, a habit is a mere two-place relation that coordinates a stimulus and a response. Like instinct, law, and matter, habit is crystallized or hardened mind or Secondness: if x then y.

Peirce's fallibilism owes to the possibility of an "outward clash," that is, a brute confrontation of an expectation and the world: x but not y ("direct consciousness of hitting and getting hit," 1885, 233 f.). No matter where we are on the doubt-belief trajectory – whether we are in the process of formulating and testing hypotheses or firmly in the realm of settled opinion and lawful habit – the world may have a surprise in store for us, frustrate our expectations, or create a novel irritation of doubt.

Quite in agreement with Peirce's dictum not "to doubt in philosophy what we do not doubt in our hearts" (1868b, 29), the mere possibility of this outward clash does not qualify our belief and cannot qualify what has become habitual and routine. Nothing would be gained if we were to attach as a footnote to all our expectations the global proviso that these expectations only hold in the absence of an outward clash. "True belief" (the cessation of thought and the formation of habit) arises as we exit the domain of epistemology. Inversely, the possibility of an outward clash and thus the in-principle fallibility of all consciously held or unconsciously embodied expectations obtains before we enter that domain. In the middle of things, however, and where we engage in mental activity, the hypothesis of reality serves the construction of the reality of hypotheses.[24]

Bibliography

Esposito, Joseph (1980), *Evolutionary Metaphysics: The Development of Peirce's Theory of Categories*. Athens: Ohio University Press.

Fisch, Max (1986), *Peirce, Semeiotic and Pragmatism*. Bloomington: Indiana University Press.

Galison, Peter (2006), "The Pyramid and the Ring", presentation at the conference of the Gesellschaft für analytische Philosophie (GAP). Berlin.

24 The final version of this paper benefits from critical commentaries by the editors and referees, but also from Helmut Pape and Andreas Hetzel. Unfortunately, it could not address all their objections or meet all requests for elaboration. In particular, Pape's incisive critique needs to be taken up elsewhere.

Hacking, Ian (1980), "The Theory of Probable Inference: Neyman, Peirce and Braithwaite," in: D. H. Mellor (ed.), *Science, Belief and Behavior: Essays in Honor of R. B. Braithwaite.* Cambridge: Cambridge University Press, 141–160.

— (2002) *Historical Ontology.* Cambridge: Harvard University Press.

Hoffmann, Michael (2004), "Peirces Philosophie der Wissenschaft, Logik und Erkenntnistheorie," *Philosophische Rundschau* 51: 193–211, 296–313.

Houser, Nathan and Kloesel, Christian (1992), "Introduction," in: Nathan Houser and Christian Kloesel (eds.), *The Essential Peirce*, vol. 1. Bloomington: Indiana University Press, xix–xli.

James, William (1907), *Pragmatism, a new name for some old ways of thinking.* London: Longman, Green.

Kuhn, Friedrich (1996), *Ein anderes Bild des Pragmatismus: Wahrscheinlichkeitstheorie und Begründung der Induktion als maßgebliche Einflußgrößen in den "Illustrations of the Logic of Science" von Charles Sanders Peirce.* Frankfurt: Klostermann.

Latour, Bruno (1990), "The Force and the Reason of Experiment," in: Homer LeGrand (ed.) *Experimental Inquiries.* Dordrecht: Kluwer, 49–80.

— (1996a), *Der Berliner Schlüssel: Erkundungen eines Liebhabers der Wissenschaften.* Berlin: Akademie Verlag.

— (1996b), "Do Scientific Objects Have a History? Pasteur and Whitehead in a Bath of Lactic Acid", *Common Knowledge* 5(1): 76–91.

— (1993), *We Have Never Been Modern.* Cambridge: Harvard University Press.

— (1999), *Pandora's Hope.* Cambridge: Harvard University Press.

Levi, Isaac (1980), "Induction as Self-Correcting According to Peirce", in: D. H. Mellor (ed.), *Science, Belief and Behavior: Essays in Honor of R. B. Braithwaite.* Cambridge: Cambridge University Press, 127–140.

Lichtenberg, Georg Christoph (1784–1788), "Selections from Waste-Book H", in: R. J. Hollingdale (ed.), *Georg Christoph Lichtenberg: Aphorisms.* London: Penguin, 1990, 111–116.

Nordmann, Alfred (2004), "Was ist TechnoWissenschaft? – Zum Wandel der Wissenschaftskultur am Beispiel von Nanoforschung und Bionik", in: T. Rossmann and C. Tropea (eds.), *Bionik: Aktuelle Forschungsergebnisse in Natur-, Ingenieur- und Geisteswissenschaften.* Berlin: Springer, 2004, 209–218.

— (2006a), "From Metaphysics to Metachemistry", in: Davis Baird, Eric Scerri, Lee McIntyre (eds.), *Philosophy of Chemistry: Synthesis of a New Discipline*, in the series *Boston Studies in the Philosophy of Science.* Dordrecht: Springer, 347–362.

— (2006b), "Critical Realism, Critical Idealism, Critical Common-Sensism: The School and World Philosophies of Riehl, Cohen, and Peirce", in: Michael Friedman and Alfred Nordmann (eds.), *The Kantian Legacy in Nineteenth Century Science.* Cambridge: MIT Press, 249–274.

— (2008), "Philosophy of Nanotechnoscience", in: G. Schmid, H. Krug, R. Waser, V. Vogel, H. Fuchs, M. Grätzel, K. Kalyanasundaram, L. Chi (eds.), *Nanotechnology,* vol. 1: G. Schmid (ed.), *Principles and Fundamentals.* Weinheim: Wiley, 217–244.

Pape, Helmut (1991), *Charles S. Peirce: Naturordnung und Zeichenprozeß*. Frankfurt: Suhrkamp.
— (1997), "Abduction and the topology of human Cognition: Review of Ansgar Richter, Der Begriff der Abduktion bei Charles S. Peirce", *Modern Logic* 7(2): 199–221.
Peirce, Charles Sanders (1868a), "Questions Concerning Certain Faculties Claimed for Man", in: Nathan Houser and Christian Kloesel (eds.), *The Essential Peirce*, vol. 1. Bloomington: Indiana University Press, 1992, 11–27, compare *Collected Papers* 5.213–63.
— (1868b), "Some Consequences of Four Incapacities", in: Nathan Houser and Christian Kloesel (eds.), *The Essential Peirce*, vol. 1. Bloomington: Indiana University Press, 1992, 28–55, compare *Collected Papers* 5: 264–317.
— (1871), "Fraser's 'The works of George Berkeley'", in: Nathan Houser and Christian Kloesel (eds.), *The Essential Peirce*, vol. 1. Bloomington: Indiana University Press, 1992, 83–105, compare *Collected Papers* 8: 7–38.
— (1877), "The Fixation of Belief", in: Nathan Houser and Christian Kloesel (eds.), *The Essential Peirce*, vol. 1. Bloomington: Indiana University Press, 1992, 109–123, compare *Collected Papers* 5: 358–387.
— (1878), "How to Make our Ideas Clear", in: Nathan Houser and Christian Kloesel (eds.), *The Essential Peirce*, vol. 1. Bloomington: Indiana University Press, 1992, 124–141, compare *Collected Papers* 5: 388–410.
— (1885), "An American Plato", in: Nathan Houser and Christian Kloesel (eds.), *The Essential Peirce*, vol. 1. Bloomington: Indiana University Press, 1992, 229–241, compare *Collected Papers* 8: 39–54).
— (1891), "The Architecture of Theories", in: Nathan Houser and Christian Kloesel (eds.), *The Essential Peirce*, vol. 1. Bloomington: Indiana University Press, 1992, 285–297, compare *Collected Papers* 6: 7–34.
— (1892), "Man's Glassy Essence", in: Nathan Houser and Christian Kloesel (eds.), *The Essential Peirce*, vol. 1. Bloomington: Indiana University Press, 1992, 334–351, compare *Collected Papers* 6: 238–271.
— (1895), "Of Reasoning in General", in: Nathan Houser and Christian Kloesel (eds.), *The Essential Peirce*, vol. 2. Bloomington: Indiana University Press, 1998, 11–26.
— (1903a), "The Nature of Meaning (Sixth Harvard Lecture)", in: Nathan Houser and Christian Kloesel (eds.), *The Essential Peirce*, vol. 2. Bloomington: Indiana University Press, 1998, 208–225, compare *Collected Papers* 5: 151–179.
— (1903b), "Pragmatism as the Logic of Abduction (Seventh Harvard Lecture)", in: Nathan Houser and Christian Kloesel (eds.), *The Essential Peirce*, vol. 2. Bloomington: Indiana University Press, 1998, 226–241, compare *Collected Papers* 5: 180–212.
— (1904), "New Elements", in: Nathan Houser and Christian Kloesel (eds.), *The Essential Peirce*, vol. 2. Bloomington: Indiana University Press, 1998, 300–324.
— (1905a), "What Pragmatism Is", in: Nathan Houser and Christian Kloesel (eds.), *The Essential Peirce*, vol. 2. Bloomington: Indiana University Press, 1998, 331–345, compare *Collected Papers* 5: 411–437.

— (1905b), "Issues of Pragmaticism", in: Nathan Houser and Christian Kloesel (eds.), *The Essential Peirce*, vol. 2. Bloomington: Indiana University Press, 1998, 346–359, compare *Collected Papers* 5: 438–463.

— (1908), "A Neglected Argument for the Reality of God", in: Nathan Houser and Christian Kloesel (eds.), *The Essential Peirce*, vol. 2. Bloomington: Indiana University Press, 1998, 434–450, compare *Collected Papers* 6: 452–491.

— (1913), "An Essay toward Improving Our Reasoning in Security and in Uberty", in: Nathan Houser and Christian Kloesel (eds.), *The Essential Peirce*, vol. 2. Bloomington: Indiana University Press, 1998, 463–474.

— (1931–35), *Collected Papers*, Charles Hartshorne and Paul Weiss (eds.), 6 vols. Cambridge: Harvard University Press (cited by volume and paragraph numbers).

— (1991), "Weitere selbständige Ideen und der Streit zwischen Nominalisten und Realisten", in: Helmut Pape (ed.), *Naturordnung und Zeichenprozeß: Schriften über Semiotik und Naturphilosophie*. Frankfurt: Suhrkamp, 378–399.

Popper, Karl Raimund (1968), *The Logic of Scientific Discovery*. New York: Harper and Row.

— (1972), *Objective Knowledge*. Oxford: Clarendon.

Reck, Andrew (1994),"Peirce's Conception of Philosophy and Its Place within His Classification of the Sciences", in: Edward Moore and Richard Robin (eds.), *From Time and Chance to Consciousness: Studies in the Metaphysics of Charles Peirce*. Oxford: Berg, 115–131.

Reichenbach, Hans (1961), *Experience and Prediction*. Chicago: University of Chicago Press.

Reynold, Andrew (2002), *Peirce's Scientific Metaphysics*. Nashville: Vanderbilt University Press.

Rosenthal, Sandra (1994),"Charles Peirce: Scientific Method and Worldly Pluralism", in: Edward Moore and Richard Robin (eds.), *From Time and Chance to Consciousness: Studies in the Metaphysics of Charles Peirce*. Oxford: Berg, 133–142.

Schiemann, Gregor (1997), *Wahrheitsgewißheitsverlust*. Darmstadt: Wissenschaftliche Buchgesellschaft.

Hypothetical Metaphysics of Nature

Michael Esfeld

Abstract:

The paper sketches out a reply to the underdetermination challenge and the incommensurability challenge that is sufficient to rebut the skeptical conclusions from these challenges and to lay the ground for the project of metaphysics of nature. Such metaphysics is as hypothetical as are our scientific theories. The paper then explains how one can argue for certain views in the metaphysics of nature based on our current fundamental physical theories, namely the commitments to a tenseless theory of time and existence instead of a tensed one, to events instead of substances, and to relations instead of intrinsic properties. Finally, the paper considers the themes of causation, laws and dispositions.

1. Introduction

Whereas the philosophy of science concentrated for a long time on epistemological issues concerning the justification of knowledge claims contained in scientific theories, the project of a metaphysics of nature (*Naturphilosophie* in German) or metaphysics of science has gathered new momentum in recent years. The new metaphysics of nature distinguishes itself from the older essays in speculative metaphysics by being close to science: metaphysical claims are based on scientific theories. Consequently, the metaphysical claims about nature are as hypothetical as our scientific theories: there is no more certainty to be gained in metaphysics than there is in science. In other words, scientific knowledge claims are fallible and metaphysics, insofar as it draws on those claims, is as fallible as science. The idea of a metaphysics of nature close to science is, however, not without forerunners among 19th and 20th century philosophers. Gustav Theodor Fechner, Hermann Lotze, Eduard von Hartmann, Erich Becher, Aloys Wenzl and others have tried to overcome the speculative metaphysics of German Idealism by clinging to the science of their day. The advent of modern physics has profoundly changed the outlook of this enterprise.

Pursuing the project of a metaphysics of nature evidently presupposes the view that the pretensions to knowledge contained in our scientific theories are not baseless. In other words, some sort of scientific realism is presupposed (cf. Hawley 2006). Accordingly, in this paper, I will first sketch out a reply to scepticism about scientific knowledge that is sufficient as a basis for a metaphysics of nature. I will then discuss a few examples of positions that it is reasonable to endorse when it comes to constructing a metaphysics of nature grounded on our current fundamental physical theories. The thesis that this paper seeks to illustrate is that there is a mutual dependence between science and philosophy: philosophy in the sense of metaphysics needs science to know about what there is in the real world, and science needs philosophy in the sense of epistemology when it comes to developing criteria for the justification of science's pretensions to knowledge in general and the assessment of knowledge claims contained in specific scientific theories.

Following David Papineau, one can divide the philosophy of science into two broad areas:

> The *epistemology* of science deals with the justification of claims to scientific knowledge. The *metaphysics* of science investigates philosophically puzzling features of the world described by science. In effect, the epistemology of science asks whether scientific theories are true, whereas the metaphysics of science considers what it would tell us about the world if they were. (Papineau 1996, 1)

In the next two sections, I shall outline a reply to the two main challenges to scientific realism stemming from underdetermination and from incommensurability. The reply to be developed here is sufficient to justify engaging in the project of a metaphysics of nature. The second half of the paper goes into that project, sketching out three metaphysical claims grounded on our current fundamental physical theories, namely a tenseless view of time and existence (section 4), a metaphysics of events (section 5) and a metaphysics of relations (section 6). Finally, I shall briefly touch upon the topics of causation, laws and dispositions (section 7).

2. The Underdetermination Challenge

If one pursues the project of a metaphysics of nature based on science, one presupposes that science is in principle capable of providing a cognitive access to the constitution of nature. Since there are different scientific theories – or different models or interpretations of a given scientific theory – that are incompatible with each other, that presupposition amounts to the thesis that we have methods at our disposition which enable a rational evaluation of the claims about the constitution of nature contained in different and incompatible scientific theories (or incompatible models or interpretations of a given theory).

That presupposition faces a challenge from confirmation holism and its consequence, namely that scientific theories are underdetermined by experience. Confirmation holism is the thesis that experience does not confirm or refute single propositions taken in isolation, but always a whole network of propositions or a whole theory. Confirmation holism goes back to Duhem (1914) – (English translation Duhem 1954; see part 2, chapter 6) – and Quine (1953). It is known as the Duhem–Quine thesis. The main argument is that whenever a single proposition is taken to be refuted by experience, there are background propositions available. There is always the logical possibility to change the truth-values attributed to some of these background propositions such that one can continue to hold on to the original proposition despite the recalcitrant experience. Confirmation holism therefore implies that theory is underdetermined by experience: for any given body of experience, it is logically possible to construct an indefinite number of theories that are incompatible with each other, but that are all in accordance with the experience – that is, it is possible to infer from each of these theories the observational propositions describing the experience in question.

Logical possibility, however, does not mean that there always are at least two rival theories that are equally credible. It is trivial that one can always construct a rival theory that is in accordance with the experimental data by *ad hoc* modifications of a given theory. This has no implications for the issue of which theory it is *rational* to accept. One may construe the claims Quine makes in "Two dogmas" as meaning, among other things, that for any proposition, circumstances are conceivable in which it is *rational* to abandon the proposition in question in order to adapt a given system of propositions to new experimental data. The issue of whether or not it is rational to renounce certain propositions regarded as logical laws – such as the law of the excluded middle – to accommodate experience in the domain of quantum physics is a case in point (see Quine 1953, 43). However, this does not imply that for any given body of experience we can conceive at least two rival theories that are both in accordance with the experience, without there being any further rational criteria that distinguish one of these theories as a better candidate for a true theory than the other one(s) (see

Laudan 1996, chapter 2, against any such claim based on underdetermination).

Whereas the history of science provides many examples of successor theories being incompatible with their predecessors, it is rare to find an example of two or more contemporary theories that contradict themselves in their ontological commitments, that make exactly the same experimental predictions and none of which is an *ad hoc* modification of a given theory. Quantum physics is one such rare example. Standard quantum mechanics as set out, for instance, in von Neumann (1932) – (English translation von Neumann 1955) – and Bohm's rival theory – (first version Bohm 1952, last major publication by Bohm himself Bohm & Hiley 1993) – make exactly the same empirical predictions. Nonetheless, there are at least two criteria that one can invoke in favour of a rational evaluation of any two such theories. First, if they agree in their experimental predictions at a given time, this does not imply that it is impossible to conceive of an area in which the two theories may distinguish themselves by different predictions of experimental results (see Laudan & Leplin 1991, 451–455, reprinted in Laudan 1996, 56–59). For instance, in the case of standard quantum mechanics vs. Bohm's theory, the domain of quantum fields may constitute such an area: it is an open question whether there can be a Bohmian theory that is equivalent to quantum field theory (but see Bohm, Hiley & Kaloyerou 1987 and Huggett & Weingard 1994, 382–387).

Moreover, if there are two rival theories that agree in their predictions of experimental results, they are rival theories because their ontological commitments are different. It is always possible to evaluate these commitments. Confirmation holism speaks against foundationalism in epistemology: experience is not a foundation of knowledge in the sense of being able on its own to justify scientific theories, for the same experience can be accounted for by rival theories. The logical possibility of such rival theories is sufficient to refute such a foundationalism. Confirmation holism therefore is linked to justification holism, that is, the view that a proposition is justified if and only if it is coherent with other propositions in a whole network of propositions (a whole theory). We can always employ the coherence criterion of justification holism to assess the ontological commitments of rival theories that agree in their experimental predictions. The coherence criterion examines two or more such rival theories against the background of our other knowledge, seeking to establish which one of these rival theories fits better into a coherent system of our knowledge as a whole. The idea behind that criterion is that explanation is linked to coherence: each of the rival theories seeks to explain a certain range of phenomena. These explanations, in turn, can be evaluated as to which one is more likely to contain a true description of entities and their connections in the world, given what else we know about the world.

In this vein, the adherent to standard quantum mechanics can point out against the Bohmian that the quantum potential that Bohm's theory has to stipulate in order to reproduce the experimental predictions of standard quantum mechanics is a very implausible assumption against the background of our other physical knowledge: it is entirely different from a potential in classical physics, and the only argument available for its introduction is the intention to reproduce the experimental predictions of standard quantum mechanics. To counter that objection, the Bohmian, in turn, can point to the measurement problem into which standard quantum mechanics runs. (The question, however, is whether ontological assumptions such as those ones contained in Bohm's theory are necessary in order to solve that problem).

But what about two or more rival accounts, each of which is conceived as a complete theory of the world? One may think of an interpretation of quantum theory in the style of Everett (1957) according to which everything is a quantum object, subject to quantum entanglement, and no quantum state reductions occur, not even in measurement. There exists only one entangled quantum state of the world at a time that includes all possible values that all properties can take as really existing in infinitely many different branches of the world. On the other hand, there is an interpretation of quantum theory that includes state reductions and thus a dissolution of entanglement (the most precise proposal in that respect goes back to Ghirardi, Rimini & Weber 1986). Consequently, there really are classical objects and classical properties that have precise numerical values in the world. Again, it is possible to evaluate and to compare the ontological commitments of any such rival complete worldviews. For instance, the worldview of universal quantum entanglement commits us in the last resort to the view that each person has infinitely many minds, existing in different branches of the universe (see Albert & Loewer 1988 and Lockwood 1989, chapters 12 & 13). In any case, it implies that each single physical object exists infinitely many times in different branches of the universe, having different values of its time-dependent properties in these branches. One may voice with good reason reservations about such extravagant commitments.

Such an assessment of ontological commitments does not have to lead to an uncontroversial result that is accepted by all parties. There rarely are such results in philosophical debates. The interpretation of quantum mechanics again is a case in point. But this does not hinder there being sound arguments which suggest that certain theories – or models or interpretations of a given theory – are the best hypotheses about the constitution of nature or a certain domain of nature that we can put forward given the state of our scientific knowledge. Hence, the underdetermination challenge does not undermine the claim that science is in principle capable of providing a cognitive access to the constitution of nature.

3. The Incommensurability Challenge

Apart from confirmation holism and justification holism, there is a further form of holism that leads to another challenge to the knowledge claims contained in scientific theories: semantic holism. The basic idea is that the content (meaning) of a concept is given by its position in a whole network of concepts, that is, a whole theory. In short, the content of a concept consists in its inferential relations to other concepts (cf. Esfeld 2001, chapter 2). Semantic holism is widely accepted at least with respect to theoretical concepts. For instance, the concept "electron" cannot be introduced by pointing to electrons. It is introduced by indicating its inferential relations to other concepts in a whole theory. The theory as a whole then has certain observational consequences.

Semantic holism implies that there is no separation between issues concerning factual matters and issues concerning conceptual content (meaning). Changes in our views as to what the world is like go together with changes in conceptual content. The reason is that a theory change entails a change in some inferential relations – that is, according to semantic holism, a change in conceptual content. For instance, in Newton's theory, gravitation is conceived of as a force that acts at a distance. In other words, there is an inferential link between the concept of gravitation and the concept of action at a distance. In general relativity, by contrast, gravitation is conceived of as an effect of the curvature of space-time. Hence, the inferential link from gravitation to action as a distance is cut off and replaced with a link from gravitation to space-time curvature and effects that the latter has. Another prominent example is the change in the content (inferential relations) of the concept "electron" from prescientific theories *via* classical field theory to quantum mechanics and quantum field theory. Examples such as these ones are taken to show that there can be a radical conceptual change when one theory is replaced with another.

Such examples drive the claim of semantic incommensurability put forward by Kuhn (1962, chapters 9 & 10) and Feyerabend (1962), which is based on semantic holism. For instance, the concept of gravitation as action at a distance is incommensurable with the concept of gravitation as an effect of space-time curvature, because these two concepts do not have any significant scientific inferences in common. There is a radical conceptual change from Newton's theory of gravitation to Einstein's theory of gravitation: the language of general relativity does not even have the conceptual means at its disposal to express the idea of an action at a distance. In this sense, Newton's concept of gravitation cannot be translated into general relativity.

The thesis of semantic incommensurability calls into question whether there is progress on the level of our views about the constitution of nature: if the proper concepts of a new theory are incommensurable with those of its predecessor, then there is no basis for claiming that the new theory makes cognitive progress in revealing the constitution of nature. Kuhn goes as far as suggesting the replacement of the idea of science making cognitive progress with the idea of changes of scientific theories not being directed to any goal (1962, chapter 13). Incommensurability is thus grist to the mills of what is known as the argument from pessimistic meta-induction (Laudan 1981 and 1984, in particular 157), although that argument is not committed to the incommensurability thesis: our past theories have turned out to be false. There is no reason to suppose that our current theories will endure a better fate. We are thus not justified in supposing that there is cognitive progress in the history of science.

There is an easy Popperian answer to the argument from pessimistic meta-induction: most of our past theories have indeed been falsified. But falsification is the means to make cognitive progress. By replacing our old theories with new ones we correct the errors that the old theories contain. We make cognitive progress in the sense that we come closer and closer to the truth about the constitution of nature by falsifying our old theories and replacing them with new ones (see Popper 1959, in particular chapters 4 & 10). This Popperian reply presupposes, however, that the newer theories do not only improve on the older theories as regards the range and the accuracy of predictions, but also that the theoretical views which the newer theories offer are comparable with the ones of the older theories. The views about the constitution of nature contained in the newer theories are surely incompatible with the ones of the older theories: they contradict them. Nonetheless, it is possible to compare the proper concepts of the newer theories with the proper concepts of the older ones. Being in a position to carry out such a comparison is a necessary condition for being able to argue that we come closer and closer to the truth through falsification. Otherwise, there would only be theory replacement, but no continuous progress towards truth. It is here that the argument from semantic incommensurability comes in: that argument seeks to establish that this presupposition of comparability is not satisfied. In that sense, the thesis of semantic incommensurability is much more radical than the conclusion of the argument from pessimistic meta-induction.

However, in order to counter that thesis, one can point out that there are precise logical relations between the formalisms – that is, the mathematical equations – of any major mature physical theories that succeed each other in the history of science. It is well known that the formalisms of the main mature classical physical theories before the 20th century can be reconstructed from the formalisms of their successor theories in relativity physics and quantum physics. More precisely, it is possible to derive within the formalisms of the newer theories an image of the formalisms of the older theories that reproduces the predictions of the older theories in the cases where one can neglect certain quantities: if one carries out a mathematical operation that lets the quantum of action approach zero ($h \rightarrow 0$) or a mathematical operation that lets the velocity of light approach infinity (so that $1/c \rightarrow 0$) one can reproduce within quantum mechanics or special relativity the predictions of classical mechanics. The situations in which classical mechanics is applied successfully are such that one can for all practical purposes proceed as if the quantum of action were zero or as if the velocity of light were infinite; the objects dealt with are very big in comparison to the quantum of action, and the velocities are very small in comparison to the velocity of light. In fact, the matter is more complicated and requires a detailed mathematical examination. These examples are only meant to give a rough idea of how one seeks to reconstruct the formalism of classical mechanics from the formalisms of its successor theories. This possibility to reconstruct the formalism of classical mechanics explains why the predictions of classical mechanics are successful in certain areas.

Some philosophers take the possibility of such a reconstruction of the formalism of the older theory from the formalism of the newer theory to be a sufficient basis for claiming that the concepts of the older theory can be integrated into the concepts of the newer theory. They go as far as maintaining that the older theories are reduced to the newer ones (see, in particular, Schaffner 1967, Hooker 1981, part I, § 3, and 2004, sections 1–3). However, that conclusion cannot be warranted on this basis alone. The formal structure of the newer theories is very different from the one of the older theories. Special relativity, for instance, is fundamentally distinct from its predecessor theories in posing the principle that the velocity of light is finite and constant, independently of any reference frame. The same goes for quantum mechanics in comparison to classical mechanics; that is why quantum mechanics, in distinction to classical mechanics, raises a number of notorious problems of interpretation. The fact that the formal structure is considerably different means that there is a great difference on the level of the views of the constitution of nature that both theories offer. What is therefore needed is not only a reconstruction of the formalism, but also a reconstruction of the concepts of the predecessor theories from the concepts of the successor theories.

There is no general rule regarding how to carry out such a reconstruction of concepts. To my mind, the examples of major theory replacements from classical to contemporary physics allow us to distinguish three cases – a smooth case of reduction (1), an intermediate case of reconstruction without reduction (2) and the radical case of incommensurability (3).

(1) Consider the example of classical mechanics and quantum mechanics. Assume a version of quantum mechanics that includes a dynamics which leads to definite numerical values as measurement outcomes (the most elaborate proposal in that respect goes back to Ghirardi, Rimini & Weber 1986, although there still are a number of physical problems with that proposal). In that case, quantum theory includes the description of a transition from quantum states, characterized by superpositions and entanglement, to classical states. There is, of course, a considerable conceptual change from classical mechanics to quantum mechanics. But this change does not amount to semantic incommensurability. The change concerns the view of what is fundamental in nature. Quantum mechanics shows that the assumptions of classical mechanics about what is fundamental in nature are false; nonetheless, classical concepts can be derived within quantum mechanics, having a limited do-

main of applicability: they apply as soon as quantum superpositions and entanglements are dissolved, and quantum mechanics, on the version proposed by Ghirardi, Rimini & Weber (1986), includes a dynamics for this dissolution (state reductions). Generally speaking, this is a case where the concepts of the older theory can be integrated into the concepts of the newer theory, having a limited domain of applicability within the domain that the newer theory covers.

(2) The case is different if one maintains a version of quantum mechanics that does not acknowledge definite numerical values as measurement outcomes so that there are no state reductions. Instead, everything that there is in the physical world is subject to quantum entanglement, and there is only a universal quantum state of the world developing in time (that version, which is based exclusively on the Schrödinger dynamics, goes back to Everett 1957). Definite numerical values of physical quantities as measurement outcomes are merely the way in which the world appears to local observers whose cognitive access is confined to a particular branch of the universe. Definite numerical values as measurement outcomes – and classical properties in general – are relative to a certain state of consciousness of a local observer. Apart from that particular branch of the universe, there are infinitely many other branches of the universe in which other numerical values of the physical properties in question exist relative to other states of mind of the observer. The universal quantum state of the world includes all these branches.

In that case, there is no domain of the world to which the concepts of classical physics apply, for there is nothing that is not subject to quantum entanglement. What classical physics says about the world is false and beside the point. Nonetheless, one can reconstruct the concepts of classical physics on the basis of the concepts of quantum theory. Given the universal quantum state of the world that includes infinitely many different branches of the world in which the same physical objects exist in different relative states, one can reconstruct the concepts of classical physics as describing how the world appears to local observers. The mistake of classical physics is to take that appearance for the physical reality. Generally speaking, given the wider conceptual framework of the successor theory, one can reconstruct the concepts of the predecessor theory within that framework, although these concepts do not yield any true or approximately true descriptions of what there is in the world. Instead, they describe how the world appears to a local observer with limited cognitive access.

The case of classical mechanics and special relativity is in a certain manner conceptually similar. If the velocity of light were infinite, the distinction between events that are separated from each other by a spacelike interval, by a timelike interval and by a lightlike interval would collapse. Spatial and temporal distances, including notably the relation of similarity between events, would no longer depend on a frame of reference. However, given that the velocity of light is finite and constant, there is no domain to which the spatio-temporal concepts of classical mechanics apply. Nonetheless, one can reconstruct these concepts within the framework that is given by the principle of the velocity of light being finite and constant, showing how these concepts describe how the world appears to an observer within a given frame of reference. The observer takes these concepts to be objective, not realizing that they depend on a frame of reference (that is privileged for her, but not privileged in a general manner).

(3) Consider the case of Newton's theory of gravitation and general relativity. Although the predictions nearly coincide within a certain domain, there is no possibility to reconstruct within general relativity the concept of an action at a distance. It is not possible to reconstruct that concept as describing how the world appears to a local observer. That concept points out a consequence to which Newton's theory of gravitation is committed; it does not belong to the repertoire of a local observer. In short, the view of there being an action at a distance is simply abandoned and replaced with a theory of curved space-time according to which gravitation is an effect of space-time curvature.

In none of the three types of cases considered, do the concepts of the predecessor theory come close to or approximate the concepts of the successor theory. Furthermore, all these cases are much more complicated than is suggested by the mathematical operation of letting a particular quantity approach zero in order to derive an image of the formalism of the old theory within the formalism of the new theory. The reason is the difference in the formal structure of the successor theories in comparison to their predecessors. Even in the case of quantum mechanics and classical mechanics, there is no question of a simple transformation of the reconstruction of the formalism into a reconstruction, or even a reduction, of the concepts. Everything depends on the way in which one interprets quantum mechanics.

Nonetheless, only the third type of case is a case of semantic incommensurability. The thesis that semantic incommensurability is widespread is unfounded. The first type of case is a case of theory reduction:

if quantum mechanics includes a dynamics for a transition from superposed and entangled quantum states to classical states, then it is possible to reduce classical mechanics to that version of quantum mechanics; for it is then possible to derive within quantum mechanics concepts that match the role (inferential relations) of the proper concepts of classical mechanics within a certain domain. The second type of case is situated somewhere in between: there is no domain in which the concepts of the predecessor theory truly apply or are approximately true, but it is possible to reconstruct the concepts of the old theory within the framework of the new theory, namely as describing how the world appears to observers with a limited cognitive access.

Already the second type of case is sufficient to ground the claim that there is not a simple theory replacement in the history of science, but that the successor theories come closer to the truth than their predecessor ones (although we do not have at our disposal a universally accepted, precise definition of the notion of coming close to the truth). One can maintain that there is an approximation to a true and complete theory of the domain in question – or even nature as a whole – in the following sense: the common cases of replacing the old theories with new ones amount to an enlargement of our point of view, because it is possible to reconstruct within the new theory the proper concepts of the old theory as describing the way in which the world appears to a local observer with a limited cognitive access. In replacing our old theories with better ones, we decrease the cognitive limits of our point of view, coming closer to an objective description, that is, a description and explanation of the world from a point of view of nowhere. Science is the ambitious project of transgressing our cognitive limits, given that we are finite thinking beings confined to a very small spatio-temporal region of the universe, seeking to reach an objective description of the world.

To sum up, the incommensurability challenge and the underdetermination challenge can be countered in such a way that they do not undermine the project of a metaphysics of nature based on science. The second type of case considered in this section is already sufficient as a basis for that project. In the remaining sections of the paper, I shall consider four claims within the scope of that project and their link with our current scientific theories.

4. The Tenseless Theory of Time and Existence

There are two rival theories of time as well as of existence. According to the tensed theory of existence, existence is relative to time in the sense that only that what is present – or only that what is present and that what is past – exists. What is in the future does not exist as yet, and, according to some versions of this theory, what is past does not exist any more. The tensed theory of existence implies the tensed theory of time according to which the past, the present and the future are objective modes of time, being out there in the world. The tensed theory of time, however, does not imply the tensed theory of existence.

Opposed to the tensed theory of time is the tenseless theory of time which claims that there are only temporal relations of being earlier than, simultaneous with and later than among events, but no objective modes of past, present and future. The tenseless theory of time implies the tenseless theory of existence according to which existence is not relative to a location in time in the same way as it is not relative to a location in space: everything that there is in space and time simply exists. The tenseless theory of existence, however, does not imply the tenseless theory of time.

There are philosophical arguments in favour of both these theories of time and existence. The case can be settled by taking science into account. The relevant scientific theory is special relativity. Special relativity shows that there is no simultaneity between events independently of a reference frame. Any event – consisting in physical properties that occur at a space-time point – that is supposed to be simultaneous with other events is so only relative to a reference frame, and there is no globally preferred reference frame. Thus, there is no objective "now" – in the same way as there is no objective "here." For any space-time point, it can be claimed that it is "present" in the same way as it can be claimed that it is "here" (see, for instance, Dorato 1995, chapters 11 to 13, in particular pp. 186–187, 210). The reason is that, according to special relativity, spatial as well as temporal distances between events are relative to a reference frame. Invariant with respect to the choice of a reference frame is only the four-dimensional, spatio-temporal distance between any two events (or points of space-time). That is the reason why special relativity is taken to show that space and time are united in a four-dimensional entity, space-time.

Special relativity hence makes a case for the tenseless theories of time and existence. Since spatial and temporal distances are relative to a reference frame, there is no basis in the physical world for upholding a tensed theory of time or existence (see Saunders 2002). Special relativity does not exclude that there is a direction of time in the universe accounting for irreversible processes (see Maudlin 2007, 108–110, 115–117); but it contains a decisive argument against the assumption that there are objective modes of time. Nonetheless, the tenseless theories of time and existence belong to the metaphysics of nature. It is logically possible to rescue the idea of an objective present by introducing the notion of one globally privileged reference frame. That notion does not contradict special relativity. The point is that it is entirely *ad hoc.*

Furthermore, basing a tenseless theory of time and existence on special relativity is a hypothetical claim, since it depends on one particular scientific theory that may be superseded in the future. General relativity – and notably its application in cosmology – goes beyond special relativity; but it does not change anything with respect to what special relativity says about the relativity of spatial and temporal distances given the principle of general covariance that excludes that there is a globally privileged coordinate system. However, it cannot be ruled out that a future theory that transcends general relativity may change that matter.

5. Events Instead of Substances

The tenseless theories of time and existence, based on the physics of special relativity, result in what is known as the view of the world as a block universe: the whole of four-dimensional space-time is a single block so to speak, including time; everything that there is exists at a space-time point or region. What is the content of the block universe?

In metaphysics, it is common to draw a distinction between substances and events. *Substances* persist as a whole for a certain time. They have spatial parts (unless they are atoms in a literal sense), but they do not have any temporal parts. Relying on physics, an *event* can be conceived as consisting in the physical properties that occur at a space-time point. Let us leave open whether these physical properties literally are properties of space-time points or whether they are properties of matter (e.g. fields) located at space-time points. Continuous sequences of events are *processes*. Processes have spatial as well as temporal parts. Four-dimensional objects such as processes are commonly conceived of as *perdurants*, persisting by having spatial as well as temporal parts, whereas three-dimensional objects such as substances are conceived as *endurants*, persisting as a whole for a certain time, having no temporal parts.

Common sense admits both substances and processes. A volcano, for instance, is regarded as a substance, persisting for a certain time by having no temporal parts, but only spatial parts. The eruption of a volcano, by contrast, is a process, persisting for a time by having temporal as well as spatial parts. The eruption can, for instance, be first mild and then heavy.

Events and processes cannot be dispensed with in metaphysics. Even if it were possible to conceive all events as consisting in changes of the properties of substances, there would be a dualism of substances and events qua changes in the properties of substances. However, it may be possible to do without substances, recognizing only events and processes. (There is an ambiguity in the notion of a substance: if one regards space-time points as substances, they are not substances in the sense of endurants, but four-dimensional objects that have neither spatial nor temporal parts; they are not in space and time, but they are what makes up space-time). Again, there is a philosophical dispute as to whether or not one should admit substances in addition to events. Again, science is relevant to that dispute.

If one switches from a physics of three-dimensional space to a physics of four-dimensional space-time (block universe), there no longer is any need to admit substances as the objects that are the enduring foundation of change, change being the change of properties of substances, motion being change in the location of substances. Moreover, there is no need to conceive of the transtemporal identity of objects as the identity of substances in time, because substances do not have temporal parts.

In the metaphysics of the block universe, transtemporal identity can be accounted for in terms of genidentity, that is, sequences of events that have the same or similar properties. In other words, the transtemporal identity of any physical object is explained by the fact that the object is a process whose temporal parts form a continuous sequence, exhibiting similar physical properties. As regards motion, what common sense considers as the motion of a substance through three-dimensional space is explained as a continuous sequence of space-time points or regions that possess a similar physical content (a world line). Change is different physical properties instantiated at points or regions of space-time forming such a continuous sequence.

Hence, against the background of special relativity, it is in the first place the philosophical arguments of coherence and parsimony that speak in favour of a metaphysics of events – and processes (perdurants) –, admitting no substances (endurants): if one makes the step to a metaphysics of a four-dimensional block universe, it is simply not coherent to recognize three-dimensional substances among the content of the block universe. Four-dimensional events and their sequences (processes) have to be accepted anyway, and they are sufficient as the furniture of the universe (see Sider 2001 as regards the philosophical arguments).

Moreover, in recent years, arguments have been developed to the effect that admitting three-dimensional substances with spatial, but no temporal parts, is not consistent with special relativity. According to special relativity, the spatial distances between points depend on a reference frame. Consequently, if one subscribes to an ontology according to which there are three-dimensional macroscopic substances, their spatial figure varies from one frame of reference to another one, because the spatial distances between the points that the substance in question occupies depend on a reference frame. If, by contrast, physical objects are four-dimensional perdurants, their figure in four-dimensional space-time is not relative to a reference frame (see Balashov 1999 as well as Hales & Johnson 2003). A further argument makes the following point: since simultaneity is relative to a reference frame, a metaphysics of enduring three-dimensional objects cannot come up with a convincing theory of the coexistence (co-presence) of objects. By contrast, a metaphysics of perduring four-dimensional objects, which have temporal parts, can easily include a theory of coexistence: any two four-dimensional objects coexist if and only if they have parts that are separated by a space-like interval (see Balashov 2000 and the discussion between Gilmore 2002 and Balashov 2005).

These arguments put the case against three-dimensional substances on a par with the case against objective simultaneity based on special relativity. Nonetheless, in any case, a metaphysics of events based on special and general relativity is as hypothetical as the tenseless theory of time and existence based on these scientific theories. They both presuppose that there is no globally preferred reference frame, and it cannot be excluded that the future development of science may introduce notions that rehabilitate the notion of a single objective temporal order in the universe.

6. Relations Instead of Intrinsic Properties

Up to now, I have argued in favour of the view of the world consisting in the last resort in the distribution of fundamental physical properties at space-time points over the whole of space-time, forming continuous sequences that are processes and that can be regarded as physical objects (albeit no substances in the sense of things that do not have temporal parts). Quantum physics can be seen in the first place as adding something to this view concerning the physical properties. Whereas special and general relativity can be conceived as theories about space-time, quantum physics is concerned with matter.

It is often taken for granted that the fundamental physical properties, instantiated at space-time points, are intrinsic properties. Intrinsic are all and only those properties that an object has irrespective of whether or not there are other contingent objects; in brief, having or lacking an intrinsic property is independent of accompaniment or loneliness (see Langton & Lewis 1998). All other properties are extrinsic, consisting in the object bearing certain relations to other objects.

Quantum physics is usually conceived of in terms of states of physical objects. The state of an object at a time can be regarded as encapsulating the properties that the object has at that time. The most striking feature of quantum theory is that the states of several objects can be entangled. In fact, starting from the formalism of quantum theory, it is to be expected that whenever one considers a complex object that consists of two or more quantum objects, the states of these objects are entangled. Entanglement is to say that it is not the case that each of the objects has a state separately. On the contrary, only the whole, that is, the complex object composed of two or more objects, is in a precise state (called a "pure state"). Philosophers of physics therefore speak of non-separability (Howard 1989) or relational holism (Teller 1986), since entanglement consists in certain relations among quantum objects.

These relations cannot be traced back to intrinsic properties of the physical objects in question. There are no intrinsic properties of the related quantum objects on which the relations of entanglement could supervene. Quantum physics can therefore be taken to suggest a metaphysics of relations, known as structural realism: insofar as quantum physics is concerned, the fundamental physical properties consist in certain relations instead of being intrinsic properties (see French & Ladyman 2003, Esfeld 2004 and Ladyman & Ross 2007, chapter 3; see furthermore Lyre 2004 as regards gauge theories and Esfeld & Lam 2008 as regards structural realism with respect to general relativity).

Again, this is a conclusion belonging to the metaphysics of nature. This conclusion could be avoided by postulating intrinsic properties in the form of hidden variables that restore separability (that is, properties of quantum objects that are there, but whose value we cannot know). However, since the discovery of the theorem of John Bell (1964), it is clear that one would have to pay a high metaphysical as well as physical price for admitting hidden variables of that kind. (The only elaborate account of quantum physics in terms of hidden variables, Bohm's theory, does not fall within the scope of that criticism, for Bohm himself interprets his theory in terms of relations and holism rather than in terms of intrinsic properties; see Bohm & Hiley 1993).

Nonetheless, again, a metaphysics of relations based on quantum physics is hypothetical. Quantum physics as it stands is definitely not the last word on matter. There is the notorious measurement problem, and there is the problem of the unification of quantum field theory and general relativity. If the project of unifying quantum field theory and general relativity succeeds, the resulting scientific theory will have important repercussions for our view of nature. Nonetheless, whatever may be the future fundamental physical theory that achieves a unified treatment of the phenomena that are currently considered by two different theories, it would be unreasonable to expect that future theory to go back behind the unification of space and time as considered by general relativity or the holism that quantum entanglement manifests.

7. Causation, Laws and Dispositions

One of the main themes of the metaphysics of nature is causation. What, if any, are the physical foundations of causation? The most developed proposal for a theory of causation that seeks to be close to physics is the transference theory according to which, in brief, a causal process consists in the exchange of a conserved physical quantity such as energy (Salmon 1998, chapters 1, 12, 16 and 18, Dowe 2000). However, that theory is modelled along the lines of Newtonian mechanics and special relativity. It faces serious difficulties from general relativity (see Rueger 1998, Curiel 2000, and Lam 2005), not to mention the holism of quantum theory.

Moreover, even if a proposal for a physical account of causation could be developed with success, it would not automatically settle the main issue in the metaphysics of causation, that is, the opposition between a Humean view of causation and a power view of causation. According to the Humean view, causation is nothing but a contingent regular pattern of co-instantiation of events of the same types. The Humean could employ a physical account of causation in order to distinguish the regularity patterns that are causal from the regularity patterns that are not causal. According to the power view of causation, by contrast, there are causal powers in the world that establish a necessary connection between causes and their effects (see notably Harré & Madden 1975, in particular chapter 5, and Shoemaker 1980). The friend of causal powers could interpret a physical account of causation as identifying the fundamental manifestations of such causal powers in the world.

The view that one takes on causation has implications for one's views about laws and dispositions: for the Humean, laws are contingent regularity patterns, whereas for the friend of causal powers, they are metaphysically necessary, deriving from the causal powers, which establish necessary connections. The Humean eliminates dispositions in favour of categorical properties, reducing statements about dispositions to counterfactual conditionals that are made true by the distribution of categorical properties in the world, in virtue of that distribution exhibiting certain regularity patterns (see e.g. Sparber 2006). For the friend of causal powers, by contrast, dispositions in the form of powers are fundamental physical properties.

Both sides in this debate claim support from physics. Since Russell's famous paper denouncing the notion of causation as production (Russell 1912), some Humeans maintain that the physical theories of the 20th century do not admit of any other sort of causation than Humean regularity patterns in the distribution of the fundamental physical properties (see e.g. Field 2003, section 1, and Norton 2007). However, the friends of powers retort that physics cannot specify the properties it deals with beyond causal-structural equivalence, thus suggesting the conclusion that the physical properties are themselves causal properties and hence powers (see notably Shoemaker 1980 and Mumford 2006). Moreover, talk in dispositional terms is widespread in fundamental physics, notably with respect to the objective probabilities of quantum theory (see Suárez 2007 for interpreting these probabilities in terms of propensities), but also applicable to general relativity, since space-time includes gravitation and thus a physical force (see e.g. Bartels 1996, 37–38, Bartels 2009 and

Bird 2009, section 2.3). Nonetheless, the interpretation of quantum probabilities in terms of propensities is by no means mandatory, and the fact that physics cannot specify the properties it deals with beyond causal-structural equivalence can also be accommodated within a structural realism that commits itself to categorical structures only (cf. e. g. Psillos 2006, section 6, and Sparber 2009, chapter 4).

Hence there is a difference between themes such as the tenseless theory of time and existence vs. the tensed theory, events vs. substances, relations vs. intrinsic properties on the one hand, and causation, disposition and laws on the other. As regards the former themes, our current fundamental physical theories establish certain metaphysical commitments on all reasonable standards of interpretation – namely the commitments to a tenseless theory of time and existence, to events and to relations. As regards the latter themes, there is no such direct route from physics to metaphysical conclusions. Nonetheless, physics is pertinent when it comes to arguments for the one or the other position on these issues, and it may well turn out that one can make a cogent case for one particular position on causation, laws and dispositions by drawing on physics and science in general (see Esfeld 2008, chapter 5, for an argument for the power view of causation based on physics as well as the overall coherence of our scientific knowledge and Esfeld 2009 for the relationship with physical structures).

In conclusion, there is, as stated in the introduction, a mutual dependence between science and philosophy: science needs philosophy in the sense of epistemology in order to assess the knowledge claims contained in scientific theories (as illustrated by the discussion of the underdetermination and the incommensurability challenges in sections 2 and 3). Philosophy in the sense of metaphysics needs science in order not to make claims about the world that do not stand up to scientific scrutiny. Nonetheless, science does not directly imply a particular metaphysics of nature. However, there are clear cases in which scientific theories strongly suggest on all reasonable criteria of interpretation certain metaphysical positions – such as the cases of the tenseless theory of time and existence vs. the tensed theory, events vs. substances and relations vs. intrinsic properties. Dependence on science makes the metaphysics of nature as fallible and hypothetical as scientific theories.

Bibliography

Albert, David Z. and Loewer, Barry (1988), "Interpreting the many worlds interpretation", *Synthese* 77: 195–213.

Balashov, Yuri (1999), "Relativistic objects", *Noûs* 33: 644–662.

— (2000), "Enduring and perduring objects in Minkowski space-time", *Philosophical Studies* 99: 129–166.

— (2005), "Special relativity, coexistence and temporal parts: a reply to Gilmore", *Philosophical Studies* 124: 1–40.

Bartels, Andreas (1996), "Modern essentialism and the problem of individuation of spacetime points", *Erkenntnis* 45: 25–43.

— (2009), "Dispositionen in Raumzeit-Theorien", in: C. F. Gethmann (ed.), *Lebenswelt und Wissenschaft. XXI. Deutscher Kongress für Philosophie. Kolloquien*. Hamburg: Meiner.

Bell, John S. (1964), "On the Einstein–Podolsky–Rosen-paradox", *Physics* 1: 195–200.

Bird, Alexander (2009): "Structural properties revisited", in: T. Handfield (ed.), *Dispositions and causes*. Oxford: Oxford University Press, 215–241.

Bohm, David (1952), "A suggested interpretation of the quantum theory in terms of 'hidden' variables", *Physical Review* 85: 166–193.

Bohm, David & Hiley, Basil (1993), *The undivided universe. An ontological interpretation of quantum theory*. London: Routledge.

Bohm, David, Hiley, Basil & Kaloyerou, P. N. (1987), "A causal interpretation of quantum fields", *Physics Reports* 144: 349–375.

Curiel, Erik (2000), "The constraints general relativity places on physicalist accounts of causality", *Theoria* 15: 33–58.

Dorato, Mauro (1995), *Time and reality. Spacetime physics and the objectivity of temporal becoming*. Bologna: Cooperativa Libraria Universitaria Editrice Bologna.

Dowe, Phil (2000), *Physical causation*. Cambridge: Cambridge University Press.

Duhem, Pierre (1914), *La théorie physique. Son objet – sa structure*. Paris: Vrin. 2nd edition.

— (1954), *The aim and structure of physical theory,* translated by Philip P. Wiener. Princeton: Princeton University Press.

Esfeld, Michael (2001), Holism in philosophy of mind and philosophy of physics. Dordrecht: Kluwer.

— (2004), "Quantum entanglement and a metaphysics of relations", *Studies in History and Philosophy of Modern Physics* 35B: 601–617.

— (2008), *Naturphilosophie als Metaphysik der Natur*. Frankfurt (Main): Suhrkamp.

— (2009), "The modal nature of structures in ontic structural realism", *International Studies in the Philosophy of Science* 23: 179–194.

Esfeld, Michael & Lam, Vincent (2008), "Moderate structural realism about space-time", *Synthese* 160: 27–46.

Everett, Hugh (1957), "'Relative state' formulation of quantum mechanics", *Reviews of Modern Physics* 29: 454–462. Reprinted in B. S. DeWitt and

N. Graham (eds.), *The many-worlds interpretation of quantum mechanics*. Princeton: Princeton University Press 1973, 141–149.

Feyerabend, Paul K. (1962), "Explanation, reduction, and empiricism", in: H. Feigl & G. Maxwell (eds.), *Scientific explanation, space, and time. Minnesota Studies in the philosophy of science. Volume 3*. Minneapolis: University of Minnesota Press, 28–97.

Field, Hartry H. (2003), "Causation in a physical world", in: M. Loux & D. Zimmerman (eds.), *The Oxford handbook of metaphysics*. Oxford: Oxford University Press, 435–460.

French, Steven & Ladyman, James (2003), "Remodelling structural realism: quantum physics and the metaphysics of structure", *Synthese* 136: 31–56.

Ghirardi, Giancarlo, Rimini, Alberto & Weber, Tullio (1986), "Unified dynamics for microscopic and macroscopic systems", *Physical Review* D34: 470–491.

Gilmore, Cody S. (2002), "Balashov on special relativity and temporal parts", *Philosophical Studies* 109: 241–263.

Hales, Steven D. & Johnson, Timothy A. (2003), "Endurantism, perdurantism, and special relativity", *Philosophical Quarterly* 53: 524–539.

Harré, Rom & Madden, E. H. (1975), *Causal powers. A theory of natural necessity*. Oxford: Blackwell.

Hawley, Katherine (2006), "Science as a guide to metaphysics?", *Synthese* 149: 451–470.

Hooker, Clifford A. (1981), "Towards a general theory of reduction. Part I: Historical and scientific setting. Part II: Identity in reduction. Part III: Cross-categorial reduction", *Dialogue* 20: 38–60, 201–236, 496–529.

—— (2004), "Asymptotics, reduction and emergence", *British Journal for the Philosophy of Science* 55: 435–479.

Howard, Don (1989), "Holism, separability, and the metaphysical implications of the Bell experiments", in: J. T. Cushing & E. McMullin (eds.), *Philosophical consequences of quantum theory. Reflections on Bell's theorem*. Notre Dame: University of Notre Dame Press, 224–253.

Huggett, Nick & Weingard, Robert (1994), "Interpretations of quantum field theory", *Philosophy of Science* 61: 370–388.

Kuhn, Thomas S. (1962), *The structure of scientific revolutions*. Chicago: University of Chicago Press.

Ladyman, James & Ross, Don with Spurrett, David & Collier, John (2007), *Every thing must go. Metaphysics naturalised*. Oxford: Oxford University Press.

Lam, Vincent (2005), "Causation and space-time", *History and Philosophy of the Life Sciences* 27: 457–470.

Langton, Rae & Lewis, David (1998), "Defining 'intrinsic'", *Philosophy and Phenomenological Research* 58: 333–345. Reprinted in David Lewis (1999), *Papers in metaphysics and epistemology*. Cambridge: Cambridge University Press, 116–132.

Laudan, Larry (1981), "A confutation of convergent realism", *Philosophy of Science* 48: 19–49. Reprinted in D. Papineau (ed.) (1996), *The philosophy of science*. Oxford: Oxford University Press, 107–138.

— (1984), "Realism without the real", *Philosophy of Science* 51: 156–162.
— (1996), *Beyond positivism and relativism. Theory, method, and evidence.* Boulder: Westview Press.
Laudan, Larry & Leplin, Jarrett (1991), "Empirical equivalence and underdetermination", *Journal of Philosophy* 88: 449–472.
Lockwood, Michael (1989), *Mind, brain and the quantum. The compound 'I'.* Oxford: Blackwell.
Lyre, Holger (2004), *Lokale Symmetrien und Wirklichkeit. Eine naturphilosophische Studie über Eichtheorien und Strukturenrealismus.* Paderborn: Mentis.
Maudlin, Tim (2007), *The metaphysics in physics.* Oxford: Oxford University Press.
Mumford, Stephen (2006), "The ungrounded argument", *Synthese* 149: 471–489.
Norton, John (2007), "Causation as folk science", in: H. Price & R. Corry (eds.) (2007), *Causation, physics, and the constitution of reality. Russell's republic revisited.* Oxford: Oxford University Press, 11–44.
Papineau, David (1996), "Introduction", in: D. Papineau (ed.), *The philosophy of science.* Oxford: Oxford University Press, 1–20.
Popper, Karl R. (1959), *The logic of scientific discovery.* London: Hutchinson.
Psillos, Stathis (2006), "What do powers do when they are not manifested?", *Philosophy and Phenomenological Research* 72: 137–156.
Quine, Willard Van Orman (1953), "Two dogmas of empiricism", in: W. V. O. Quine, *From a logical point of view.* Cambridge (Massachusetts): Harvard University Press, 20–46.
Rueger, Alexander (1998), "Local theories of causation and the a posteriori identification of the causal relation", *Erkenntnis* 48: 25–38.
Russell, Bertrand (1912), "On the notion of cause", *Proceedings of the Aristotelian Society* 13: 1–26.
Salmon, Wesley C. (1998), *Causality and explanation.* Oxford: Oxford University Press.
Saunders, Simon (2002), "How relativity contradicts presentism", in: C. Callender (ed.), *Time, reality & experience.*, Cambridge: Cambridge University Press, 277–292.
Schaffner, Kenneth F. (1967), "Approaches to reduction", *Philosophy of Science* 34: 137–147.
Shoemaker, Sydney (1980), "Causality and properties", in: P. van Inwagen (ed.), *Time and cause.* Dordrecht: Reidel, 109–135. Reprinted in: Sydney Shoemaker (1984), *Identity, cause, and mind. Philosophical essays.* Cambridge: Cambridge University Press, 206–233.
Sider, Theodore R. (2001), *Four-dimensionalism.* Oxford: Clarendon Press.
Sparber, Georg (2006), "Powerful causation", in: M. Esfeld (ed.), *John Heil. Symposium on his ontological point of view.* Frankfurt (Main): Ontos, 123–137.
— (2009), *Unorthodox Humeanism.* Frankfurt (Main): Ontos.
Suárez, Mauricio (2007), "Quantum propensities", *Studies in History and Philosophy of Modern Physics* 38B: 418–438.

Teller, Paul (1986), "Relational holism and quantum mechanics", *British Journal for the Philosophy of Science* 37: 71–81.

von Neumann, Johann (1932), *Mathematische Grundlagen der Quantenmechanik*. Berlin: Springer.

— (1955), *Mathematical foundations of quantum mechanics,* translated by R. T. Beyer. Princeton: Princeton University Press.

Contributors

Andreas Bartels has been professor for Philosophy of Science at the University of Bonn since 2000. He taught before at the Universities of Giessen, Heidelberg, Berlin, Munich, Jena, Erfurt and Paderborn. His major publications include: *Kausalitätsverletzungen in allgemein-relativistischen Raumzeiten* (1986), *Bedeutung und Begriffsgeschichte* (1994), *Grundprobleme der modernen Naturphilosophie* (1996), *Strukturale Repräsentation* (2005). He now leads the "Knowing how and Knowing that" research group, supported by the Volkswagen Foundation. The group focuses upon the study of cognitive and, in particular, conceptual abilities in humans and animals.

Christophe Bouriau is Associate Professor of Philosophy and Chair of the Philosophy Department at the University of Nancy. He studies the role of imagination in knowledge and thought from the 17th century to the present, with emphasis on Descartes, Kant, the neo-Kantians, Poincaré, Vaihinger, and Binet. He recently finished a French translation of Vaihinger's popular edition (1923) of *Die Philosophie des Als Ob* (2008), and is writing a book on Vaihinger's pragmatism. He also studies the debates between Pierre Janet and Henri Bergson on the operation of memory and its relation to imagination. In November 2008, he defended his habilitation on the subject of "neo-Kantian readings: from imagination to fiction."

Michael Esfeld is full professor of philosophy of science at the University of Lausanne (Switzerland). His main areas of research are the metaphysics of science (particularly with regard to physics) as well as the philosophy of mind. His books include *Mechanismus und Subjektivität in der Philosophie von Thomas Hobbes* (1995), *Holism in philosophy of mind and philosophy of physics* (2001/2002 in German), *Einführung in die Naturphilosophie* (2002), *La philosophie de l'esprit. De la relation entre l'esprit et la nature* (2005) and *Naturphilosophie als Metaphysik der Natur* (2008).

Gad Freudenthal is Senior Research Fellow at the CNRS (Paris). He has written on the history of science in Antiquity and in the Middle Ages, especially in Jewish cultures. His books include: *Aristotle's Theory* *of Material Substance: Form and Soul, Heat and Pneuma* (1995) and *Science in the Medieval Hebrew and Arabic Traditions* (2005). He edited *Studies on Gersonides – A Fourteenth-Century Jewish Philosopher-Scientist* (1992) and (with S. Kottek), *Mélanges d'histoire de la médecine hébraïque: Études choisies de la* Revue de l'histoire de la médecine hébraïque, *1948–1985* (2003). The edited volume *Science in Medieval Jewish Cultures* is forthcoming. He also is the editor of the journal *Aleph: Historical Studies in Science and Judaism*, established in 2001.

Michael Heidelberger holds the chair for Logic and Philosophy of Science at the University of Tübingen. At the centre of his interest are topics related to causality and probability, measurement and experiment. He specializes in the history of the philosophy of science, mainly of the late nineteenth and early twentieth century and focuses on philosophy and history of psychology and physics and related subjects in this period. He is the author of *Nature from Within: Gustav Theodor Fechner's Psychophysical Worldview* (2004) and of many other works on a wide variety of subjects.

Gerhard Heinzmann is Professor in the Department of Philosophy at the University of Nancy (France) and Director of the *Social Sciences and Humanities Research Institute of Lorraine (MSH Lorraine)*. He is the editor of the *Publications of the Henri Poincaré Archives* and of the journal *Philosophia Scientiae*. Specializing in philosophy of mathematics and logic, he is the author of numerous articles and books on the philosophy of Ferdinand Gonseth, Jean Cavaillès, Paul Bernays and Poincaré, among them *Zwischen Objektkonstruktion und Strukturanalyse: Zur Philosophie der Mathematik bei Henri Poincaré* (1995). Current projects are a book on cognitive intuition in mathematics and two textbooks on the foundations of geometry and arithmetic around 1900.

Andreas Hüttemann is Professor of Philosophy at the Westfälische Wilhelms-Universität Münster. His areas of specialization are philosophy of science and early modern philosophy. He is particularly interested in the relation of science and metaphysics. Among his recent publications are *What's wrong with Microphysicalism?* (2004), "Explanation, Emergence and Entanglement" (*Philosophy of Science*, 2005), "Physicalism Decomposed" (with David Papineau in *Analysis*, 2005). He also edited *Zur Deutungsmacht der Biowissenschaften* (2008).

Ernan McMullin is Professor Emeritus of Philosophy and Director Emeritus of the Program in History and Philosophy of Science at the University of Notre Dame. His areas of interest are contemporary philosophy of science, the history of the philosophy of science, and the interactions both historical and philosophical between religious belief and the natural sciences. He is a Fellow of the American Academy of Arts and Sciences, the International Academy of the History of Science, and the American Association for the Advancement of Science. He served at different times as President of the four major philosophical associations in the United States. He is the author or editor of 14 books and over 200 articles. Among the books: *The Concept of Matter* (1963), *Galileo: Man of Science* (1967), *Newton on Matter and Activity* (1978), *The Inference That Makes Science* (1992), and most recently *The Church and Galileo* (2005).

Alfred Nordmann teaches philosophy of science at the Technische Universität Darmstadt, Germany. His historical interests concern conceptions of science and objectivity in the formation of new fields of scientific knowledge such as theories of electricity and chemistry in the 18th century (Lichtenberg, Priestley, Lavoisier), mechanics (Hertz), evolutionary biology (Darwin, Bateson), and sociology in the 19th century, nursing science and nanoscale research in the 20th century (Gleiter). His epistemological interests concern the trajectory that leads from Immanuel Kant *via* Heinrich Hertz and Ludwig Wittgenstein to contemporary analyses of models, simulations, and visualizations. His publications include *Wittgenstein's Tractatus: An Introduction* (2005), a German-language *Introduction to the Philosophy of Technology* (2008) and he edited several volumes on the philosophy of Ludwig Wittgenstein, of Heinrich Hertz, of neo-Kantianism and on nanoscience.

Helmut Pulte is full professor for philosophy with special respect to history and philosophy of science at Ruhr-Universität Bochum. He served as co-editor of the multivolumed *Historisches Wörterbuch der Philosophie* (finished in 2006) and is co-editor of the *Journal for General Philosophy of Science*. His main research areas are history of philosophy of science, history of modern mathematics and physics and current philosophy of the exact sciences. His publications include the monograph *Axiomatik und Empirie: Eine wissenschaftstheoriegeschichtliche Untersuchung von Newton bis Neumann* (2005), which analyses the theory change of mathematical physics from the late 17th to the late 19th century. He is currently preparing two volumes on *The Reception of Isaac Newton in Europe* with Scott Mandelbrote.

Gregor Schiemann is Professor of Philosophy at the Department of Philosophy at the University of Wuppertal. Having become a toolmaker in 1978, he received his physics diploma in Zürich (1988), a PhD in philosophy from Darmstadt (1995) and his habilitation in philosophy at the University of Tübingen (2003). His areas of expertise include philosophy of science, technology, the relation of mind and nature and the history of science. He specializes particularly in philosophy and history of physics in the 19th and 20th century and in the history of the concept of nature. His publications include: *Hermann von Helmholtz's Mechanism: The Loss of Certainty* (2009, German ed. 1997), *Natur, Technik, Geist. Kontexte der Natur nach Aristoteles und Descartes in lebensweltlicher und subjektiver Erfahrung* (2005) and an intellectual biography of Heisenberg (2007), as well as many articles in philosophy and history of science.

Laura J. Snyder is Associate Professor of Philosophy at St. John's University (New York) and 2009–2010 President of the International Society for the History of Philosophy of Science (HOPOS). Her work focuses on the history of inductive reasoning, especially ways that discussions of induction have been embedded in broader debates about science and its role in society. Her *Reforming Philosophy: A Victorian Debate on Science and Society* (2006) explores the way that the controversy between John Stuart Mill and William Whewell over induction was set in the context of their divergent views of politics, morality and economics, as well as their notions of the best way to reform society. In addition, she has published nearly twenty articles in the history and philosophy of science.

Rainer Specht is Professor Emeritus of the University of Mannheim. He habilitated in philosophy at Hamburg University in 1964 and taught in the philosophy department at Mannheim from 1967 until his retirement in 1995. He was visiting professor at the University of Bochum, the Universidad de San Marcos (Lima, Peru) and the Collège de France (Paris). He has published widely, especially in the philosophy of the 16th, 17th and 18th centuries. Among his books are *Commercium mentis et corporis* (1966), *Innovation und Folgelast* (1972), *René Descartes* (10th ed. 2006), *John Locke* (2nd ed. 2007) and a translation and commentary of Francisco Suárez's 5th metaphysical disputation (1976). He also translated a number of literary Spanish authors.

Scott Walter is Associate Professor in the Department of Philosophy at the University of Nancy (France). He studies the history and philosophy of mathematics and physics from 1850 to 1930, with a focus on the interplay and shifting boundaries of these disciplines. His recent publications include a critical edition of Henri Poincaré's correspondence with physicists, chemists and engineers (2007), and a history of the emergence of four-dimensional approaches to physics, in *The Genesis of General Relativity*, edited by Jürgen Renn (2007).

Index

Bei Fragen zur Produktsicherheit wenden Sie sich bitte an:
If you have any questions regarding product safety,
please contact:

Walter de Gruyter GmbH
Genthiner Straße 13
10785 Berlin
productsafety@degruyterbrill.com